BIOHYDROGEN

Woodhead Series in Bioenergy

The *Woodhead Series in Bioenergy* publishes titles on the topic of bioenergy as it relates to renewable sources of energy. The series notably covers biological resources, chemical and biological processes, and biomass products for energy production, along with the environmental, social, management, and economic aspects of bioenergy.

Series Editor

Prof. Yves Gagnon

Université de Moncton, Moncton, Canada

Previous Volumes in the Series

- Kishor Kumar Sadasivuni and Mithra Geetha, *Biofuel Cells and Energy Generation*, 2025, 9780443216022
- Krishna Kumar Jaiswal, Bhaskar Singh and Amit K. Jaiswal, *Microalgal Biofuels: Sustainable Production, Conversion, and Life Cycle Assessment*, 2025, 9780443241109
- Jeyabalan Sangeetha and Devarajan Thangadurai, *Microalgal Biomass for Bioenergy Applications*, 2024, 9780443139277
- Lina María Grajales Agudelo, Juan Carlos Valdés Serra and Erich Collicchio, *Agroenergy*, 2024, 9780443214301
- Stefano Ferrari Interlenghi, José Luiz de Medeiros and Ofélia de Queiroz Fernandes Araújo, *Sustainability of Methylic and Ethylic Biodiesel Production Routes*, 2024, 9780443219429
- Hwai Chyuan Ong, Islam Md Rizwanul Fattah and Indra Mahlia, *Waste Valorization for Bioenergy and Bioproducts*, 2024, 9780443191718

Woodhead Series in Bioenergy

BIOHYDROGEN

Advances in Theory and Production

Edited by

Danping Jiang

Professor, Key Laboratory of New Materials and Facilities for Rural Renewable Energy, Ministry of Agriculture and Rural Affairs of China, Henan Agricultural University, Zhengzhou, Henan, China

Chaoyang Lu

Professor, Key Laboratory of New Materials and Facilities for Rural Renewable Energy, Ministry of Agriculture and Rural Affairs of China, Henan Agricultural University, Zhengzhou, Henan, China

Tian Zhang

Lecturer, Key Laboratory of New Materials and Facilities for Rural Renewable Energy, Ministry of Agriculture and Rural Affairs of China, Henan Agricultural University, Zhengzhou, Henan, China

ELSEVIER

WP WOODHEAD PUBLISHING

An imprint of Elsevier

elsevier.com/books-and-journals

Woodhead Publishing is an imprint of Elsevier
50 Hampshire Street, 5th Floor, Cambridge, MA 02139, United States

ISBN: 978-0-443-27686-6

For Information on all Woodhead Publishing publications visit our website at https://www.elsevier.com/books-and-journals

Publisher: Priscilla Braglia
Acquisitions Editor: Ruchi Bhardwaj
Editorial Project Manager: Ashavari Bhattacharya
Production Project Manager: Maria Shalini A
Cover Designer: Raman Kumar

Typeset by Aptara, New Delhi, India

Contents

4. Dark fermentation biohydrogen production from organic matters

Yang Zhang, Wachiranon Chuenchart, Shaoqing Cui, Yu Tang, and Jianzhi Yue

5. Photo-fermentation biohydrogen production from organic matters

Yanyan Jing, Wen Cao, Jiyan Lu, Chaoyang Lu, Eric Lichtfouse, and Wenling Yang

6. Combined dark–photo fermentation biohydrogen production from organic matters

Yameng Li, Zhiping Zhang, Hang Yan, Lu Feng, Ikumi Umetani, Zhihan Zhang, and Junyu Tao

7. Economic analysis of production and sustainability of biohydrogen

Yangyang Li, Peiyu Tian, Beier Shang, Nan An, Bingyan Wu, Yuao Su, Hui Sun, Tian Yuan, and Tingting Hou

Contributors

Mujeeb Alam Henan Agricultural University, Zhengzhou, China
Nan An China Agricultural University, Beijing, China
Wen Cao Xi'an Jiaotong University, Xi'an, China
Haixing Chang Wuhan University, Wuhan, China
Wachiranon Chuenchart Quasar Energy Group, Independence, United States
Shaoqing Cui Zhejiang University, Hangzhou, China
Lu Feng Norwegian Institute of Bioeconomy Research (NIBIO), Ås, Norway
Tingting Hou Henan Agricultural University, Zhengzhou, China
Yun Huang Chongqing University, Chongqing, China
Danping Jiang Henan Agricultural University, Zhengzhou, China
Yanyan Jing Henan Agricultural University, Zhengzhou, China
Yameng Li Henan Agricultural University, Zhengzhou, China
Yangyang Li China Agricultural University, Beijing, China
Eric Lichtfouse Xi'an Jiaotong University, Xi'an, China
Hong Liu Henan Agricultural University, Zhengzhou, China
Shan Liu China Agricultural University, Beijing, China
Chaoyang Lu Henan Agricultural University, Zhengzhou, China
Jiyan Lu Xi'an Jiaotong University, Xi'an, China
Beier Shang China Agricultural University, Beijing, China
Zhipeng Song Henan Agricultural University, Zhengzhou, China
Yuao Su Seoul National University of Science & Technology, Seoul, South Korea
Hui Sun Henan Agricultural University, Zhengzhou, China
Nadeem Tahir Henan Agricultural University, Zhengzhou, China
Yu Tang Zhejiang University, Hangzhou, China
Junyu Tao Henan Rural Social Development Service Center, Zhengzhou, China
Peiyu Tian China Agricultural University, Beijing, China
Ikumi Umetani Norwegian Institute of Bioeconomy Research (NIBIO), Ås, Norway
Chao Wang Henan University of Science and Technology, Luoyang, China
Bingyan Wu China Agricultural University, Beijing, China
Fuqing Xu Xi'an Jiaotong University, Xi'an, China
Hang Yan Henan Agricultural University, Zhengzhou, China

Wenling Yang Henan Academy of Sciences, Institute of Biology Co. Ltd., Zhengzhou, China

Tian Yuan University of Tsukuba, Tsukuba, Japan

Jianzhi Yue Henan Agricultural University, Zhengzhou, China

Huan Zhang Henan Agricultural University, Zhengzhou, China

Quanguo Zhang Henan Agricultural University, Zhengzhou, China; Huanghe Science and Technology University, Zhengzhou, China

Tian Zhang Henan Agricultural University, Zhengzhou, China

Xiuqiang Zhang Henan Academy of Sciences, Energy Research Institute Co. Ltd., Zhengzhou, China

Xueting Zhang Huanghe Science and Technology University, Zhengzhou, China

Yang Zhang Henan Agricultural University, Zhengzhou, China

Zhihan Zhang Zhengzhou University of Aeronautics, Zhengzhou, China

Zhiping Zhang Henan Agricultural University, Zhengzhou, China

Yaping Zheng Henan Agricultural University, Zhengzhou, China

Shengnan Zhu Henan Agricultural University, Zhengzhou, China

Foreword

The energy sector is undergoing an important transition to meet the world's energy needs, to reduce greenhouse gas emissions, to support climate change mitigation, and to protect our planet for future generations. A major element of the energy transition is the integration of large-scale renewable energy sources, including bioenergy, in the energy mix of countries.

Bioenergy is the transformation of biomass, derived from recently living organic materials, into electricity, heat, and transportation fuels. Along with wind and solar energy, bioenergy is poised to play a significant role in the energy transition.

In support of the global research community and the commercial bioenergy sector, the *Woodhead Series in Bioenergy* publishes research and application-oriented book titles on the topic of bioenergy as it relates to renewable sources of energy. The series covers biological resources, chemical and biological processes, and biomass products for energy production, along with the sustainability of bioenergy, namely the environmental, social, management, and economic aspects of bioenergy.

Building on their extensive research and editorship experiences, Prof Ong, Dr. Fattah, and Prof Mahlia have edited a book that documents, in a single volume, the various technologies to transform biological waste streams in bioenergy production.

Well anchored in the UN's Sustainable Development Goal 7, Affordable and Clean Energy, this book presents a broad perspective of bioenergy processes in the valorization of biological wastes. Besides documenting the scientific and technological elements of transforming wastes into bioenergy, this book introduces the challenges and opportunities in the production of waste-to-energy processes, along with addressing the life cycle perspective of these processes.

Considering the scope of the topics covered, *Waste Valorization for Bioenergy and Bioproducts: Biofuels, Biogas, and Value-Added Products* is a book of interest for a broad range of readers, including postgraduate students, academic and industrial researchers, and industrial practitioners working on the valorization of biological waste feedstocks for bioenergy and bio-product applications.

Prof Yves Gagnon PEng, DSc
Université de Moncton, Moncton, Canada
Series Editor, Woodhead Series in Bioenergy

Acknowledgments

We would like to extent our gratitude to the National Key Research and Development Program of China (2023YFE0106000), the China Postdoctoral Science Foundation (2024T170240), and the National Natural Science Foundation of China (52306238 and 52506230) for providing solid foundation throughout the writing of this book. We also thank individuals and institutions who have walked alongside and provided support for us throughout the journey of writing of this book. We wish to express our deepest gratitude to each and every one of them.

CHAPTER

1

Introduction

Tian Zhang[a], *Hong Liu*[a,*], *Fuqing Xu*[b], *Nadeem Tahir*[a], *and Danping Jiang*[a]

[a]Henan Agricultural University, Zhengzhou, China, [b]Xi'an Jiaotong University, Xi'an, China

*Corresponding author. E-mail address: liuhong@henau.edu.cn

1.1 Hydrogen energy

Hydrogen is the most abundant substance in the world, accounting for 75% of the total mass of the universe (Gielen et al., 2019). On Earth, hydrogen energy primarily exists in the form of chemical compounds. Unlike primary energy sources such as coal, petroleum, and natural gas, which can be directly extracted, hydrogen needs to be produced using other energy sources (Sazali, 2020). In its elemental state, hydrogen energy exists as hydrogen gas (H_2), a colorless, odorless gas that is poorly soluble in water and highly flammable at room temperature and atmospheric pressure (Niermann et al., 2019; Wichman, 2003). Hydrogen energy can be sourced from virtually limitless seawater, renewable plants, coal, natural gas, and more. It is ubiquitous on Earth, representing an inexhaustible and ubiquitous energy resource (Abdin et al., 2020; Chemat et al., 2019).

Hydrogen energy is regarded as an ideal alternative energy source because of the following advantages:

1. High calorific value: The heat generated by the combustion of hydrogen per unit mass is more than other fuels, approximately 3 times that of gasoline, 3.9 times that of alcohol, and 4.5 times that of coke (Kamal et al., 2019; Liu et al., 2017).
2. Excellent combustion performance: Hydrogen requires very little ignition energy, which makes hydrogen-fueled internal combustion engines almost never misfire during operation and have good starting performance. Moreover, hydrogen has a fast combustion speed and

Biohydrogen: Advances in Theory and Production
DOI: https://doi.org/10.1016/B978-0-443-27686-6.00008-6

flame propagation speed, making it more effective in utilizing heat and improving thermal efficiency (Najjar, 2013).

3. Clean and environmentally friendly: The sole product of hydrogen combustion is water vapor. It produces no greenhouse gases or other pollutants, making it an environmentally friendly and ideal clean energy source (Fallah Vostakola et al., 2022).
4. Diverse production methods: Hydrogen can be produced through various methods, including water electrolysis, extraction from fossil fuels, recovery from industrial by-products, methanol reforming, ammonia decomposition, and biological production (biohydrogen) (Fallah Vostakola et al., 2022).
5. Versatile utilization: Hydrogen energy finds wide application across numerous sectors, including industry, power generation, transportation, aerospace, and construction. It can serve as a fuel providing thermal energy for industrial processes like steelmaking, petroleum refining, and metallurgy, and it can also generate electricity via fuel cells to power vehicles such as cars (Fayaz et al., 2012; Show et al., 2012).

Hydrogen plays a triple core role in the energy supply system as an efficient and low-carbon secondary energy, a flexible and intelligent energy carrier, and a green and low-carbon industrial raw material. Its importance is becoming increasingly prominent as the global energy transition and low-carbon process (Sgobbi et al., 2016).

1.1.1 Efficient and low-carbon secondary energy

Hydrogen is a secondary energy source with high energy density, remarkable cleanliness, and high efficiency. It has strong electrochemical activity and can directly generate electricity through fuel cells. Its power generation efficiency is generally higher than that of traditional energy conversion methods, and it has broad application prospects in many fields.

Zero emissions in end-use applications: When hydrogen is used to generate electricity through fuel cells, only water is produced, with no emissions of pollutants such as carbon dioxide or particulate matter, making it an ideal clean energy source. Its energy density per unit mass is 3.3 times that of gasoline, significantly improving energy utilization efficiency.

Substitution in the transportation sector: Hydrogen fuel cell vehicles can replace traditional fuel vehicles, reducing oil consumption and exhaust emissions. Hydrogen powertrains have advantages in long-distance driving and rapid refueling, and the hydrogen storage system has a higher weight and energy density than batteries, making it suitable for heavy loads and long-distance transportation (Chanda & Bose, 2020; Xia & Niu, 2020).

Cogeneration: Hydrogen can be used in domestic cogeneration systems to provide both electricity and heat, reducing dependence on traditional energy sources (Hydrogen Council, 2017).

Backup power application: Hydrogen fuel cells can be used as backup emergency power supplies to ensure power supply security in scenarios such as 5G base stations and big data centers. As the technology matures, its application scenarios will further expand.

Blending into natural gas pipeline network: Hydrogen can be blended into natural gas pipelines (at a ratio of 5%–20%) and burned directly, serving as an alternative energy source to natural gas and reducing carbon emissions (Cristello et al., 2023; Messaoudani et al., 2016).

1.1.2 Flexible and intelligent energy carrier

The conversion of hydrogen produced through water electrolysis technology into other energy forms can enhance the utilization of renewable energy, provide long-term energy storage, optimize regional material and energy flows, and establish a new energy development model with multi-energy complementary.

Renewable energy integration: Hydrogen can be produced through water electrolysis technology, converting intermittent renewable energy sources such as wind power and photovoltaics into chemical energy for storage. Hydrogen is produced and stored during periods of excess electricity and then used to generate electricity, provide combined heat and power, or directly burn for energy through electrochemical reactions during periods of insufficient electricity and heat supply, thereby balancing the supply and demand of the power grid and heat network (Shabaneh et al., 2024).

Long-term energy storage: Hydrogen is the only medium that can achieve cross-seasonal and large-scale energy storage. Underground hydrogen storage technology is mature and can solve the problem of seasonal imbalances in renewable energy, meeting seasonal demand differences. Compared with other energy storage methods such as batteries, supercapacitors, and compressed air, hydrogen has a larger storage capacity and longer storage time (Abdin et al., 2021; Sharma & Ghoshal, 2015).

Multienergy complementary hub: Hydrogen can react with carbon dioxide to produce methane or coproduce with oxygen (as a by-product of water electrolysis), thereby constructing an "electricity-hydrogen-heat-gas" multienergy complementary system.

Energy redistribution: Hydrogen and its compounds have high energy density and are easy to transport, which helps to redistribute energy effectively and flexibly (Moradi & Groth, 2019).

1.1.3 Green and low-carbon industrial raw materials

Hydrogen has broad application prospects in the industrial sector, serving as a green and low-carbon industrial raw material to promote deep decarbonization in the industrial sector.

Deep decarbonization of industry: Hydrogen can replace reducing agents such as coke and natural gas, eliminating carbon emissions from processes such as iron and steel production.

Chemical raw material substitution: In the production of synthetic ammonia and methanol, replacing "gray hydrogen" (hydrogen produced from fossil fuels) with "green hydrogen" (hydrogen produced from renewable energy) can achieve deep decarbonization in the chemical industry (Qazi, 2022).

Industrial heating: Hydrogen can be burned in hydrogen burners or used in fuel cells, providing a zero-emission alternative for industrial heating. Whether it is the decarbonization treatment of low-grade or high-grade heat, hydrogen has significant advantages.

Improved flexibility of energy system: Hydrogen can be quickly started and stopped through electrochemical reactions to respond to fluctuations in renewable energy supply. In addition, electrolysis cells can also provide auxiliary services such as frequency regulation for the power grid.

Under the trend of global energy development toward cleanliness, low-carbonization and intelligence, the development of the hydrogen energy industry has become a key direction in the current global energy technology revolution (Dincer, 2002). Countries, such as Japan and Germany, have elevated hydrogen energy to a national energy strategy, issuing documents such as the "Hydrogen energy basic strategy" and the "Hydrogen energy economic development roadmap" clearly establishing hydrogen energy's central role in the energy transition. China has released the "Long-term development plan for the hydrogen energy industry (2021–2035)," proposing to establish a diversified hydrogen energy application ecosystem by 2035, with hydrogen energy technology fully adopted across industries, transportation, construction, and other fields (European Commission, 2020; Iida & Sakata, 2019; Solar Now, 2023).

1.2 Hydrogen production technology

Hydrogen energy is a secondary energy source that can be generated from renewable energy sources (such as hydropower, wind power, wave

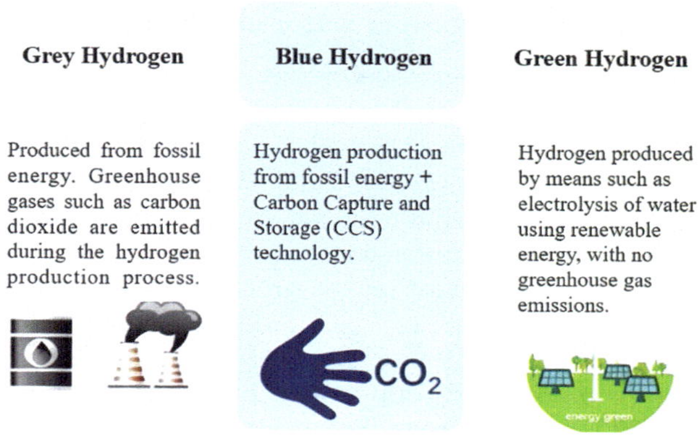

FIGURE 1.1 Hydrogen energy is divided into *gray hydrogen, blue hydrogen,* and *green hydrogen* according to the production method.

power, solar energy, biomass, and geothermal energy) as well as nonrenewable energy sources (such as coal, natural gas, and nuclear energy) (Muradov, 2017).

There are mainly three mature technologies for hydrogen production:

1. Hydrogen production from fossil fuels such as coal and natural gas, known as "gray hydrogen," which has a relatively low cost but high carbon emissions during the process (Kang et al., 2017).
2. Hydrogen production from industrial by-products such as coke oven gas, chlor-alkali tail gas, and propane dehydrogenation (PDH), known as "blue hydrogen," which can capture greenhouse gases and reduce carbon emissions during the production process.
3. Hydrogen production through electrolysis of water using alkaline electrolysis of water and proton exchange membrane (PEM) electrolysis of water, known as "green hydrogen," which has high purity and zero pollution, representing an important development direction for hydrogen production (Guban et al., 2020).

Technologies, such as solar photocatalytic water splitting and biological hydrogen production, are still in the experimental and development stages and have not yet reached the requirements for industrial-scale hydrogen production (Fig. 1.1).

Current mainstream hydrogen production technologies include water electrolysis, fossil fuel-based hydrogen production, industrial by-product hydrogen production, methanol reforming, ammonia decomposition, biological hydrogen production (Zhang et al., 2024a).

1.2.1 Hydrogen production via water electrolysis

The phenomenon of hydrogen generation via water electrolysis was first observed in 1978. In 1948 Zdansk and Lonza constructed the first pressurized water electrolysis system, which was the rudiment of industrial-scale water electrolysis plant (Leroy, 1983). Water electrolysis technology has evolved into a process characterized by technical reliability, operational simplicity, and low maintenance requirements. It operates without generating pollutants or CO_2 emissions, producing hydrogen with purity up to 99.8% (primarily containing H_2O and O_2 impurities). Furthermore the raw materials for system fabrication are readily accessible (Yu et al., 2018). The only drawback lies in its substantial electricity consumption, accounting for 85% of total production costs, which directly elevates the cost of hydrogen (Ding & Yi, 2006).

Electrolysis of water to produce hydrogen is a process in which water molecules are dissociated into hydrogen gas and oxygen gas through an electrochemical process under the action of direct current, which are, respectively, deposited at the anode and cathode poles. Based on electrolyte differentiation the technology divides into three categories: alkaline water electrolysis (AWE), PEM electrolysis, and solid oxide electrolysis cell (SOEC) (Raveendran et al., 2023). The basic principle of the three is the same, which is to prevent the free exchange of electrons in the oxidation–reduction reaction process and decompose the charge transfer process into electron transfer in the external circuit and ion transfer in the internal circuit, thereby achieving the generation and utilization of hydrogen gas. However, the electrode materials and electrolysis reaction conditions of the three are different (Yu et al., 2022).

1.2.1.1 Alkaline water electrolysis (AWE)

AWE for hydrogen production is the longest developed and most mature technology in industrialization at present. This process utilizes an aqueous potassium hydroxide (KOH) solution as the electrolyte with an asbestos separator membrane. Upon application of direct current, water molecules undergo electrochemical decomposition into hydrogen and oxygen gases. Critically the alkaline solution solely enhances ionic conductivity without participating in the electrode reactions (Zeng & Zhang, 2010). The operating principle is illustrated in Fig. 1.2.

The alkaline electrolyzer serves as the core equipment in AWE technology, comprising the cell stack, anode, cathode, and separator. Dominating China's electrolytic hydrogen market, this technology has achieved indigenous manufacturing capability for megawatt-scale units with hydrogen production capacities up to 1000 Nm^3/h (David et al., 2019). It offers process simplicity, operational convenience, carbon-free operation

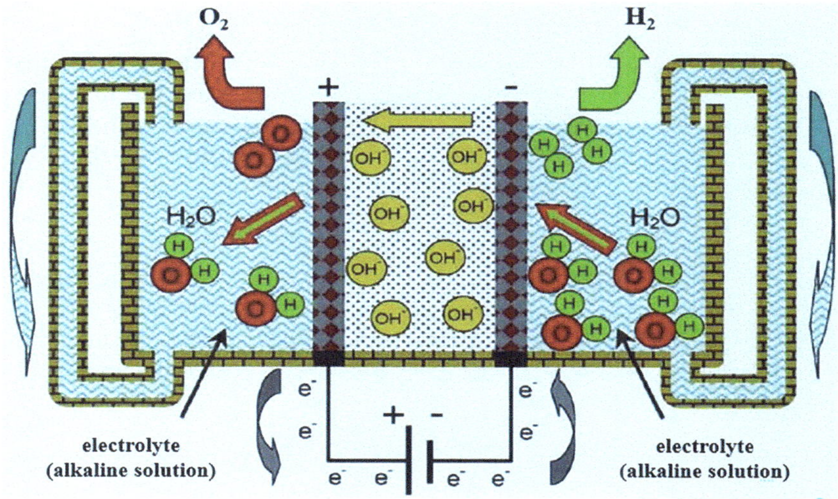

FIGURE 1.2 Mechanism of hydrogen production by alkaline water electrolysis.

(zero CO_2 emissions), and high-purity hydrogen output (typically 99%–99.99%) (Brauns & Turek, 2020). However, limitations include sluggish start–stop cycles and limited dynamic response, hindering compatibility with intermittent renewable sources such as photovoltaic and wind power. Additional purification stages (alkali removal and dehumidification) increase energy consumption and pose environmental risks. Nevertheless, its technological maturity and cost competitiveness render alkaline electrolysis ideal for large-scale stable power systems integrated with renewables, securing significant development potential.

1.2.1.2 Proton exchange membrane (PEM) electrolysis

PEM water electrolysis employs a proton-conductive polymer membrane as the solid electrolyte, eliminating liquid alkali solutions. Within the PEM electrolyzer, water undergoes electrochemical oxidation at the anode to generate oxygen (O_2), electrons (e^-), and protons (H^+). Oxygen is discharged at the anode, and electrons travel via an external circuit to the cathode, whereas protons migrate through the membrane to the cathode compartment. At the cathode, two protons combine with two electrons to form hydrogen gas (H_2) (Barbir, 2005). This operational principle is illustrated in Fig. 1.3.

The main components of the PEM electrolytic cell include membrane electrodes (PEM, catalytic layer, and diffusion layer), bipolar plate, epoxy resin plate, and end plate (Shiva Kumar & Himabindu, 2019). Among them

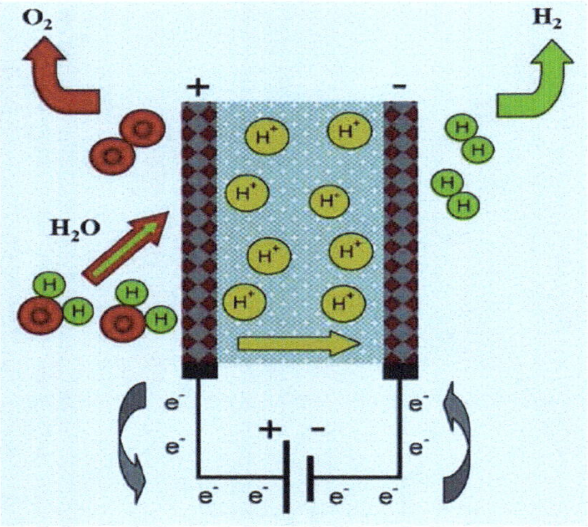

FIGURE 1.3 Mechanism of hydrogen production by proton exchange membrane (PEM) water electrolysis.

the end plate plays a fixed electrolytic cell assembly, guiding the transmission of electricity and water and gas distribution, etc.; the diffusion layer plays a role in collecting current and promoting the transmission of gas and liquid; the core of the catalytic layer is a three-phase interface composed of a catalyst, an electron-conducting medium, and a proton-conducting medium, which is the core place for the occurrence of electrochemical reactions; a perfluorosulfonic acid film is generally used as the PEM (a solid electrolyte) to isolate the gas generated by the anode and prevent the transmission of electrons and the transfer of protons (Carmo et al., 2013). PEM electrolytic cells have the advantages of high current density, high hydrogen purity, and fast response speed and are more suitable for combining with wind and light storage technology. However, as PEM electrolytic cells need to operate in a working environment with strong acids and high oxidation, PEM electrolytic cells are more dependent on precious metal materials such as iridium, platinum, and titanium, resulting in the current PEM electrolytic cells' cost. Improving the current density and reducing the load of precious metals iridium the localization and scale of PEMs, titanium felts, titanium plates, and balance of plant (BOP) auxiliary systems are also important ways to reduce costs in PEM electrolytic cell systems. Although PEM electrolytic cells account for a small proportion in the short term and have a smaller advantage over alkaline electrolytic cells, in the long term, there is a huge room for cost reduction in PEM electrolytic cells, and the proportion of applications in the future will expand rapidly.

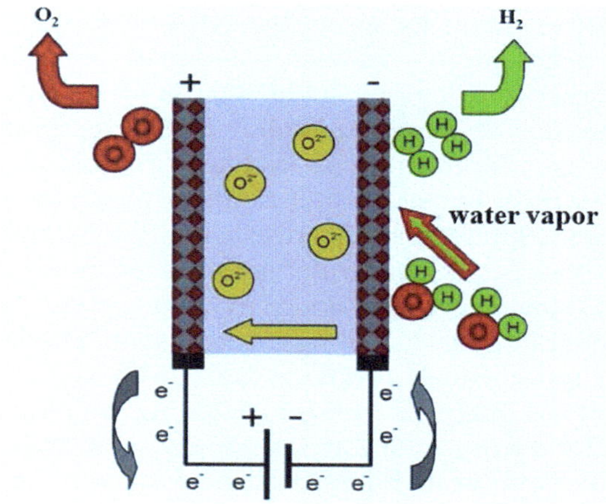

FIGURE 1.4 Mechanism of hydrogen production by SOEC water electrolysis (oxygen-ion-conducting).

1.2.1.3 *Solid oxide electrolysis cell (SOEC) technology for hydrogen production*

Solid oxide electrolysis cell (SOEC) technology is classified into two types based on operational principles: oxygen-ion-conducting and proton conducting (Kamlungsua et al., 2020). Because of higher technical requirements for proton-conducting systems, particularly significant material selection barriers, their development progress lags substantially behind oxygen-ion conducting types (Laguna-Bercero, 2012; Nechache & Hody, 2021). Commercialization efforts currently focus predominantly on oxygen-ion-conducting SOEC. In oxygen-ion-conducting systems, oxygen ions migrate through the electrolyte. Unlike proton-conducting variants, these systems introduce water vapor at the cathode (hydrogen electrode), where water molecules accept electrons to generate hydrogen gas while releasing oxygen ions. These ions transport through the electrolyte to the anode, oxidizing to form oxygen gas (Kim et al., 2018). This operational principle for oxygen-ion-conducting SOEC is illustrated in Fig. 1.4.

Solid oxide electrolysis cell (SOEC) technology fundamentally differs from AWE and PEM electrolysis by utilizing solid oxides as the electrolyte material and operating at 600°C~1000°C, significantly enhancing electrochemical performance and production efficiency (Table 1.1). SOEC electrolyzers employ nonprecious metal catalysts: the cathode material uses porous metal–ceramic Ni/YSZ, the anode material adopts perovskite oxides, and the electrolyte employs YSZ oxygen-ion conductors, with the all-ceramic structure preventing material corrosion issues (Zhang et al., 2020a). The high-temperature and high-humidity operating environment restricts material selection for electrolyzers to those with high stability, durability, and degradation resistance, consequently limiting application scenarios and large-scale deployment of SOEC hydrogen production technology (Chelmehsara & Mahmoudimehr, 2018). Durability remains the foremost challenge; thermochemical cycling, particularly during system shutdown and startup, accelerates aging and reduces operational lifespan. SOEC technology demonstrates the lowest energy consumption and highest conversion efficiency among water-splitting technologies. Research in science confirms stable operation under dynamic power conditions without significant degradation, indicating strong potential for large-scale, low-cost hydrogen supply (Fallah Vostakola et al., 2023).

Water electrolysis offers advantages including simple operation, high reliability, small footprint with low investment, and high hydrogen purity. However, because of its high electricity consumption, it is typically adopted for applications with moderate hydrogen demand, especially in sectors like semiconductors requiring ultrahigh purity. Among the three electrolysis technologies (AWE, PEM, SOEC), AWE remains the most mature and widely used method. Yet it employs highly corrosive chemicals (KOH/NaOH), posing safety risks and relatively high operational costs. Consequently, PEM and SOEC featuring enhanced stability and efficiency are undergoing rapid development. Practical hydrogen production units utilizing both are already operational, though high electrolyzer costs currently limit large-scale industrial deployment. Nevertheless, PEM and SOEC represent the definitive future direction for electrolytic hydrogen production (Guo et al., 2019; Sebbahi et al., 2022).

1.2.2 Hydrogen production from fossil fuels

Among the various methods of hydrogen production, fossil fuels, such as coal and natural gas, are still the most economically viable and dominant hydrogen production materials that can be scaled up in industry at present (Zhang et al., 2016). The proportion of hydrogen sources in China in 2020 is shown in Fig. 1.5.

Although fossil fuel reserves are limited and the hydrogen production process can cause environmental pollution, the technology and processes for hydrogen production from fossil fuels are relatively mature and cost-effective, making it the preferred method for large-scale industrial hydrogen production at present. The representative fossil energy hydrogen production processes include coal gasification and natural gas reforming.

1.2.2.1 Coal to hydrogen

Coal-based hydrogen production involves the primary use of coal in the form of a coal–water slurry or coal powder, which reacts with a gasifying agent (steam/O_2) in a gasifier at temperatures above 1000°C to

TABLE 1.1 Comparison of hydrogen production technologies by water electrolysis.

Item	Alkaline water electrolysis	PEM water electrolysis	SOEC water electrolysis
Maturity	Large-scale application	Small-scale application	Not yet commercialized
Operating temperature (°C)	70–90	70–80	600–1000
Current density (A/cm^2)	0.2–0.4	1.0–2.0	1.0–10.0
Hydrogen production scale of single device (Nm^3/h)	0.5–1000	0.01–500	/
Energy consumption of electrolyzer (kWh/Nm^3)	4.5–5.5	3.8–5.0	2.6–3.6
Conversion efficiency (%)	60–75	70–90	85–100
System lifespan	Reached 10–20 years	Reached 10–20 years	/
Start-stop speed	Hot start-stop: minute level Cold start-stop: >60 minutes	Hot start-stop: second level Cold start-stop: 5 minutes	Slow start-stop
Dynamic response capability	Relatively strong	Strong	Relatively weak
Power quality requirement	Stable power	Stable or fluctuating power	Stable power

(continued on next page)

TABLE 1.1 Comparison of hydrogen production technologies by water electrolysis—cont'd

Item	Alkaline water electrolysis	PEM water electrolysis	SOEC water electrolysis
Load regulation range	15%–100% rated load	0%–160% rated load	/
System operation and maintenance	Complex, high cost	Simple, low cost	No demand
Floor area	Large	Small	/
Electrolyzer price	280–420 USD/kW (China) 840–1120 USD/kW (other countries)	980–1680 USD/kW	/
Characteristics	Mature technology, low cost, easy to realize large-scale application, large power consumption, stable power supply	Small floor area, high adaptability to intermittent power sources, easy to combine with renewable energy, high equipment cost	High-temperature electrolysis has low energy consumption, and nonnoble metal catalysts can be used, but there are problems with the stability of electrode materials, and additional heating is required
Combination with renewable energy	Suitable for power systems with large installed capacity of stable power sources	Suitable for renewable energy power generation systems with large volatility	Suitable for solar thermal power generation systems that produce high-temperature and high-pressure steam

PEM, *proton exchange membrane.*

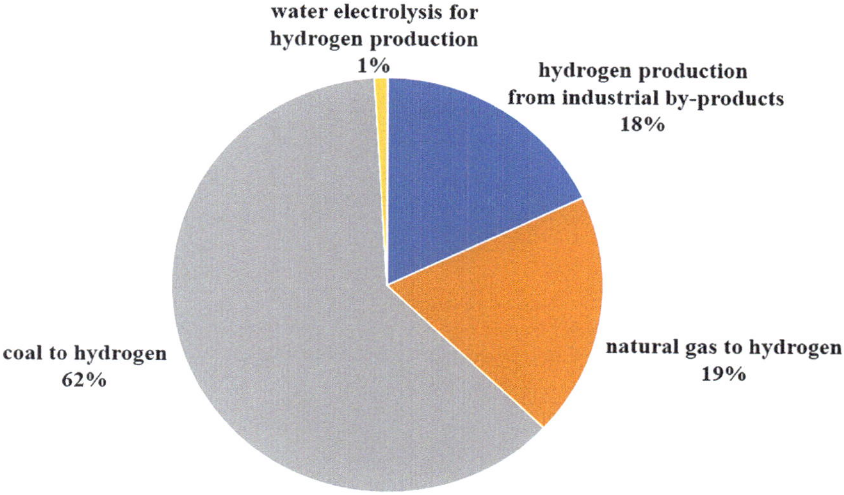

FIGURE 1.5 The proportion of hydrogen sources in China in 2020.

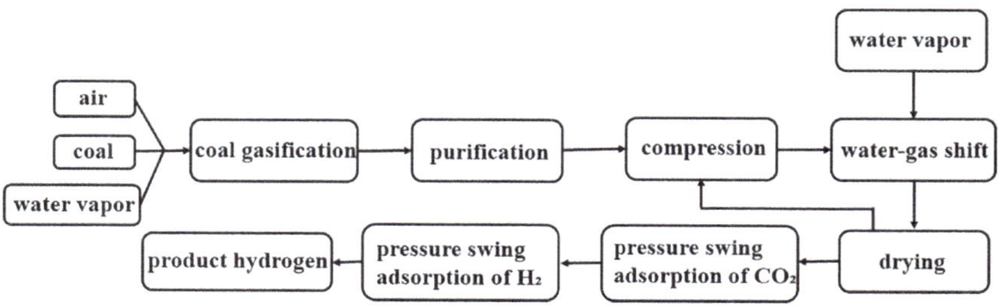

FIGURE 1.6 Coal-to-hydrogen process flow.

generate syngas (H_2 + CO). After separating CO from H_2, CO is converted into H_2 and CO_2 through water–gas shift (WGS) reaction. The process further includes the removal of acidic gases (CO_2 + SO_2) and hydrogen purification using pressure swing adsorption (PSA) technology, resulting in high-purity hydrogen (Muellerlanger et al., 2007). The typical coal-to-hydrogen process flow is illustrated in Fig. 1.6. The traditional coal-based hydrogen production process is characterized by its mature technology, low raw material costs, and large-scale facilities. However, it also has many shortcomings, including a complicated production process and complex equipment. Moreover, the hydrogen production process emits ash, sulfur-containing substances, and a significant amount of greenhouse gas carbon dioxide, thereby impacting the environment (Holladay et al., 2009; Leroy, 1983).

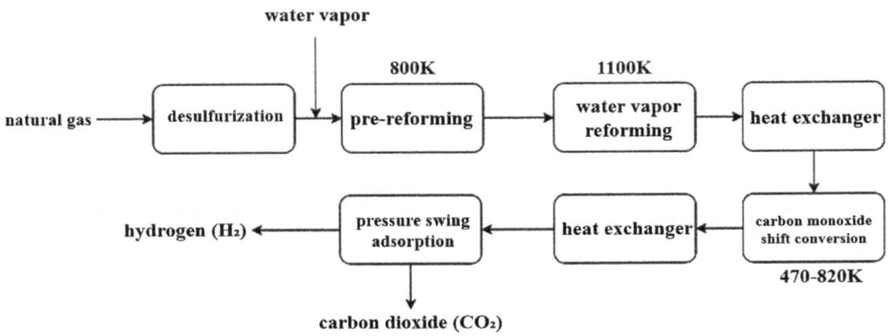

FIGURE 1.7 The process flow of hydrogen generation from natural gas (SMR). *SMR*, Steam methane reforming.

1.2.2.2 Hydrogen production from natural gas

The predominant pathways for hydrogen generation from natural gas include steam methane reforming (SMR), partial oxidation (POX) of natural gas, and autothermal reforming (ATR) (Boretti & Banik, 2021). The industry consensus maintains that SMR represents the optimal method for industrial hydrogen production using natural gas feedstock.

The steam reforming process involves the reforming reaction of methane and other hydrocarbons with steam in the presence of a catalyst under high-temperature conditions, producing a mixture of H_2, CO, and other gases. This reaction is strongly endothermic, requiring external heat supply (typically from natural gas combustion) (Lee et al., 2021; Shagdar et al., 2020). The SMR reaction requires operation under elevated temperatures (750°C–920°C) and moderate pressures (2–3 MPa), with Ni/Al_2O_3 being the catalyst. In industrial production the molar ratio of steam to methane is typically 3–5, and the resulting H_2/CO ratio is approximately 3. The syngas produced from methane steam reforming is then fed into a WGS reactor, where high- and low-temperature shift reactions convert CO into CO_2 and additional H_2, thereby increasing the hydrogen yield. The basic process flow is illustrated in Fig. 1.7. Natural gas steam reforming represents a mature hydrogen production technology that is widely deployed for industrial-scale generation of synthesis gas, high-purity hydrogen, and ammonia feedstock. As the predominant commercial hydrogen production method, SMR nevertheless faces the key technical challenge of high separation costs associated with hydrogen purification from CO-rich gas mixtures.

Although substantial investments are being made by numerous countries and oil companies in renewable energy-based green hydrogen production and traditional hydrogen production coupled with carbon capture

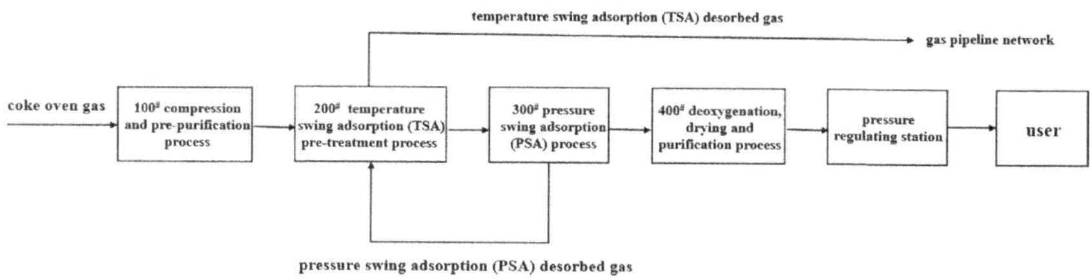

FIGURE 1.8 Diagram of hydrogen production process from coke oven gas.

and storage (CCS) to produce blue hydrogen, these alternatives currently remain more costly than fossil fuel-derived hydrogen production. Until more advanced and mature hydrogen production methods emerge, fossil fuel-derived hydrogen will continue to play a significant role in the coming decades (Sadeq et al., 2024).

1.2.3 Hydrogen production from industrial by-products

Large quantities of by-products and emissions generated in major industrial processes such as oil refining, chemical engineering, and coking mostly contain hydrogen. Hydrogen production from industrial by-products mainly includes hydrogen production from coke oven gas, hydrogen production from chlor-alkali tail gas, and hydrogen production from by-products of light hydrocarbon cracking.

1. Hydrogen production from coke oven gas: Coke oven gas is a by-product of coking, and its yield and composition vary depending on the quality of coking coal and coking processes. Generally, 300–350 m^3 (standard state) of coke oven gas can be produced per ton of dry coal. The compressed coke oven gas first removes hydrocarbons with C_5 and above, naphthalene, and trace high-boiling-point impurities through a temperature-swing adsorption process to purify the coke oven gas. Then, all impurities except oxygen are removed via a pressure-swing adsorption process to obtain hydrogen with a purity of over 99.9%. Oxygen is further removed through catalytic reactions to make the oxygen content $\leq$2 ppm, and after drying, the product hydrogen with a purity of over 99.999% and a dew point ≤ -60°C is obtained (Yang et al., 2010). The entire process is divided into four steps, as shown in Fig. 1.8. Hydrogen production from coke oven gas can effectively solve the problem of coke oven gas disposal, obtain clean energy, turn waste into treasure, bring economic benefits to enterprises, reduce greenhouse gas emissions, and contribute to sustainable development (Chen et al., 2011).

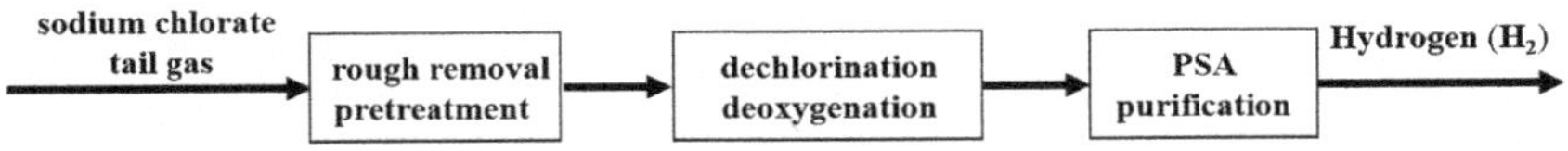

FIGURE 1.9 Hydrogen production process from by-product gas of sodium chlorate. *PSA*, Pressure swing adsorption.

2. Hydrogen production from chlor-alkali tail gas: The chlor-alkali industry uses industrial salt (sodium chloride) as raw material and electrolyzes industrial brine to produce sodium hydroxide, hydrochloric acid, chlorine gas, and hydrogen. For example, electrolytic plants for caustic soda and sodium chlorate generate a large amount of hydrogen-containing tail gas during the electrolysis process (Paulu et al., 2024). The hydrogen content in sodium chlorate tail gas is as high as over 95%, which contains oxygen, chlorine gas, water vapor, etc. As can be seen from the composition of sodium chlorate tail gas, the main impurity components in the tail gas are oxygen and chlorine gas. The key to treatment lies in deoxygenation, dechlorination, and PSA separation and purification, with the process flow shown in Fig. 1.9 (O'Brien et al., 2005). Hydrogen production from chlor-alkali tail gas has advantages such as low purification difficulty, low impurity content, and effective utilization of hydrogen. Considering that the purity of by-product hydrogen is over 99% before purification, with main impurities being oxygen, nitrogen, and water vapor, the production cost of high-purity hydrogen under this process is only 0.18–0.21 USD/Nm^3. Compared with other preparation methods, its cost and environmental protection advantages are prominent (Erden & Karakilcik, 2024).
3. Hydrogen production from by-products of light hydrocarbon cracking.

There are two main routes for hydrogen production from light hydrocarbon cracking: PDH and ethane cracking. Among them, ethane cracking technology is mature, but the incremental growth of by-product hydrogen from ethane cracking is relatively small because of factors such as limited sources of ethane resources. PDH is one of the important methods for producing propylene. Propane generates propylene through dehydrogenation under the action of a catalyst, with hydrogen produced as a by-product in the process (propane → propylene + hydrogen), and the entire process flow is shown in Fig. 1.10. The hydrogen content in PDH products ranges from 60% to 95%, and high-purity hydrogen that meets fuel cell applications can be obtained through purification technology. The by-product hydrogen from PDH plants has high purity and low purification difficulty (Akah et al., 2019). Compared with other hydrogen production methods, PDH has advantages such as a short production process, small floor area, short construction cycle, low cost, low energy consumption, low carbon emissions, high hydrogen purity, and large hydrogen production capacity, making it a

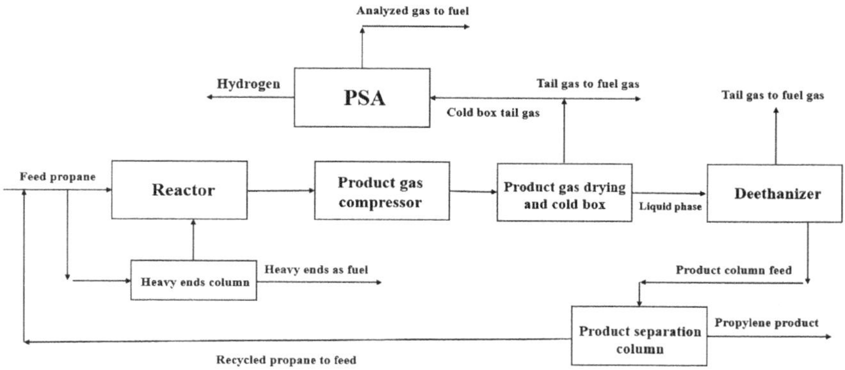

FIGURE 1.10 Process flow of propane dehydrogenation (PDH). *PSA*, Pressure swing adsorption.

method with great hydrogen supply potential in China's current hydrogen production market.

Industrial by-product gases have large emissions and wide sources and are regarded as one of the optimal ways to produce low-cost hydrogen at present. However, they have drawbacks: limited by the supply of raw materials, production sites must be close to raw material enterprises; the raw materials are polluting, and excess harmful gases need to be recovered and treated (Muradov, 2008). The cost of hydrogen production from industrial by-products is difficult to calculate separately. Compared with other methods, hydrogen production from industrial by-products should be the most price-advantaged hydrogen production method besides coal-based hydrogen production and also one of the effective ways for industrial hydrogen production at this stage (Taghvaei et al., 2013).

1.2.4 Hydrogen production via methanol reforming (methanol to hydrogen)

The primary methods for hydrogen production from methanol mainly include methanol decomposition, methanol POX, and methanol steam reforming (MSR).

1.2.4.1 Hydrogen production by methanol decomposition

The fundamental principle of methanol decomposition for hydrogen production involves the thermal vaporization of water and methanol, followed by the cracking of methanol to produce hydrogen. In this process the two most basic raw materials are demineralized water and methanol. With the hydrogen production temperature ranging between 220°C and

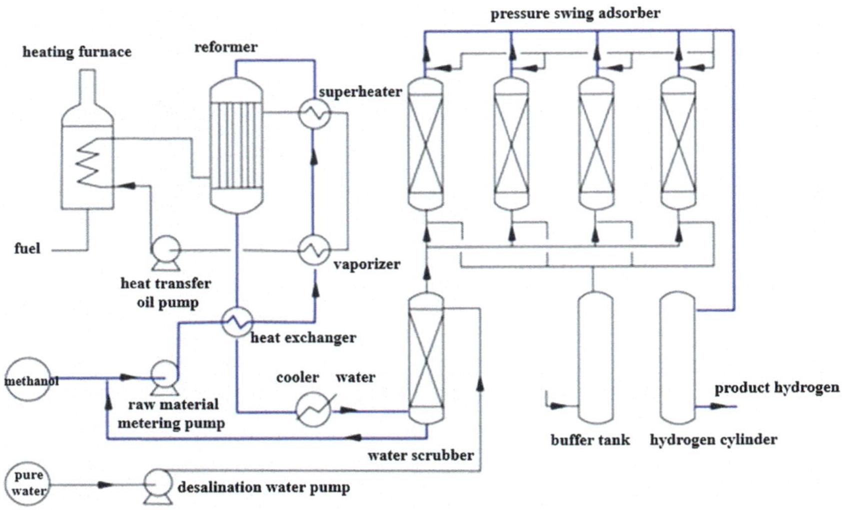

FIGURE 1.11 The process flow of methanol decomposition coupled with PSA. *PSA*, Pressure swing adsorption.

280°C, highly active and efficient catalysts are required during the reaction (Rico et al., 2010). Typically, copper-based catalysts, such as Cu–Mg–Al, are employed (Chang et al., 2009; Xu et al., 2005). Following the cracking reaction, the resultant gas mixture is processed via PSA, with process conditions being modulated as a function of the required hydrogen purity specification. This method enables hydrogen production with purity levels surpassing 99.9%. In industrial hydrogen production, methanol decomposition technology is well-established and has been scaled up, making it a common method for producing syngas and pure hydrogen. The process flow of methanol decomposition coupled with PSA is illustrated in Fig. 1.11. From an economic perspective, methanol decomposition demonstrates process simplicity, straightforward operation, and competitive capital costs. Environmentally, this technology exhibits superior performance through reduced by-product generation, material recyclability, and minimal environmental impact, eliminating the need for purification while maintaining low energy consumption. Technically the mature process architecture ensures consistent production of high-purity hydrogen (Sá et al., 2010).

1.2.4.2 Hydrogen production by partial oxidation of methanol

Hydrogen production by partial oxidation of methanol (POM) is a technology that generates hydrogen through incomplete combustion of

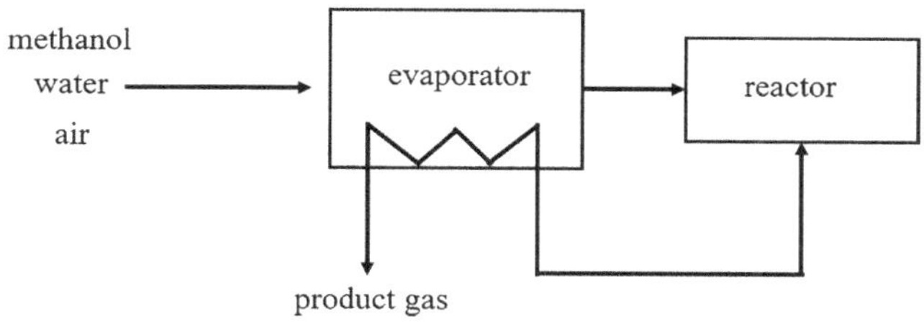

FIGURE 1.12 Process flow of hydrogen production by methanol partial oxidation.

methanol with oxygen. This process primarily involves controlled chemical reactions between methanol and oxygen to achieve POX, yielding hydrogen (Hohn & Lin, 2009). The procedure encompasses multiple reactions, including methanol decomposition, WGS, steam reforming, and POX. Notably, MSR is highly endothermic, requiring a stable external heat supply to maintain high hydrogen yield, whereas oxidative reforming is exothermic, necessitating efficient heat dissipation. As illustrated in Fig. 1.12 the POM process involves preheating methanol, water, and air to a specific temperature in an evaporator before entering the reactor. The product gas undergoes heat recovery in a heat exchanger and is purified in PSA to obtain high-purity hydrogen (Chang et al., 2006, 2008). Compared to other methanol-based hydrogen production methods, POM exhibits lower technological maturity and has yet to achieve large-scale application.

1.2.4.3 *Hydrogen production by methanol steam reforming*

MSR produces hydrogen by catalytic cracking of a methanol–water vapor mixture under pressure at 170°C–300°C, yielding hydrogen and carbon dioxide (Palo, 2007). As depicted in Fig. 1.13 the process involves mixing methanol and demineralized water in a fixed ratio, followed by vaporization and superheating in a gasification unit. The vapor undergoes catalytic cracking and conversion reactions, generating syngas (CO_2, H_2, and CO). After cooling via heat exchange and condensation, unreacted methanol is recovered, and the gas is purified in a PSA system to obtain high-purity hydrogen (Sanz et al., 2016). MSR offers advantages such as operational simplicity, mild reaction conditions, and easily separable by-products. With a flexible production capacity ranging from 10 to 10,000 Nm^3/h, it is a mature and widely adopted hydrogen production method.

1.2.5 Ammonia decomposition for hydrogen production (ammonia-hydrogen)

Ammonia decomposition for hydrogen production is a technology that utilizes the ammonia decomposition reaction to generate hydrogen. It can

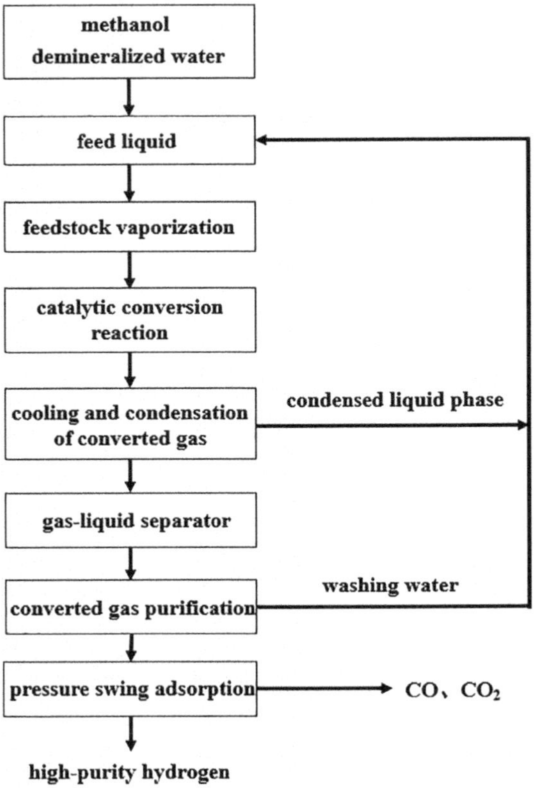

FIGURE 1.13 Process flow of methanol steam reforming for hydrogen production.

be categorized into thermal cracking of ammonia for hydrogen production technology, plasma-driven ammonia cracking technology, and electrocatalytic ammonia oxidation decomposition technology.

1.2.5.1 *Thermal cracking of ammonia for hydrogen production*

Thermal cracking of ammonia is a hydrogen production method that employs the ammonia decomposition reaction. This process involves heating liquid ammonia to 800°C–850°C in the presence of nickel-based catalysts, resulting in a hydrogen-nitrogen mixture containing 75% H_2 and 25% N_2 (Nguyen et al., 2025; Yousefi Rizi & Shin, 2022). As illustrated in Fig. 1.14 the thermal cracking process represents the reverse reaction of ammonia synthesis and is endothermic in nature. The conversion rate of ammonia is thermodynamically constrained under certain conditions. Although the

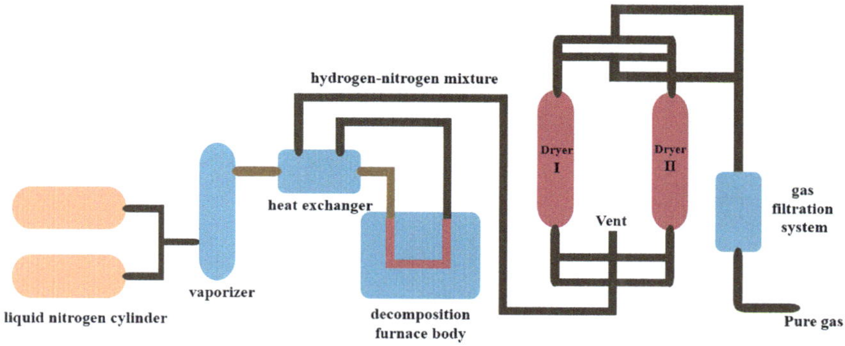

FIGURE 1.14 Process flow of hydrogen production by ammonia thermal cracking.

equilibrium conversion rate exceeds 99% at 450°C, reaction kinetics limitations restrict the reaction rate, necessitating temperatures of 700°C or higher for complete conversion. Consequently, continuous external heat supply is required to enhance thermal catalytic conversion efficiency, leading to substantial energy consumption and higher costs. Industrial applications typically employ catalytic technologies to improve production efficiency, with Ru-based catalysts demonstrating higher activity for ammonia thermal cracking, whereas transition metal Ni- and Fe-based catalysts offer better economic viability. Thermal cracking of ammonia for hydrogen production represents a significant hydrogen generation method, offering advantages in efficiency, cleanliness, and renewability. However, challenges remain in catalyst stability and selectivity, energy consumption, and cost, necessitating further research and development to enhance efficiency and economic viability, thereby promoting its application in the energy sector (Yuan et al., 2023).

1.2.5.2 Plasma-driven ammonia cracking technology

Plasma-driven ammonia cracking is an emerging technology that utilizes the high energy of plasma to decompose ammonia molecules (Spatolisano et al., 2023). Although the fundamental chemical principles of hydrogen production through plasma excitation are similar to traditional methods, the difference lies in the active species that initiate the chemical reactions: Traditional methods employ catalysts, whereas plasma methods utilize high-energy electrons and free radicals. This technology offers advantages in efficiency, low energy consumption, and pollution-free operation, though it remains in the laboratory research phase (Bang et al., 2023).

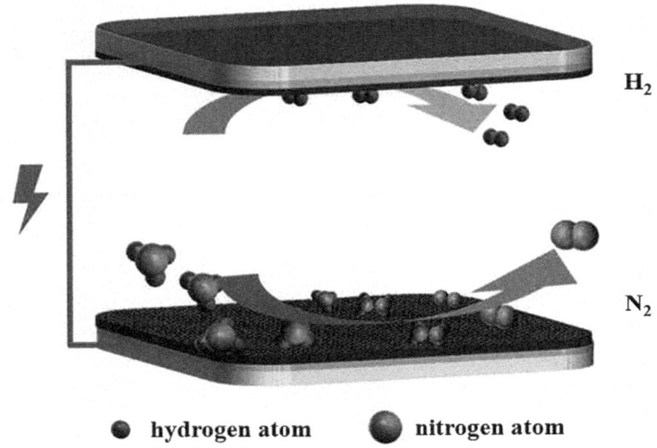

FIGURE 1.15 Schematic diagram of electrocatalytic decomposition of ammonia in alkaline aqueous electrolyte.

1.2.5.3 Electrocatalytic ammonia oxidation decomposition technology

The electrocatalytic decomposition of ammonia consists of two electrode reactions: the hydrogen evolution reaction (HER) at the cathode and the ammonia oxidation reaction (AOR) at the anode (Jing et al., 2022). Fig. 1.15 illustrates the schematic of electrocatalytic ammonia decomposition in alkaline aqueous electrolyte. Theoretically the external voltage required for ammonia electrolysis is significantly lower than that for water electrolysis. Specifically, ammonia electrolysis requires 33 MJ of energy to produce 1 kg of H_2, compared to 180 MJ for water electrolysis, though both processes require additional overpotential to overcome kinetic barriers. Electrocatalytic ammonia oxidation decomposition technology offers advantages in efficiency, low energy consumption, and pollution-free operation, though challenges remain in catalyst lifespan and cost (Zhang et al., 2021b; Zhong et al., 2013).

1.2.6 Biohydrogen production

Biohydrogen production is the process in which energy stored in organic compounds in nature is converted into hydrogen energy through microbial metabolism or photosynthesis. It can be divided into photolysis of water for hydrogen production, photo fermentation for hydrogen production, dark fermentation for hydrogen production, combined dark–photo fermentation for hydrogen production, and so on (Argun & Kargi, 2011; Azwar et al., 2014; Ren et al., 2011). The specific mechanisms, advantages, and disadvantages of each hydrogen production method are described in the following text.

1.3 Biohydrogen energy

In the context of global energy transition, hydrogen energy has become one of the most promising future energy sources because of its advantages of cleanliness, efficiency, and sustainability (Li & Zheng, 2024). Biohydrogen, as a unique member of the hydrogen energy family, is gradually emerging and receiving increasing attention (Meena & Patane, 2025). Biohydrogen, in simple terms, refers to hydrogen gas produced through biological processes or reforming processes (Agyekum & Odoi-Yorke, 2024). Unlike traditional fossil fuel-based hydrogen production the production process of biohydrogen mainly relies on the metabolic activity of microorganisms and the technology of biomethanol reforming for hydrogen production. It utilizes abundant biomass resources in nature and converts them into hydrogen under relatively mild conditions (Ananthi et al., 2024).

The production of biohydrogen mainly occurs through microbial fermentation, photosynthesis of algae, and the hydrogen production technology from biomethanol reforming (Zhao et al., 2024). Microbial fermentation refers to the decomposition and metabolism of organic substances by anaerobic bacteria or facultative anaerobic bacteria in an anaerobic or microaerobic environment, thereby producing hydrogen (Zhang et al., 2024b). These organic substances have a wide range of sources, including various agricultural wastes (such as straw and manure), industrial organic wastewater, food processing wastes, and so on (Prado & Rial, 2025). For example, in anaerobic fermentation tanks, bacteria such as *Clostridium butyricum* and *Clostridium pasteurianum* can gradually degrade complex organic macromolecules like carbohydrates, proteins, and fats, ultimately converting them into hydrogen, carbon dioxide, as well as small amounts of organic acids and alcohols. This method of producing biohydrogen from waste not only achieves energy regeneration but also effectively addresses environmental pollution issues, which can be described as killing two birds with one stone. Algae photosynthesis is the ability of some photosynthetic microorganisms, such as cyanobacteria (cyanobacteria) and green algae, to use light energy to split water to produce hydrogen and oxygen (Pathy et al., 2022). In this process, special enzyme systems in photosynthetic microorganisms, such as hydrogenase, play a key role. Hydrogenase can catalyze the reduction of protons to hydrogen and, at the same time, transfer the electrons produced by photosynthesis to protons, realizing the conversion of light energy to chemical energy. Compared to microbial fermentation, photosynthesis produces hydrogen more cleanly and directly uses solar energy, making it a promising and sustainable way to produce hydrogen. However, at present, the efficiency of hydrogen production by photosynthesis is still relatively low, and how to improve the hydrogen production efficiency of photosynthetic microorganisms is a difficult problem for researchers to overcome (Putatunda et al., 2023). Biomethanol reforming technology refers to the process of catalyzing the conversion of biomethanol into hydrogen under the action of a catalyst under certain temperature and pressure conditions (Garcia et al., 2021). This method has the advantages of fast hydrogen production, high hydrogen content, and fewer by-products, which can enable the on-demand and quantity production of hydrogen energy. However, this method requires continuous energy and is greatly affected by the catalyst and the structure of the reforming hydrogen production reactor, so the research and development of this technology mainly focuses on the structural design of the reforming reactor and the enhancement of the catalyst (Lu et al., 2022). Therefore, to improve the efficiency of hydrogen production from biomethanol catalytic reforming, it is necessary to select the appropriate reforming reactor structure and catalyst type (Tang et al., 2024). It can be seen that the advantages of biohydrogen are numerous. First of all, it is a clean energy source; the combustion product is only water. It does not produce carbon dioxide, nitrogen oxides, and other pollutants; there is almost zero burden on the environment; and it helps to alleviate global climate change and air pollution problems (Zhao et al., 2024). Second, biohydrogen has a wide range of raw sources and is renewable, and the abundant biomass resources on the earth provide a solid guarantee for its large-scale production (Kim et al., 2021). In addition, the production process of biohydrogen is mild, usually at room temperature and pressure, without the need for harsh conditions such as high temperature and high pressure, which not only reduces the cost and energy consumption of production equipment but also improves the safety and stability of the production process (Zheng et al., 2022).

Despite the many advantages of biohydrogen, there are still some challenges to its large-scale commercial application (Rani et al., 2024). For example, the efficiency of biohydrogen production needs to be further improved, and the production cost is relatively high, which limits its competitiveness in the market. In addition, the storage and transportation technology of biohydrogen is not yet perfect, and how to store and transport biohydrogen safely and efficiently is also an important problem that needs to be solved (Ariful Islam et al., 2025). However, with the continuous progress in biotechnology, materials science, and other fields, these problems are gradually being overcome. It is believed that in the near future, biohydrogen will play an important role on the global energy stage and inject new vitality into the sustainable development of mankind.

1.4 Current main preparation methods for biohydrogen

At present, the production methods of biohydrogen mainly include biohydrogen production technology and biomethanol reforming hydrogen production technology (Lu et al., 2023; Shim et al., 2024). Biohydrogen production technology is a process that uses the energy absorbed by nature to absorb solar energy and store it in its own organic compounds through the metabolic activities and photosynthesis of hydrogen-producing microorganisms, which only needs to be carried out under normal temperature and pressure conditions (Xue et al., 2024). The biomethanol reforming technology is a process that converts biomethanol into hydrogen under the action of a catalyst at a certain temperature and pressure (Hu et al., 2024). The essence of the two hydrogen production technologies is to convert the solar energy stored in the biomass into hydrogen energy, and the biological hydrogen production technology is divided into dark fermentation biological hydrogen production, photo fermentation biological hydrogen production, dark–photo combined fermentation biological hydrogen production, and so on (Yue et al., 2021). A light source is required for photo-fermentation biological hydrogen production, whereas dark fermentation biological hydrogen production does not need to provide a light source.

1.4.1 Photocatalytic water splitting for hydrogen production

Photobiological hydrogen production is a process in which algae and cyanobacteria use solar energy as the primary energy source and water as the raw material. Through photosynthesis and their unique hydrogen-producing enzyme systems, they decompose water into hydrogen and oxygen gases (Shi et al., 2019; Wang et al., 2020). The mechanism of direct biophotolysis in algae is illustrated in Fig. 1.16. Light-harvesting pigments absorb photons and transfer the energy to photosystem II (PSII), which catalyzes water splitting into protons (H^+), molecular oxygen (O_2), and released electrons. The generated oxygen diffuses into mitochondria for respiratory consumption. Protons are translocated into the stromal compartment via adenosine triphosphate (ATP) synthase, whereas electrons traverse an electron transport chain to reach photosystem I (PSI). Under light irradiation, PSI photo excites these electrons and transfers them to ferredoxin (Fd), ultimately delivering them to hydrogenase. This enzyme catalyzes the combination of stromal protons with incoming electrons to form molecular hydrogen (H_2) (Ramachandran & Menon, 1998). Gaffron & Rubin (1942) reported the phenomenon of hydrogen production by algae via photosynthesis in 1942. Antal et al. (2003) investigated the relationship between PSII of sulfur-starved *Chlamydomonas reinhardtii* and hydrogen

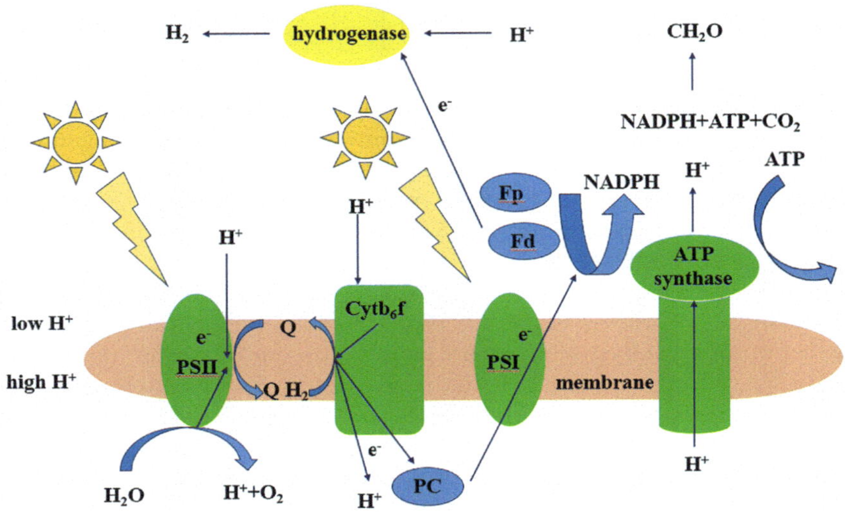

FIGURE 1.16 Mechanism of hydrogen production by photolysis of water.

production from algae. Although direct biophotolysis of water for hydrogen production enables the conversion of light energy into hydrogen using water as the sole feedstock, numerous technical challenges remain to be addressed. For instance, hydrogen generation is concomitant with oxygen evolution, and oxygen exerts an inhibitory effect on the hydrogen-producing process of algae. In addition, low light conversion efficiency and the complexity of photobioreactors are also major constraints limiting the development of microalgal hydrogen production. These existing issues render the technology of water photolysis for hydrogen production difficult to achieve large-scale promotion and application (Çelik & Yıldız, 2017).

1.4.2 Photofermentation of biological hydrogen production

Photofermentation biological hydrogen production refers to the process in which photosynthetic bacteria metabolize organic matter and release hydrogen under anaerobic and light conditions (Sagir & Alipour, 2021). Photosynthetic bacteria are prokaryotes of the primitive photosynthesis system, and the most studied photosynthetic bacteria mainly include *Rhodobacter sphaeroides, Rhodobacter capsulatus, Rhodopseudomonas palustris, and Rhodospirillum rubrum* (Fan et al., 2025; Wang et al., 2022). As early as the 40s of the 20th century, people began to carry out basic research on photosynthetic bacteria. Because of the increasingly severe

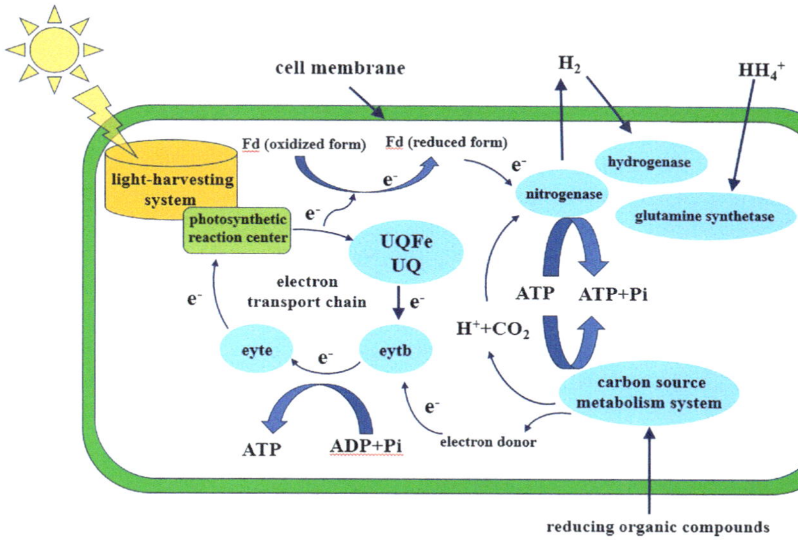

FIGURE 1.17 Mechanism of hydrogen production by photo-fermentation.

global energy and environmental problems in recent years, many experts and scholars have conducted further research on the mechanism and technical route of photofermentation biological hydrogen production and have made a series of research progress (Wei et al., 2021). Lu et al. (2016) investigated the photo-fermentative hydrogen production capacity of photosynthetic bacteria using apple waste as the substrate, and the hydrogen yield under the optimal process conditions reached 111.85 ± 1 mL H_2/g total solid (TS). Li et al. (2017) optimized the photo-fermentative hydrogen production process with phoenix tree leaves as the raw material, achieving a hydrogen yield of 64.1 mL H_2/g TS under the optimal conditions.

As illustrated in Fig. 1.17 the hydrogen production mechanism of photosynthetic bacteria involves only one photosynthetic center (PSI), which exclusively carries out anoxygenic photosynthesis (Ming, 1999). The carbon source metabolism system in photosynthetic bacteria can utilize organic matter to generate ATP, H^+, CO_2, and electron donors. Electrons are transferred from these electron donors to cytochromes and subsequently enter the electron transport chain. Upon absorption of light energy, electrons are converted into high-energy electrons. A portion of these high-energy electrons circulates within the electron transport chain, reducing their energy level to produce ATP, whereas another portion is transferred to nitrogenase via ferredoxin. Nitrogenase utilizes high-energy electrons, H^+, and ATP for nitrogen fixation and concurrent hydrogen production (Meyer & Cusanovich, 2003).

In the field of hydrogen production by photofermentation, many experts at home and abroad have conducted in-depth research. The research of foreign experts mainly focuses on the breeding of photosynthetic bacterial strains, the improvement of photoconversion efficiency, and the optimization of the hydrogen production process (Cheng et al., 2022b; Sivaramakrishnan et al., 2021; Zhang et al., 2024b). China has mainly studied the improvement of light conversion efficiency, the design and operation of large and medium-sized reactors, etc., and the research of large- and medium-sized reactors is the most representative of Henan Agricultural University, which has successively developed 5 and 10 m^3 reactors and been able to operate normally, realizing the light source from sequential batch experiments to continuous experiments, from incandescent lamps to fiber introduction of sunlight to provide light sources, and the experimental research on biological hydrogen production based on solar photo fermentation has been studied (Yin et al., 2024). At present, photofermentation biological hydrogen production technology is widely regarded as the most ideal hydrogen production method, which has the advantages of a wide source of raw materials, high substrate conversion efficiency, and the combination of organic waste treatment, solar energy utilization, and clean energy production (Zhang et al., 2021c). The disadvantage is that it requires light, and the light conversion efficiency is low (Rai & Singh, 2016; Zhang et al., 2021e).

1.4.3 Dark fermentation of biological hydrogen production

Biological hydrogen production by dark fermentation refers to the process by which dark fermentation bacteria metabolize organic compounds to release hydrogen in an anaerobic and dark environment (Brennan & Owende, 2010; Dong et al., 2025; Lin et al., 2012). According to the type of metabolites, it is divided into four categories, namely, propionic acid fermentation, butyric acid fermentation, ethanol fermentation, and mixed fermentation (Feng et al., 2023; Yin et al., 2024). The main dark fermentation bacteria include *Clostridium, Clostridium* species, *Paenibacillus*, facultative *Enterobacter* sp., *Bacillus* sp., some thermophilic bacteria, and sludge (Li et al., 2020; Mugnai et al., 2021). The metabolic pathway of dark fermentation for hydrogen production is illustrated in Fig. 1.18. Carbohydrates are first hydrolyzed into glucose; glucose undergoes glycolysis to generate pyruvate, ATP, and NADH (Reduced Nicotinamide Adenine Dinucleotide); subsequently, pyruvate produces hydrogen via formic acid cracking or pyruvate decarboxylation.

As early as the 70s of the 20th century the research and development of dark fermentation biological hydrogen production technology began,

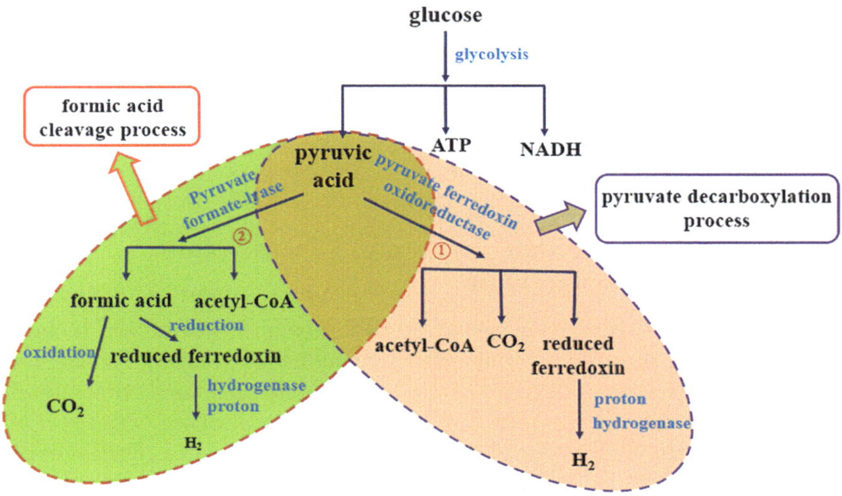

FIGURE 1.18 Metabolic pathway of hydrogen production by dark-fermentation.

and significant progress has been made in strain breeding and hydrogen production mechanism (Li et al., 2024). Chen et al. (2022a) compared the hydrogen production efficiency of rice straw as the substrate for dark fermentation under normal and high-temperature conditions and found that the hydrogen yield was higher at high temperatures. Li et al. (2018) compared the hydrogen production capacities of agricultural waste under synchronous saccharification dark fermentation (SSDF) and asynchronous saccharification dark fermentation (ASDF) modes, concluding that SSDF achieved better hydrogen production performance for agricultural waste in dark fermentation.

Dark fermentation biological hydrogen production has the characteristics of a fast rate, short cycle, no need for light, and a wide source of raw materials (Kannah et al., 2020). At the same time, dark fermentation also has some disadvantages regarding hydrogen production; it can only convert organic matter into small molecule acids and alcohols, but these small molecule substances cannot be fully utilized. In addition, more by-products will be produced during the process of biological hydrogen production via dark fermentation, and the biochemical oxygen demand (BOD) content of organic matter is high after the end of dark fermentation. The accumulation of organic acids will occur if the raw materials cannot be fully utilized, thereby destroying the environment for fermentation and hydrogen production, resulting in the termination of the fermentation process (Martinez-Burgos et al., 2021). Because by-products such as organic acids and alcohols will be produced in the process of hydrogen production, the fermentation broth cannot be directly discharged, and the fermentation broth needs to be rendered harmless before discharge, which is also an important issue restricting the development of hydrogen production by dark fermentation (Yin et al., 2022). Therefore improving substrate utilization rates and hydrogen production efficiency is an urgent issue to be addressed in dark fermentation for hydrogen production (Sivagurunathan et al., 2016).

1.4.4 Biological hydrogen production by dark–photo combined fermentation

Dark–photo combined fermentation biohydrogen production technology refers to a bioprocess that efficiently converts organic waste into hydrogen through two stages of tandem anaerobic dark fermentation and photofermentation, using microbial synergistic metabolism (Zhang et al., 2020b). According to the system configuration and metabolic process, it can be divided into two categories: Split tandem system, dark fermentation, and photo fermentation are carried out in an independent reactor, and the dark fermentation broth is input into the photo fermentation unit after pH adjustment; in the integrated coupling system, dark fermentation bacteria and photosynthetic bacteria coexist in the same reactor, and the in situ metabolism relay is realized through light partitioning (Zhang et al., 2021a). The core breakthrough of this technology lies in the deep transformation of dark fermentation by-products in the photo fermentation stage, which converts the organic acids that are inhibitory in traditional dark fermentation into additional hydrogen. In the dark fermentation stage, anaerobic bacteria such as *Clostridium* and *Enterobacter* decompose organic matter in a dark environment to produce hydrogen and, at the same time, produce by-products such as propionic acid, butyric acid, and ethanol. In the photofermentation stage, photosynthetic bacteria (such as *Rhodobacter sphaeroides, Rhodopseudomonas palustris*) are driven by near-infrared light (800–1000 nm) and use the aforementioned organic acids/alcohols as substrates to further produce hydrogen through nitrogenase (Elsharkawy et al., 2020; Policastro et al., 2021). Therefore it is an effective way to improve the substrate conversion utilization rate and hydrogen production efficiency by combining the advantages of anaerobic bacteria and photosynthetic bacteria to form an efficient hydrogen production system (Ramprakash & Muthukumar, 2018; Zhang et al., 2017).

The core advantage of this technology is to break through the bottleneck of hydrogen production in a single dark fermentation and convert the residual organic acids from dark fermentation into hydrogen in the photofermentation stage, so that the total hydrogen production can be increased to two to six times that of dark fermentation and the substrate utilization rate is significantly improved. The raw materials

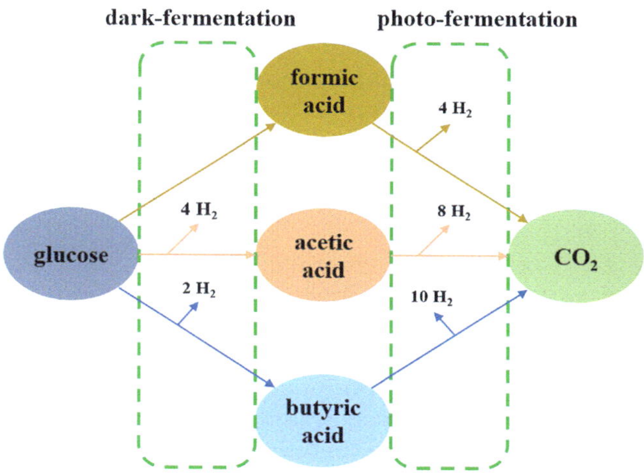

FIGURE 1.19 Roadmap of hydrogen production by combined dark–photo fermentation.

have wide applicability (wastewater, straw, food waste, etc.), and solar energy is used to drive photo fermentation to reduce energy consumption (Sivaramakrishnan et al., 2021). Results show that the hydrogen production and chemical oxygen demand (COD) removal rate of the combined hydrogen production system are significantly increased by the combined dark–photo fermentation mode with starch and glucose as the matrix (Hitit et al., 2017). Another study showed that photo-fermentation bacteria continued to produce hydrogen using the residual liquid after dark fermentation, and the combination of dark–photo fermentation increased hydrogen production from 1.59 to 5.48 mol H_2/mol glucose, and the energy conversion efficiency increased from 13.3% to 46% (Su et al., 2009). However, there are also some shortcomings, such as limited light transfer, serious light attenuation of deep liquid in the photofermentation stage, and complex design of large-scale reactors. The system is difficult to control, and the two-stage microbiota needs to be optimized independently, such as dark fermentation pH 5.5–6.5 and photofermentation pH 7.0–7.5. The purity of the product is low, and the hydrogen content is only 50%–60%, which requires additional purification (Meky et al., 2021). Although the accumulation of organic acids was alleviated by photofermentation, microbial community interaction (such as competition inhibition) and long-term operational stability were still the key constraints for industrialization (Sargsyan et al., 2015) (Fig. 1.19).

1.4.5 Biomethanol reforming for hydrogen production

Biomethanol reforming to hydrogen production is a process in which biomethanol is fully mixed with deionized water, preheated, superheated to a gaseous state, and converted into a hydrogen-based mixture under the action of a catalyst (Gürgan et al., 2020). Among them, biomethanol is obtained by catalytic reforming of biohydrogen and carbon dioxide; biohydrogen is obtained from microbial anaerobic fermentation of renewable resources such as agricultural and forestry waste. The efficiency of hydrogen production from biomethanol reforming is affected by many aspects, mainly including reactor structure and hydrogen production process (Garcia et al., 2021).

In the research of methanol reforming hydrogen production technology, methanol–water vapor reforming hydrogen production technology is considered to be the most efficient and widely used hydrogen production method. Methanol–water vapor reforming technology refers to the process in which methanol and water vapor undergo methanol cracking reaction and water–gas displacement reaction under the action of catalyst under certain temperature and pressure conditions and generate hydrogen, carbon dioxide, and carbon monoxide (Luo et al., 2025). The reaction is a complex system of multicomponent, multireaction gas–solid catalyzed reactions. The total reaction of methanol–water vapor reforming to hydrogen is shown in Eq. (1.1).

$$CH_3OH + H_2O \rightarrow 3H_2 + CO_2 \ \Delta H = +49.5\ KJ/mol \quad (1.1)$$

There are three different views on the specific catalytic reaction process in the process of hydrogen production by methanol–water vapor reforming. That is, parallel reaction mechanism, decomposition transformation mechanism, and decomposition inverse transformation mechanism (Ranjekar & Yadav, 2023).

1.4.5.1 *Parallel reaction mechanism*

$$CH_3OH + H_2O \rightarrow 3H_2 + CO_2 \quad (1.2)$$

$$CH_3OH \rightarrow 2H_2 + CO$$

$$H_2O + CO \rightarrow H_2 + CO_2$$

Regarding the research on the mechanism of methanol–water vapor reforming hydrogen production, early researchers believe that methanol–water vapor reforming hydrogen production is a process in which methanol and water go through different catalytic reactions to finally generate a mixture containing H_2, CO_2, and CO, as shown in Eq. (1.2). However, with the continuous in-depth research of researchers, some people proposed different reforming hydrogen production mechanisms.

1.4.5.2 Decomposition and transformation mechanism

$$CH_3OH \rightarrow 2H_2 + CO \tag{1.3}$$

$$CO + H_2O \rightarrow H_2 + CO_2$$

The theory is proposed that methanol is first catalyzed to produce a mixture of H_2 and CO, and then the CO in the product undergoes a water vapor shift reaction with H_2O to form H_2 and CO_2, as shown in Eq. (1.3). The proposal of this theory has attracted the attention of many scholars, and it has been found that the experimental results are inconsistent with the theory in a large number of subsequent targeted experimental studies. Therefore the theory of hydrogen production by methanol–water vapor reforming is controversial.

1.4.5.3 Decomposition of the inverse transformation mechanism

$$2CH_3OH \rightarrow CH_3OCHO + 2H_2 \tag{1.4}$$

$$CH_3OCHO + H_2O \rightarrow CH_3OH + HCOOH$$

$$HCOOH \rightarrow H_2 + CO_2$$

Later, some researchers proposed the decomposition inverse transformation theory of methanol–water vapor reforming to hydrogen, which believed that methanol first decomposes into CH_3OCHO and H_2, then decomposes CH_3OCHO into HCOOH under the action of water vapor, and finally HCOOH undergoes a series of catalytic conversions to generate H_2 and CO_2, as shown in Eq. (1.4). In this theory the reverse reaction of the water vapor shift also occurs to generate a small amount of CO. So far, there is no universally recognized mechanism for hydrogen production from methanol–water vapor reforming. However, in the continuous research on methanol–water vapor reforming for hydrogen production, it is found that the decomposition inverse transformation mechanism of methanol–water vapor reforming for hydrogen production is being verified more and more (Ranjekar & Yadav, 2021).

It can be seen that the reaction in the process of hydrogen production by methanol–water vapor reforming is shown in Eqs. (1.5) and (1.6). By comparing the three major reaction formulas of methanol–water vapor reforming to hydrogen, it can be seen that Eqs. (1.1) and (1.5) are endothermic reactions, and Eq. (1.5) is a strong endothermic reaction, whereas Eq. (1.6) is an exothermic reaction. Therefore, in the process of reforming hydrogen production, the reforming temperature has a greater influence on the whole reforming hydrogen production process. When the reforming temperature is high, although the activity of the catalyst will be increased, it will also promote the reaction of Eqs. (1.1) and (1.5), and the reaction of Eq. (1.1) will be intensified, whereas the reaction of Eq. (1.6) will be inhibited. Therefore in the process of hydrogen production by methanol–water vapor reforming, it is necessary to select the appropriate catalytic reforming temperature to ensure the catalytic reforming efficiency while ensuring the activity of the catalyst (Cheng et al., 2022).

$$CH_3OH \rightarrow 2H_2 + CO \tag{1.5}$$

$$CO + H_2O \rightarrow H_2 + CO_2 \tag{1.6}$$

At present, methanol–water vapor reforming is considered to be one of the most promising hydrogen production methods to directly provide hydrogen energy for fuel cells. From the reaction formula of methanol–water vapor reforming to hydrogen production, it can be seen that from a theoretical point of view, each mole of methanol can finally produce three moles of hydrogen, and the hydrogen content in the product is high. The methanol–water vapor reforming reaction is relatively mild and simple to operate, and there is no very strict requirement for reforming temperature, and the CO content in the product is low. However, the overall reaction process of methanol–water vapor reforming to hydrogen production is an endothermic process, so it needs to continuously provide energy from the outside, which will consume a part of the raw materials to ensure the normal progress of the reforming reaction, which will affect the performance of methanol reforming hydrogen production (Saravanan et al., 2021).

Comparing several biological hydrogen production methods, it is found that each hydrogen production method has its own advantages and disadvantages. Although photolysis of water to produce hydrogen can convert solar energy into hydrogen energy only by using water as raw material, the oxygen generated during hydrogen production will inhibit the activity of hydrogenase. Although photofermentation hydrogen production can rely on unlimited solar energy resources to produce hydrogen from various organic matter as raw materials, it has high requirements for photo biochemical reactors. Although the process of dark fermentation hydrogen production is simple and easy to operate and can decompose complex organic matter for hydrogen production, the utilization rate of raw materials is low, and the organic acids generated during hydrogen production will affect the activity of hydrogen production enzymes. Although combined dark–photo fermentation hydrogen production can improve the utilization rate of raw materials and hydrogen production efficiency to a certain extent, its coupling process is complex and difficult to control (Table 1.2).

TABLE 1.2 Comparison of biological hydrogen production technologies.

Method	Microorganism	Enzyme	Electron donor	Main products	Limiting factor
Photolysis of water	Algae cyanobacteria	Hydrogenase Nitrogenase, hydrogenase	Water	H_2, O_2	Light; the generated oxygen inhibits the activity of hydrogen-producing enzymes
Photo-fermentation	Photosynthetic bacteria	Nitrogenase, hydrogenase	Organic matter	H_2, CO_2	Light; high requirements for reactors
Dark fermentation	Anaerobic bacteria	Hydrogenase	Organic matter	H_2, CO_2, organic acids	Low substrate utilization rate: the generated organic acids affect the activity of hydrogen-producing enzymes
Combined dark–photo fermentation	Anaerobic bacteria photosynthetic bacteria	Nitrogenase, hydrogenase	Organic matter	H_2, CO_2, organic acids	Complex process and difficult to regulate

1.5 Advantages of biohydrogen over other energy sources

Compared with other energy sources, biohydrogen has many significant advantages not only as an energy carrier but also as an ecological hub connecting "waste treatment carbon fixation clean energy" (Ayodele et al., 2023). Its development will reshape the underlying logic of future energy networks, shifting from centralized supply to distributed circulation. In terms of environmental protection, when biohydrogen is burned or used in fuel cells, it only produces water vapor and does not emit harmful waste or greenhouse gases. It has almost zero pollution to the environment and is a truly clean energy source; in terms of sustainability, its raw materials come from biomass, such as plants, organic waste, etc. These resources are renewable and can be continuously obtained through recycling waste, effectively reducing dependence on fossil fuels; from a convenience perspective, biohydrogen can be used on existing infrastructure without the need for large-scale construction of new compression, storage, and other facilities, reducing the cost and difficulty of energy utilization (Ariful Islam et al., 2025).

The core competitiveness of biohydrogen lies in its construction of a unique lifecycle negative-carbon closed-loop system, which is its essential characteristic that distinguishes it from fossil fuels and green hydrogen (Zech et al., 2015). Compared to gray hydrogen and green hydrogen, biohydrogen achieves active carbon removal through a triple mechanism: First, raw biomass (energy crops, microalgae) fixes atmospheric CO_2 through photosynthesis during the growth stage, absorbing 1.6–1.8 t of CO_2 per ton of dry matter; second the hydrogen production process is coupled with CCS technology (bioenergy with carbon capture and storage (BECCS)), which enables biomass gasification to achieve a net removal of 10–15 kg CO_2 per kilogram of hydrogen. Furthermore, dark fermentation of organic waste for hydrogen production can avoid methane emissions from landfill treatment. The dual environment of carbon-negative emissions and coordinated pollution control makes biohydrogen the only energy carrier that combines deep decarbonization and ecological restoration functions, providing an irreplaceable ecological solution for carbon neutrality goals.

Biohydrogen exhibits significant regional universality and cost competitiveness in terms of resource resilience and economic viability (Chen et al., 2009). Its raw material system forms a three-level resource resilience structure: the global annual production of agricultural and forestry waste exceeds 10 billion t, and only 30% can produce 180 million t of hydrogen annually, with almost zero raw material costs; planting drought-tolerant *Miscanthus floridulus* on barren marginal land can increase land energy output efficiency by three times compared to photovoltaic hydrogen production; this resource structure breaks through the rigid dependence of green hydrogen on renewable electricity (requiring 50–55 kWh of electricity per kilogram of green hydrogen) and gives rise to a unique economic model. The synergistic effect of the broad-spectrum, negative carbon financialization, and coproduction value-added of this raw material creates a cost depression for biohydrogen in agricultural and industrial areas, laying the foundation for a decentralized energy network.

With the continuous advancement of technology, biohydrogen will play an important role in the future energy field and has extremely broad application prospects (Garcia et al., 2021). It can serve as a gas supply source for hydrogen fuel cells, providing clean and efficient energy for electric vehicles, distributed power generation systems, etc., and promoting green transformation in the fields of transportation and energy supply. In the field of environmental protection, biohydrogen technology can convert organic waste into valuable hydrogen resources, achieve resource recycling, reduce pressure on natural resources, and reduce greenhouse gas emissions such as carbon dioxide. Although biohydrogen has many advantages and potential, the current biohydrogen production technology is not fully mature and still faces some challenges before large-scale application (Tang et al., 2024). For example, the efficiency of the biological hydrogen production process is relatively low, resulting in higher production costs. The infrastructure for the collection, storage, and distribution of biohydrogen is still incomplete, which limits its large-scale promotion and application. However, with the continuous efforts of researchers and the continuous innovation of technology, these problems are expected to be gradually solved. Biohydrogen, as a clean and sustainable energy source, provides a new direction for humanity to address energy crises and environmental challenges. Although there are still some technical and economic challenges at present, with the deepening of research and technological breakthroughs, it is believed that in the near future, biohydrogen will shine on the energy stage and make important contributions to the sustainable development of humanity.

1.6 Potential to contribute to the energy transition to achieve CO_2 reduction targets

Biohydrogen, as a clean hydrogen produced from biomass resources through thermochemical or biological conversion processes, has shown irreplaceable potential in promoting the low-carbon transformation of the global energy system and achieving the goal of deep decarbonization (Feng et al., 2023). Its core competitiveness stems from its unique negative carbon emission mechanism and organic waste resource utilization capacity. On the one hand, biomass raw materials (agricultural and forestry wastes, energy crops, algae) absorb CO_2 from the atmosphere through photosynthesis during the growth process, forming a natural carbon sink. On the other hand the coupling of CCS technology (BECCS) in the process of gasification or pyrolysis hydrogen production can permanently sequester biosource CO_2 and achieve net negative carbon emissions throughout the life cycle, with 10–20 t of CO_2 per 1 t of hydrogen produced (Rani et al., 2024). Compared with the "zero-carbon" characteristics of green hydrogen by electrolysis of water the active carbon removal ability of biohydrogen makes it a key tool for deep decarbonization in hard-to-abate areas. Studies have shown that using 10% of the world's 150 billion t of biomass resources per year to produce hydrogen could meet 30% of the world's current hydrogen demand and contribute 800 million t of annual emission reductions by 2050. In addition, microbial hydrogen production technology further amplifies its climate benefits by producing hydrogen while suppressing methane emissions. By building a closed-loop system of "biomass growth, hydrogen production, and carbon sequestration," biohydrogen provides a path for the carbon neutrality goal with both emission reduction intensity and resource recycling advantages (Ahmad et al., 2024).

In addition, biohydrogen has shown significant emission reduction efficiency in three major areas: industry, transportation, and energy systems (Zhang et al., 2021d). In the industrial sector, biohydrogen can replace coke as a reducing agent in steel smelting, reducing CO_2 emissions by 1.5 t per tonne of steel produced, and replacing natural gas reforming processes in ammonia synthesis and methanol, reducing CO_2 emissions by 1.8 t per tonne of ammonia. In the transportation sector, biojet fuel (sustainable aviation fuel (SAF)) derived from biohydrogen can reduce aviation carbon emissions by 50%–80% and can also produce "zero-carbon steel" from hydrogen production from forestry waste, which also verifies the feasibility of industrial decarbonization. In the energy system, biohydrogen can enhance the flexibility of the power grid, and as an energy storage medium, it can reduce the intermittency of wind and solar power generation, whereas the distributed hydrogen production model can promote the local transformation of rural biomass resources and help the equitable transition of regional energy. These applications not only directly replace fossil fuels but also create net environmental benefits through carbon-negative properties.

However, the large-scale development of biohydrogen still faces the triple bottleneck of technology, resources, and policies (Puricelli et al., 2021). At the technical level the efficiency of microbial hydrogen production is low (only 5%–15% of photo fermentation), and it is necessary to modify hydrogen-producing strains through synthetic biology. In the process of biomass gasification, tar pollution and catalyst deactivation have pushed up the treatment cost by more than 30%. In terms of resources, energy crops may occupy 5%–7% of the world's arable land, and conflicts need to be mitigated through marginal land planting and waste-first strategies. The cost of biomass collection and transportation accounts for 30%–50% of the final hydrogen cost, which restricts the economic radius of the industry. In terms of policy mechanism, there is a lack of unified carbon accounting standards for the whole life cycle of biohydrogen, which makes it difficult to quantify the negative carbon value, and the existing subsidy framework does not fully cover the cost of BECCS storage. To solve these problems, we need to make breakthroughs in three aspects. In terms of technology, we have developed high-efficiency anaerobic fermentation reactors and anticarbon gasification catalysts; in terms of resources, we need to build a regional ecological industrial chain of "agriculture and forestry-hydrogen production-carbon sequestration"; in terms of policy, biohydrogen will be included in the carbon market trading system, and negative hydrocarbon premium subsidies will be formulated. Only through interdisciplinary technology iteration and industrial chain integration can biohydrogen become a strategic pillar supporting the goal of carbon neutrality after 2035 (Pandey et al., 2025).

In summary, biohydrogen provides a unique path for the deep decarbonization of the energy system by virtue of the dual advantages of negative carbon closed-loop and waste recycling. By overcoming efficiency bottlenecks, optimizing resource allocation, and innovating policy tools, it is expected to occupy more than 15% of the energy landscape dominated by green hydrogen and become a key negative carbon carrier to achieve the 1.5°C temperature control goal.

References

Abdin, Z., Tang, C., Liu, Y., & Catchpole, K. (2021). Large-scale stationary hydrogen storage via liquid organic hydrogen carriers. *Iscience, 24*(9). doi:10.1016/j.isci.2021.102966.

Abdin, Z., Zafaranloo, A., Rafiee, A., Mérida, W., Lipiński, W., & Khalilpour, K. R. (2020). Hydrogen as an energy vector. *Renewable and Sustainable Energy Reviews, 120*. doi:10.1016/j.rser.2019.109620.

Agyekum, E. B., & Odoi-Yorke, F. (2024). Review of over two decades of research on dark and photo fermentation for biohydrogen production—A combination of traditional, systematic, and bibliometric approaches. *International Journal of Hydrogen Energy, 91*, 1149–1169. doi:10.1016/j.ijhydene.2024.10.218.

Ahmad, A., R. , K., Hasan, S. W., Show, P. L., & Banat, F. (2024). Biohydrogen production through dark fermentation: Recent trends and advances in transition to a circular bioeconomy. *International Journal of Hydrogen Energy, 52*, 335–357. doi:10.1016/j.ijhydene.2023.05.161.

Akah, A., Williams, J., & Ghrami, M. (2019). An overview of light olefins production via steam enhanced catalytic cracking. *Catalysis Surveys from Asia, 23*(4), 265–276. doi:10.1007/s10563-019-09280-6.

Ananthi, V., Ramesh, U., Balaji, P., Kumar, P., Govarthanan, M., & Arun, A. (2024). A review on the impact of various factors on biohydrogen production. *International Journal of Hydrogen Energy, 52*, 33–45. doi:10.1016/j.ijhydene.2022.08.046.

Antal, T. K., Krendeleva, T. E., Laurinavichene, T. V., Makarova, V. V., Ghirardi, M. L., Rubin, A. B., Tsygankov, A. A., & Seibert, M. (2003). The dependence of algal H_2 production on Photosystem II and O_2 consumption activities in sulfur-deprived *Chlamydomonas reinhardtii* cells. *Biochimica et Biophysica Acta (BBA)—Bioenergetics, 1607*(2–3), 153–160. doi:10.1016/j.bbabio.2003.09.008.

Argun, H., & Kargi, F. (2011). Bio-hydrogen production by different operational modes of dark and photo-fermentation: An overview. *International Journal of Hydrogen Energy, 36*(13), 7443–7459. doi:10.1016/j.ijhydene.2011.03.116.

Ariful Islam, M., Chowdhury, A., Jahan, I., & Farrok, O. (2025). Mitigation of environmental impacts and challenges during hydrogen production. *Bioresource Technology, 415*. doi:10.1016/j.biortech.2024.131666.

Ayodele, D. T., Ogunbiyi, O. D., Akamo, D. O., Otun, K. O., Akinpelu, D. A., Adegoke, J. A., Fapojuwo, D. P., & Oladoye, P. O. (2023). Factors affecting biohydrogen production: Overview and perspectives. *International Journal of Hydrogen Energy, 48*(71), 27513–27539. doi:10.1016/j.ijhydene.2023.04.001.

Azwar, M. Y., Hussain, M. A., & Abdul-Wahab, A. K. (2014). Development of biohydrogen production by photobiological, fermentation and electrochemical processes: A review. *Renewable and Sustainable Energy Reviews, 31*, 158–173. doi:10.1016/j.rser.2013.11.022.

Bang, S., Snoeckx, R., & Cha, M. S. (2023). Kinetic study for plasma assisted cracking of NH_3: Approaches and challenges. *Journal of Physical Chemistry A, 127*(5), 1271–1282. doi:10.1021/acs.jpca.2c06919.

Barbir, F. (2005). PEM electrolysis for production of hydrogen from renewable energy sources. *Solar Energy, 78*(5), 661–669. doi:10.1016/j.solener.2004.09.003.

Boretti, A., & Banik, B. K. (2021). Advances in hydrogen production from natural gas reforming. *Advanced Energy and Sustainability Research, 2*(11). doi:10.1002/aesr.202100097.

Brauns, J., & Turek, T. (2020). Alkaline water electrolysis powered by renewable energy: A review. *Processes, 8*(2). doi:10.3390/pr8020248.

Brennan, L., & Owende, P. (2010). Biofuels from microalgae—A review of technologies for production, processing, and extractions of biofuels and co-products. *Renewable and Sustainable Energy Reviews, 14*(2), 557–577. doi:10.1016/j.rser.2009.10.009.

Carmo, M., Fritz, D. L., Mergel, J., & Stolten, D. (2013). A comprehensive review on PEM water electrolysis. *International Journal of Hydrogen Energy, 38*(12), 4901–4934. doi:10.1016/j.ijhydene.2013.01.151.

Çelik, D., & Yıldız, M. (2017). Investigation of hydrogen production methods in accordance with green chemistry principles. *International Journal of Hydrogen Energy, 42*(36), 23395–23401. doi:10.1016/j.ijhydene.2017.03.104.

Chanda, C. K., & Bose, D. (2020). Challenges of employing renewable energy for reducing greenhouse gases (GHGs) and carbon footprint. *Encyclopedia of Renewable and Sustainable Materials, 1–5*, 346–365. doi:10.1016/B978-0-12-803581-8.11170-1.

Chang, F. W., Ou, T. C., Roselin, L. S., Chen, W. S., Lai, S. C., & Wu, H. M. (2009). Production of hydrogen by partial oxidation of methanol over bimetallic Au-Cu/TiO_2-Fe_2O_3 catalysts. *Journal of Molecular Catalysis A: Chemical, 313*(1–2), 55–64. doi:10.1016/j.molcata.2009.08.002.

Chang, F. W., Roselin, L. S., & Ou, T. C. (2008). Hydrogen production by partial oxidation of methanol over bimetallic Au-Ru/Fe_2O_3 catalysts. *Applied Catalysis A: General, 334*(1–2), 147–155. doi:10.1016/j.apcata.2007.10.003.

Chang, F. W., Yu, H. Y., Roselin, L. S., Yang, H. C., & Ou, T. C. (2006). Hydrogen production by partial oxidation of methanol over gold catalysts supported on TiO_2-MO_x (M = Fe, Co, Zn) composite oxides. *Applied Catalysis A: General, 302*(2), 157–167. doi:10.1016/j.apcata.2005.12.028.

Chelmehsara, M. E., & Mahmoudimehr, J. (2018). Techno-economic comparison of anode-supported, cathode-supported, and electrolyte-supported SOFCs. *International Journal of Hydrogen Energy, 43*(32), 15521–15530. doi:10.1016/j.ijhydene.2018.06.114.

Chemat, F., Abert-Vian, M., Fabiano-Tixier, A. S., Strube, J., Uhlenbrock, L., Gunjevic, V., & Cravotto, G. (2019). Green extraction of natural products. Origins, current status, and future challenges. *TrAC—Trends in Analytical Chemistry, 118*, 248–263. doi:10.1016/j.trac.2019.05.037.

Chen, F., Chang, M. H., Kuo, C. Y., Hsueh, C. Y., & Yan, W. M. (2009). Analysis of a plate-type microreformer for methanol steam reforming reaction. *Energy and Fuels, 23*(10), 5092–5098. doi:10.1021/ef9005252.

Chen, H., Wu, J., Huang, R., Zhang, W., He, W., Deng, Z., Han, Y., Xiao, B., Luo, H., & Qu, W. (2022). Effects of temperature and total solid content on biohydrogen production from dark fermentation of rice straw: Performance and microbial community characteristics. *Chemosphere, 286*(Pt 1), 131655. doi:10.1016/j.chemosphere.2021.131655.

Chen, W. H., Lin, M. R., Leu, T. S., & Du, S. W. (2011). An evaluation of hydrogen production from the perspective of using blast furnace gas and coke oven gas as feedstocks. *International Journal of Hydrogen Energy, 36*(18), 11727–11737. doi:10.1016/j.ijhydene.2011.06.049.

Cheng, D., Ngo, H. H., Guo, W., Chang, S. W., Nguyen, D. D., Bui, X. T., Wei, W., Ni, B., Varjani, S., & Hoang, N. B. (2022b). Enhanced photo-fermentative biohydrogen production from biowastes: An overview. *Bioresource Technology, 357*. doi:10.1016/j.biortech.2022.127341.

Cheng, D., Ngo, H. H., Guo, W., Chang, S. W., Nguyen, D. D., Deng, L., Chen, Z., Ye, Y., Bui, X. T., & Hoang, N. B. (2022a). Advanced strategies for enhancing dark fermentative biohydrogen production from biowaste towards sustainable environment. *Bioresource Technology, 351*. doi:10.1016/j.biortech.2022.127045.

Cristello, J. B., Yang, J. M., Hugo, R., Lee, Y., & Park, S. S. (2023). Feasibility analysis of blending hydrogen into natural gas networks. *International Journal of Hydrogen Energy, 48*(46), 17605–17629. doi:10.1016/j.ijhydene.2023.01.156.

David, M., Ocampo-Martínez, C., & Sánchez-Peña, R. (2019). Advances in alkaline water electrolyzers: A review. *Journal of Energy Storage, 23*, 392–403. doi:10.1016/j.est.2019.03.001.

Dincer, I. (2002). Technical, environmental and exergetic aspects of hydrogen energy systems. *International Journal of Hydrogen Energy, 27*(3), 265–285. doi:10.1016/s0360-3199(01)00119-7.

Ding, F., & Yi, Y. (2006). *Hydrogen production and storage technologies*. Chemical Industry Press.

Dong, Z., Cao, S., Zhao, B., Wang, G., duan, J., & Yuan, A. (2025). Enhancement of hydrogen production in dark fermentation of corn stover hydrolysates through composite electric field pretreatment: Improving enzymatic efficiency and regulating microbial metabolic balance. *Bioresource Technology, 415*, 131721. doi:10.1016/j.biortech.2024.131721.

Elsharkawy, K., Gar Alalm, M., Fujii, M., Afify, H., Tawfik, A., & Elsamadony, M. (2020). Paperboard mill wastewater treatment via combined dark and LED-mediated fermentation in the absence of external chemical addition. *Bioresource Technology, 295*. doi:10.1016/j.biortech.2019.122312.

Erden, M., & Karakilcik, M. (2024). Experimental investigation of hydrogen production performance of various salts with a chlor-alkali method. *International Journal of Hydrogen Energy, 52*, 546–560. doi:10.1016/j.ijhydene.2023.08.049.

European Commission. (2020). *A hydrogen strategy for a climate-neutral Europe*. Global CCS Institute A. VIC.

Fallah Vostakola, M., Ozcan, H., El-Emam, R. S., & Horri, B. A. (2023). Recent advances in high-temperature steam electrolysis with solid oxide electrolysers for green hydrogen production. *Energies, 16*(8). doi:10.3390/en16083327.

Fallah Vostakola, M., Salamatinia, B., & Amini Horri, B. (2022). A review on recent progress in the integrated green hydrogen production processes. *Energies, 15*(3), 1209. doi:10.3390/en15031209.

Fan, Y., Zhang, F., He, K., Yu, D., Chen, H., Tian, D., Shi, Y., Li, Z., & Wang, X. (2025). Functional microorganisms in hydrogen production: Mechanisms and applications. *Bioresource Technology, 419*. doi:10.1016/j.biortech.2024.132007.

Fayaz, H., Saidur, R., Razali, N., Anuar, F. S., Saleman, A. R., & Islam, M. R. (2012). An overview of hydrogen as a vehicle fuel. *Renewable and Sustainable Energy Reviews, 16*(8), 5511–5528. doi:10.1016/j.rser.2012.06.012.

Feng, S., Ngo, H. H., Guo, W., Chang, S. W., Nguyen, D. D., Bui, X. T., Zhang, X., Ma, X. Y., & Hoang, B. N. (2023). Biohydrogen production, storage, and delivery: A comprehensive overview of current strategies and limitations. *Chemical Engineering Journal, 471*, 144669.

Gaffron, H., & Rubin, J. (1942). Fermentative and photochemical production of hydrogen in algae. *Journal of General Physiology, 26*(2), 219–240. doi:10.1085/jgp.26.2.219.

Garcia, G., Arriola, E., Chen, W. H., & De Luna, M. D. (2021). A comprehensive review of hydrogen production from methanol thermochemical conversion for sustainability. *Energy, 217*, 119384. doi:10.1016/j.energy.2020.119384.

Gielen, D., Boshell, F., Saygin, D., Bazilian, M. D., Wagner, N., & Gorini, R. (2019). The role of renewable energy in the global energy transformation. *Energy Strategy Reviews, 24*, 38–50. doi:10.1016/j.esr.2019.01.006.

Guban, D., Muritala, I. K., Roeb, M., & Sattler, C. (2020). Assessment of sustainable high temperature hydrogen production technologies. *International Journal of Hydrogen Energy, 45*(49), 26156–26165. doi:10.1016/j.ijhydene.2019.08.145.

Guo, Y., Li, G., Zhou, J., & Liu, Y. (2019). Comparison between hydrogen production by alkaline water electrolysis and hydrogen production by PEM electrolysis. *IOP Conference Series: Earth and Environmental Science, 371*(4), 042022. doi:10.1088/1755-1315/371/4/042022.

Gürgan, M., Koku, H., Eroglu, I., & Yücel, M. (2020). Transcriptome analysis of the effects of light and dark cycle on hydrogen production metabolism of *Rhodobacter capsulatus* DSM1710. *International Journal of Hydrogen Energy, 45*(60), 34707–34719. doi:10.1016/j.ijhydene.2020.03.108.

Hitit, Z. Y., Zampol Lazaro, C., & Hallenbeck, P. C. (2017). Increased hydrogen yield and COD removal from starch/glucose based medium by sequential dark and photo-fermentation using *Clostridium butyricum* and *Rhodopseudomonas palustris*. *International Journal of Hydrogen Energy, 42*(30), 18832–18843. doi:10.1016/j.ijhydene.2017.05.161.

Hohn, K. L., & Lin, Y. C. (2009). Catalytic partial oxidation of methanol and ethanol for hydrogen generation. *ChemSusChem, 2*(10), 927–940. doi:10.1002/cssc.200900104.

Holladay, J. D., Hu, J., King, D. L., & Wang, Y. (2009). An overview of hydrogen production technologies. *Catalysis Today, 139*(4), 244–260. doi:10.1016/j.cattod.2008.08.039.

Hu, B., Shu, R., Khairun, H. S., Tian, Z., Wang, C., & Gupta, N. K. (2024). Methanol steam reforming for hydrogen production over Ni-based catalysts: State-of-the-art review and future prospects. *Chemistry—An Asian Journal, 19*(16). doi:10.1002/asia.202400217.

Hydrogen Council. (2017). *Roadmap towards a hydrogen economy*. Hydrogen Council.

Iida, S., & Sakata, K. (2019). Hydrogen technologies and developments in Japan. *Clean Energy, 3*(2), 105–113. doi:10.1093/ce/zkz003.

Jing, H., Yang, H., Yu, X., Hu, C., Li, R., & Li, H. (2022). Treatment of organic matter and ammonia nitrogen in wastewater by electrocatalytic oxidation: A review of anode material preparation. *Environmental Science: Water Research and Technology, 8*(2), 226–248. doi:10.1039/d1ew00695a.

Kamal, A., Al-Ghamdi, S. G., & Koc, M. (2019). Revaluing the costs and benefits of energy efficiency: A systematic review. *Energy Research and Social Science, 54*, 68–84. doi:10.1016/j.erss.2019.03.012.

Kamlungsua, K., Su, P.-C., & Chan, S. H. (2020). Hydrogen generation using solid oxide electrolysis cells. *Fuel Cells, 20*(6), 644–649. doi:10.1002/fuce.202070602.

Kang, P., Morrow, G., Zhang, X. L., Wang, T. P., Tan, Z. F., & Agarwal, J. (2017). Systematic comparison of hydrogen production from fossil fuels and biomass resources. *International Journal of Agricultural and Biological Engineering, 10*(6), 192–200. doi:10.25165/j.ijabe.20171006.2990.

Kannah, R. Y., Kavitha, S., Preethi, O. P. K., Kumar, G., Dai-Viet, N. V., & Banu, J. R. (2020). Techno-economic assessment of various hydrogen production methods—A review. *Bioresource Technology, 319*, 124175.

Kim, J., Jun, A., Gwon, O., Yoo, S., Liu, M., Shin, J., Lim, T. H., & Kim, G. (2018). Hybrid-solid oxide electrolysis cell: A new strategy for efficient hydrogen production. *Nano Energy, 44*, 121–126. doi:10.1016/j.nanoen.2017.11.074.

Kim, S. H., Kumar, G., Chen, W. H., & Khanal, S. K. (2021). Renewable hydrogen production from biomass and wastes (ReBioH$_2$-2020). *Bioresource Technology, 331*. doi:10.1016/j.biortech.2021.125024.

Laguna-Bercero, M. A. (2012). Recent advances in high temperature electrolysis using solid oxide fuel cells: A review. *Journal of Power Sources, 203*, 4–16. doi:10.1016/j.jpowsour.2011.12.019.

Lee, S., Kim, H. S., Park, J., Kang, B. M., Cho, C.-H., Lim, H., & Won, W. (2021). Scenario-based techno-economic analysis of steam methane reforming process for hydrogen production. *Applied Sciences, 11*(13), 6021. doi:10.3390/app11136021.

Leroy, R. (1983). Industrial water electrolysis: Present and future. *International Journal of Hydrogen Energy, 8*(6), 401–417. doi:10.1016/0360-3199(83)90162-3.

Li, H., Cheng, J., Xia, R., Dong, H., & Zhou, J. (2024). Electron syntrophy between mixed hydrogenogens and *Geobacter metallireducens* boosted dark hydrogen fermentation: Clarifying roles of electroactive extracellular polymeric substances. *Bioresource Technology, 395*. doi:10.1016/j.biortech.2024.130350.

Li, W., He, L., Cheng, C., Cao, G., & Ren, N. (2020). Effects of biochar on ethanol-type and butyrate-type fermentative hydrogen productions. *Bioresource Technology, 306*, 123088. doi:10.1016/j.biortech.2020.123088.

Li, W., & Zheng, X. (2024). Development mechanism and technological innovation of hydrogen energy: Evaluating collaborative innovation based on hydrogen patent data. *International Journal of Hydrogen Energy, 52*, 415–427. doi:10.1016/j.ijhydene.2023.05.310.

Li, Y., Zhang, Z., Jing, Y., Ge, X., Wang, Y., Lu, C., Zhou, X., & Zhang, Q. (2017). Statistical optimization of simultaneous saccharification fermentative hydrogen production from *Platanus orientalis* leaves by photosynthetic bacteria HAU-M1. *International Journal of Hydrogen Energy, 42*(9), 5804–5811. doi:10.1016/j.ijhydene.2016.11.182.

Li, Y., Zhang, Z., Zhu, S., Zhang, H., Zhang, Y., Zhang, T., & Zhang, Q. (2018). Comparison of bio-hydrogen production yield capacity between asynchronous and simultaneous saccharification and fermentation processes from agricultural residue by mixed anaerobic cultures. *Bioresource Technology, 247*, 1210–1214. doi:10.1016/j.biortech.2017.09.053.

Lin, C. Y., Lay, C. H., Sen, B., Chu, C. Y., Kumar, G., Chen, C. C., & Chang, J. S. (2012). Taiwan Fermentative hydrogen production from wastewaters: A review and prognosis 37. *International Journal of Hydrogen Energy, 20*, 15632–15642. doi:10.1016/j.ijhydene.2012.02.072.

Liu, J., Duan, X., Yuan, Z., Liu, Q., & Tang, Q. (2017). Experimental study on the performance, combustion and emission characteristics of a high compression ratio heavy-duty spark-ignition engine fuelled with liquefied methane gas and hydrogen blend. *Applied Thermal Engineering, 124*, 585–594. doi:10.1016/j.applthermaleng.2017.06.067.

Lu, C., Wang, G., Zhang, Q., Yang, X., Yu, J., Liu, T., Petracchini, F., Zhang, Z., Sun, Y., Jiang, D., Liang, X., Li, Y., Zhang, Y., Zhang, T., & Zhang, H. (2023). Comparison of biorefinery characteristics: Photo-fermentation biohydrogen, dark fermentation biohydrogen, biomethane, and bioethanol production. *Applied Energy, 347*. doi:10.1016/j.apenergy.2023.121463.

Lu, C., Zhang, Z., Ge, X., Wang, Y., Zhou, X., You, X., Liu, H., & Zhang, Q. (2016). Bio-hydrogen production from apple waste by photosynthetic bacteria HAU-M1. *International Journal of Hydrogen Energy, 41*(31), 13399–13407. doi:10.1016/j.ijhydene.2016.06.101.

Lu, W., Zhang, R., Toan, S., Xu, R., Zhou, F., Sun, Z., & Sun, Z. (2022). Microchannel structure design for hydrogen supply from methanol steam reforming. *Chemical Engineering Journal, 429*, 132286. doi:10.1016/j.cej.2021.132286.

Luo, B., Jiaqiang, E., Feng, C., Ding, J., & Yang, W. (2025). Investigation on performance optimization of a novel microreactor with multiple-pulsation combustion for methanol steam reforming to hydrogen production for proton exchange membrane fuel cell. *Applied Energy, 384*, 125488. doi:10.1016/j.apenergy.2025.125488.

Martinez-Burgos, W. J., Sydney, E. B., de Paula, D. R., Medeiros, A. B. P., de Carvalho, J. C., Molina, D., & Soccol, C. R. (2021). Hydrogen production by dark fermentation using a new low-cost culture medium composed of corn steep liquor and cassava processing water: Process optimization and scale-up. *Bioresource Technology, 320*. doi:10.1016/j.biortech.2020.124370.

Meena, P. K., & Patane, P. M. (2025). Biohydrogen: Advancing a sustainable transition from fossil fuels to renewable energy. *International Journal of Hydrogen Energy, 100*, 955–970. doi:10.1016/j.ijhydene.2024.12.396.

Meky, N., Elreedy, A., Ibrahim, M. G., Fujii, M., & Tawfik, A. (2021). Intermittent versus sequential dark-photo fermentative hydrogen production as an alternative for bioenergy recovery from protein-rich effluents. *Energy, 217*. doi:10.1016/j.energy.2020.119326.

Messaoudani, Z. l., Rigas, F., Binti Hamid, M. D., & Che Hassan, C. R. (2016). Hazards, safety and knowledge gaps on hydrogen transmission via natural gas grid: A critical review. *International Journal of Hydrogen Energy, 41*(39), 17511–17525. doi:10.1016/j.ijhydene.2016.07.171.

Meyer, T. E., & Cusanovich, M. A. (2003). Discovery and characterization of electron transfer proteins in the photosynthetic bacteria. *Photosynthesis Research, 76*(1–3), 111–126. doi:10.1023/A:1024910323089.

Ming, Z. (1999). Advances in the mechanism of hydrogen production by photosynthetic bacteria. *Chinese Journal of Applied and Environmental Biology, 5*, 25–29.

Moradi, R., & Groth, K. M. (2019). Hydrogen storage and delivery: Review of the state of the art technologies and risk and reliability analysis. *International Journal of Hydrogen Energy, 44*(23), 12254–12269. doi:10.1016/j.ijhydene.2019.03.041.

Muellerlanger, F., Tzimas, E., Kaltschmitt, M., & Peteves, S. (2007). Techno-economic assessment of hydrogen production processes for the hydrogen economy for the short and medium term. *International Journal of Hydrogen Energy, 32*(16), 3797–3810. doi:10.1016/j.ijhydene.2007.05.027.

Mugnai, G., Borruso, L., Mimmo, T., Cesco, S., Luongo, V., Frunzo, L., Fabbricino, M., Pirozzi, F., Cappitelli, F., & Villa, F. (2021). Dynamics of bacterial communities and substrate conversion during olive-mill waste dark fermentation: Prediction of the metabolic routes for hydrogen production. *Bioresource Technology, 319*. doi:10.1016/j.biortech.2020.124157.

Muradov, N. (2017). Low to near-zero CO_2 production of hydrogen from fossil fuels: Status and perspectives. *International Journal of Hydrogen Energy, 42*(20), 14058–14088. doi:10.1016/j.ijhydene.2017.04.101.

Muradov, N. Z. (2008). Production of hydrogen from hydrocarbons. *Hydrogen fuel* (pp. 45–114). CRC Press.

Najjar, Y. S. H. (2013). Hydrogen safety: The road toward green technology. *International Journal of Hydrogen Energy, 38*(25), 10716–10728. doi:10.1016/j.ijhydene.2013.05.126.

Nechache, A., & Hody, S. (2021). Alternative and innovative solid oxide electrolysis cell materials: A short review. *Renewable and Sustainable Energy Reviews, 149*, 111322. doi:10.1016/j.rser.2021.111322.

Nguyen, D. H., Nguyen, P. Q., Kim, S. M., Kim, J. H., Shim, G. H., & Ahn, H. S. (2025). Hydrogen production from ammonia: Preliminary design of portable cracking system and related micro-channel reactor. *International Journal of Hydrogen Energy, 128*, 597–612. doi:10.1016/j.ijhydene.2025.04.241.

Niermann, M., Beckendorff, A., Kaltschmitt, M., & Bonhoff, K. (2019). Liquid organic hydrogen carrier (LOHC)—Assessment based on chemical and economic properties. *International Journal of Hydrogen Energy, 44*(13), 6631–6654. doi:10.1016/j.ijhydene.2019.01.199.

O'Brien, T. F., Bommaraju, T. V., & Hine, F. (2005). Chlor-alkali technologies. *Handbook of chlor-alkali technology: Volume I: Fundamentals, Volume II: Brine Treatment and cell operation, Volume III: Facility design and product handling, Volume IV: Plant commissioning and support systems, Volume V: Corrosion, environmental issues, and future development* (pp. 387–442). Springer.

Palo, D. R., Dagle, R. A., & Holladay, J. D. (2007). Methanol steam reforming for hydrogen production. *Chemical Reviews, 107*(10), 3992–4021. doi:10.1021/cr050198b.

Pandey, A. K., Park, K. D., Morya, R., Joo, H. H., & Kim, S. H. (2025). Exploring sustainable biohydrogen production from dried food waste: Optimization strategies and environmental implications. *Journal of Environmental Management, 375*. doi:10.1016/j.jenvman.2025.124195.

Pathy, A., Nageshwari, K., Ramaraj, R., Maniam, G. P, Govindan, N., & Balasubramanian, P. (2022). Biohydrogen production using algae: Potentiality, economics and challenges. *Bioresource Technology, 360*, 127514. doi:10.1016/j.biortech.2022.127514.

Paulu, A., Matuštík, J., Trecáková, T., & Kočí, V. (2024). Overlooked source of hydrogen: The environmental potential of chlor-alkali by-product. *International Journal of Hydrogen Energy, 49*, 1437–1443. doi:10.1016/j.ijhydene.2023.10.246.

Policastro, G., Carraturo, F., Compagnone, M., Giugliano, M., Guida, M., Luongo, V., Napolitano, R., & Fabbricino, M. (2021). A preliminary study on a novel bioaugmentation technique enhancing lactic acid production by mixed cultures fermentation. *Bioresource Technology, 340*, 125595. doi:10.1016/j.biortech.2021.125595.

Prado, E. R. L., & Rial, R. C. (2025). Biohydrogen production from residual biomass: The potential of wheat, corn, rice, and barley straw—Recent advances. *Bioresource Technology, 432*. doi:10.1016/j.biortech.2025.132638.

Puricelli, S., Cardellini, G., Casadei, S., Faedo, D., van den Oever, A. E. M., & Grosso, M. (2021). A review on biofuels for light-duty vehicles in Europe. *Renewable and Sustainable Energy Reviews, 137*, 110398. doi:10.1016/j.rser.2020.110398.

Putatunda, C., Behl, M., Solanki, P., Sharma, S., Bhatia, S. K., Walia, A., & Bhatia, R. K. (2023). Current challenges and future technology in photofermentation-driven biohydrogen production by utilizing algae and bacteria. *International Journal of Hydrogen Energy, 48*(55), 21088–21109. doi:10.1016/j.ijhydene.2022.10.042.

Qazi, U. Y. (2022). Future of hydrogen as an alternative fuel for next-generation industrial applications; challenges and expected opportunities. *Energies, 15*(13), 4741. doi:10.3390/en15134741.

Rai, P. K., & Singh, S. P. (2016). Integrated dark- and photo-fermentation: Recent advances and provisions for improvement. *International Journal of Hydrogen Energy, 41*(44), 19957–19971. doi:10.1016/j.ijhydene.2016.08.084.

Ramachandran, R., & Menon, R. K. (1998). An overview of industrial uses of hydrogen. *International Journal of Hydrogen Energy, 23*(7), 593–598. doi:10.1016/s0360-3199(97)00112-2.

Ramprakash, B., & Muthukumar, K. (2018). Influence of sulfuric acid concentration on biohydrogen production from rice mill wastewater using pure and coculture of *Enterobacter aerogenes* and *Citrobacter freundii*. *International Journal of Hydrogen Energy, 43*(19), 9254–9258. doi:10.1016/j.ijhydene.2018.03.198.

Rani, P., Kumar Yadav, D., Yadav, A., Ram Bishnoi, N., Kumar, V., Ram, C., Pugazhendhi, A., & Kumar, S. S. (2024). Frontier in dark fermentative biohydrogen production from lignocellulosic biomass: Challenges and future prospects. *Fuel, 366*. doi:10.1016/j.fuel.2024.131187.

Ranjekar, A. M., & Yadav, G. D. (2021). Steam reforming of methanol for hydrogen production: A critical analysis of catalysis, processes, and scope. *Industrial and Engineering Chemistry Research, 60*(1), 89–113. doi:10.1021/acs.iecr.0c05041.

Ranjekar, A. M., & Yadav, G. D. (2023). Hydrogen production by steam reforming of methanol by Cu-Zn/$CeAlO_3$ perovskite. *New Journal of Chemistry, 47*(10), 4860–4870. doi:10.1039/d2nj05491d.

Raveendran, A., Chandran, M., & Dhanusuraman, R. (2023). A comprehensive review on the electrochemical parameters and recent material development of electrochemical water splitting electrocatalysts. *RSC Advances, 13*(6), 3843–3876. doi:10.1039/d2ra07642j.

Ren, N., Guo, W., Liu, B., Cao, G., & Ding, J. (2011). Biological hydrogen production by dark fermentation: Challenges and prospects towards scaled-up production. *Current Opinion in Biotechnology, 22*(3), 365–370. doi:10.1016/j.copbio.2011.04.022.

Rico, V. J., Hueso, J. L., Cotrino, J., & González-Elipe, A. R. (2010). Evaluation of different dielectric barrier discharge plasma configurations as an alternative technology for green C1 chemistry in the carbon dioxide reforming of methane and the direct decomposition of methanol. *Journal of Physical Chemistry A, 114*(11), 4009–4016. doi:10.1021/jp100346q.

Sá, S., Silva, H., Brandão, L., Sousa, J. M., & Mendes, A. (2010). Catalysts for methanol steam reforming—A review. *Applied Catalysis B: Environmental, 99*(1–2), 43–57. doi:10.1016/j.apcatb.2010.06.015.

Sadeq, A. M., Homod, R. Z., Hussein, A. K., Togun, H., Mahmoodi, A., Isleem, H. F., Patil, A. R., & Moghaddam, A. H. (2024). Hydrogen energy systems: Technologies, trends, and future prospects. *Science of the Total Environment, 939*. doi:10.1016/j.scitotenv.2024.173622.

Sagir, E., & Alipour, S. (2021). Photofermentative hydrogen production by immobilized photosynthetic bacteria: Current perspectives and challenges. *Renewable and Sustainable Energy Reviews, 141*. doi:10.1016/j.rser.2021.110796.

Sanz, O., Velasco, I., Pérez-Miqueo, I., Poyato, R., Odriozola, J. A., & Montes, M. (2016). Intensification of hydrogen production by methanol steam reforming. *International Journal of Hydrogen Energy, 41*(10), 5250–5259. doi:10.1016/j.ijhydene.2016.01.084.

Saravanan, A., Kumar, P. S., Shiong Khoo, K., Show, P.-L., Femina Carolin, C., Fetcia Jackulin, C., Jeevanantham, S., Karishma, S., Show, K.-Y., Lee, D.-J., & Chang, J.-S. (2021). Biohydrogen from organic wastes as a clean and environment-friendly energy source: Production pathways, feedstock types, and future prospects. *Bioresource Technology, 342*, 126021. doi:10.1016/j.biortech.2021.126021.

Sargsyan, H., Gabrielyan, L., Hakobyan, L., & Trchounian, A. (2015). Light–dark duration alternation effects on *Rhodobacter sphaeroides* growth, membrane properties and biohydrogen production in batch culture. *International Journal of Hydrogen Energy, 40*(11), 4084–4091. doi:10.1016/j.ijhydene.2015.01.163.

Sazali, N. (2020). Emerging technologies by hydrogen: A review. *International Journal of Hydrogen Energy, 45*(38), 18753–18771. doi:10.1016/j.ijhydene.2020.05.021.

Sebbahi, S., Nabil, N., Alaoui-Belghiti, A., Laasri, S., Rachidi, S., & Hajjaji, A. (2022). Assessment of the three most developed water electrolysis technologies: Alkaline water electrolysis, proton exchange membrane and solid-oxide electrolysis. *Materials Today: Proceedings, 66*, 140–145. doi:10.1016/j.matpr.2022.04.264.

Sgobbi, A., Nijs, W., De Miglio, R., Chiodi, A., Gargiulo, M., & Thiel, C. (2016). How far away is hydrogen? Its role in the medium and long-term decarbonisation of the European energy system. *International Journal of Hydrogen Energy, 41*(1), 19–35. doi:10.1016/j.ijhydene.2015.09.004.

Shabaneh, R., Roychoudhury, J., Braun, J. F., & Saxena, S. (2024). *The clean hydrogen economy and Saudi Arabia: Domestic developments and international opportunities*. Taylor & Francis.

Shagdar, E., Lougou, B. G., Shuai, Y., Ganbold, E., Chinonso, O. P., & Tan, H. (2020). Process analysis of solar steam reforming of methane for producing low-carbon hydrogen. *RSC Advances, 10*(21), 12582–12597. doi:10.1039/c9ra09835f.

Sharma, S., & Ghoshal, S. K. (2015). Hydrogen the future transportation fuel: From production to applications. *Renewable and Sustainable Energy Reviews, 43*, 1151–1158. doi:10.1016/j.rser.2014.11.093.

Shi, Y., Yang, A. F., Cao, C. S., & Zhao, B. (2019). Applications of MOFs: Recent advances in photocatalytic hydrogen production from water. *Coordination Chemistry Reviews, 390*, 50–75. doi:10.1016/j.ccr.2019.03.012.

Shim, H., Khani, Y., Valizadeh, B., Hoon Lee, S., Hyun Ko, C., Lee, D., & Park, Y. K. (2024). Improved biohydrogen production using Ni/ZrxCeyO$_2$ loaded on foam reactor through steam gasification of sewage sludge. *Bioresource Technology, 413*. doi:10.1016/j.biortech.2024.131530.

Shiva Kumar, S., & Himabindu, V. (2019). Hydrogen production by PEM water electrolysis—A review. *Materials Science for Energy Technologies, 2*(3), 442–454. doi:10.1016/j.mset.2019.03.002.

Show, K. Y., Lee, D. J., Tay, J. H., Lin, C. Y., & Chang, J. S. (2012). Biohydrogen production: Current perspectives and the way forward. *International Journal of Hydrogen Energy, 37*(20), 15616–15631. doi:10.1016/j.ijhydene.2012.04.109.

Sivagurunathan, P., Kumar, G., Bakonyi, P., Kim, S. H., Kobayashi, T., Xu, K. Q., Lakner, G., Tóth, G., Nemestóthy, N., & Bélafi-Bakó, K. (2016). A critical review on issues and overcoming strategies for the enhancement of dark fermentative hydrogen production in continuous systems. *International Journal of Hydrogen Energy, 41*(6), 3820–3836. doi:10.1016/j.ijhydene.2015.12.081.

Sivaramakrishnan, R., Shanmugam, S., Sekar, M., Mathimani, T., Incharoensakdi, A., Kim, S.-H., Parthiban, A., Edwin Geo, V., Brindhadevi, K., & Pugazhendhi, A. (2021). Insights on biological hydrogen production routes and potential microorganisms for high hydrogen yield. *Fuel, 291*, 120136. doi:10.1016/j.fuel.2021.120136.

Solar Now. (2023). *Green hydrogen: A guide to policy making*. Solar Now.

Spatolisano, E., Pellegrini, L. A., de Angelis, A. R., Cattaneo, S., & Roccaro, E. (2023). Ammonia as a carbon-free energy carrier: NH_3 cracking to H_2. *Industrial and Engineering Chemistry Research, 62*(28), 10813–10827. doi:10.1021/acs.iecr.3c01419.

Su, H., Cheng, J., Zhou, J., Song, W., & Cen, K. (2009). Combination of dark- and photo-fermentation to enhance hydrogen production and energy conversion efficiency. *International Journal of Hydrogen Energy, 34*(21), 8846–8853. doi:10.1016/j.ijhydene.2009.09.001.

Taghvaei, H., Jahanmiri, A., Rahimpour, M. R., Shirazi, M. M., & Hooshmand, N. (2013). Hydrogen production through plasma cracking of hydrocarbons: Effect of carrier gas and hydrocarbon type. *Chemical Engineering Journal, 226*, 384–392. doi:10.1016/j.cej.2013.04.035.

Tang, X., Fang, Z., Wu, Y., Yuan, Z., Deng, B., Du, Z., Sun, C., Zhou, F., Qiao, X., & Li, X. (2024). Efficiency NiCu/t-zirconia catalysts for methanol steam reforming: Experimental and DFT insights. *Energy, 297*, 131293. doi:10.1016/j.energy.2024.131293.

Wang, K., Lu, C., Zhang, H., Guo, S., Ru, G., Wang, J., Hu, J., Zhang, N., & Zhang, Q. (2022). Enhancement effect of defoamer additives on photo-fermentation biohydrogen production process. *Bioresource Technology, 352*, 127070. doi:10.1016/j.biortech.2022.127070.

Wang, Y., Yang, H., Zhang, X., Han, F., Tu, W., & Yang, W. (2020). Microalgal hydrogen production. *Small Methods., 4*(3). doi:10.1002/smtd.201900514.

Wei, X., Feng, J., Cao, W., & Guo, L. (2021). Enhanced biohydrogen production by an ammonium-tolerant *Rhodobacter capsulatus* from sugarcane bagasse. *Fuel, 300*, 121009. doi:10.1016/j.fuel.2021.121009.

Wichman, I. S. (2003). Material flammability, combustion, toxicity and fire hazard in transportation. *Progress in Energy and Combustion Science, 29*(3), 247–299. doi:10.1016/S0360-1285(03)00027-3.

Xia, J., & Niu, W. (2020). Pushing carbon footprint reduction along environment with carbon-reducing information asymmetry. *Journal of Cleaner Production, 249*, 119376. doi:10.1016/j.jclepro.2019.119376.

Xu, Y., Kameoka, S., Kishida, K., Demura, M., Tsai, A. P., & Hirano, T. (2005). Catalytic properties of alkali-leached Ni_3Al for hydrogen production from methanol. *Intermetallics, 13*(2), 151–155. doi:10.1016/j.intermet.2004.07.039.

Xue, S., Yan, J., Liang, D., Wang, F., & Lv, G. (2024). Effects of carbon source variability on enhanced bio-hydrogen production. *Bioresource Technology, 406*. doi:10.1016/j.biortech.2024.131000.

Yang, Z., Ding, W., Zhang, Y., Lu, X., Zhang, Y., & Shen, P. (2010). Catalytic partial oxidation of coke oven gas to syngas in an oxygen permeation membrane reactor combined with NiO/MgO catalyst. *International Journal of Hydrogen Energy, 35*(12), 6239–6247. doi:10.1016/j.ijhydene.2009.07.103.

Yin, T., Wang, W., Guo, W., Zhuo, S., Cao, G., Ren, H., Li, J., Xie, G., Ding, J., & Liu, B. (2024). Enhanced thermophilic hydrogen production by an enriched novel acetic-acid-type fermentative bacterium from inoculum sludge with nonheat pretreatment. *Energy and Fuels, 38*(10), 8749–8761. doi:10.1021/acs.energyfuels.3c04696.

Yin, Y., Song, W., & Wang, J. (2022). Inhibitory effect of acetic acid on dark-fermentative hydrogen production. *Bioresource Technology, 364*, 128074. doi:10.1016/j.biortech.2022.128074.

Yousefi Rizi, H., & Shin, D. (2022). Green hydrogen production technologies from ammonia cracking. *Energies, 15*(21), 8246. doi:10.3390/en15218246.

Yu, M., Budiyanto, E., & Tüysüz, H. (2022). Principles of water electrolysis and recent progress in cobalt-, nickel-, and iron-based oxides for the oxygen evolution reaction. *Angewandte Chemie—International Edition, 61*(1). doi:10.1002/anie.202103824.

Yu, P., Ma, J., Zhang, R., Zhang, J. Z., & Botte, G. G. (2018). Novel Pd-Co electrocatalyst supported on carbon fibers with enhanced electrocatalytic activity for coal electrolysis to produce hydrogen. *ACS Applied Energy Materials, 1*(2), 267–272. doi:10.1021/acsaem.7b00085.

Yuan, P., Chen, L., Liu, C., & Wang, Z. (2023). Numerical studies on hydrogen production from ammonia thermal cracking with catalysts. *Energies, 16*(13), 5196. doi:10.3390/en16135196.

Yue, T., Jiang, D., Zhang, Z., Zhang, Y., Li, Y., Zhang, T., & Zhang, Q. (2021). Recycling of shrub landscaping waste: Exploration of bio-hydrogen production potential and optimization of photo-fermentation bio-hydrogen production process. *Bioresource Technology, 331*, 125048. doi:10.1016/j.biortech.2021.125048.

Zech, K., Oehmichen, K., Grasemann, E., Michaelis, J., Funke, S., & Seiffert, M. (2015). Technical, economic and environmental assessment of technologies for the production of biohydrogen and its distribution results of the Hy-NOW study. *International Journal of Hydrogen Energy, 40*(15), 5487–5495. doi:10.1016/j.ijhydene.2015.01.177.

Zeng, K., & Zhang, D. (2010). Recent progress in alkaline water electrolysis for hydrogen production and applications. *Progress in Energy and Combustion Science, 36*(3), 307–326. doi:10.1016/j.pecs.2009.11.002.

Zhang, F., Zhao, P., Niu, M., & Maddy, J. (2016). The survey of key technologies in hydrogen energy storage. *International Journal of Hydrogen Energy, 41*(33), 14535–14552. doi:10.1016/j.ijhydene.2016.05.293.

Zhang, G., Ruan, J., & Du, Tingting (2021b). Recent advances on photocatalytic and electrochemical oxidation for ammonia treatment from water/wastewater. *ACS ES&T Engineering, 1*(3), 310–325. doi:10.1021/acsestengg.0c00186.

Zhang, H., Li, Y., Chen, L., & Zhang, Q. (2021c). Effect of zinc ion on photo-fermentative hydrogen production performance, kinetics and electronic distribution in biohydrogen production by HAU-M1. *Bioresource Technology, 324*. doi:10.1016/j.biortech.2021.124680.

Zhang, J., Lenser, C., Menzler, N. H., & Guillon, O. (2020a). Comparison of solid oxide fuel cell (SOFC) electrolyte materials for operation at 500 °C. *Solid State Ionics, 344*, 115138. doi:10.1016/j.ssi.2019.115138.

Zhang, J., Yang, M., Zhao, W., Zhang, J., & Zang, L. (2021d). Biohydrogen production amended with nitrogen-doped biochar. *Energy and Fuels, 35*(2), 1476–1487. doi:10.1021/acs.energyfuels.0c03405.

Zhang, L., Jia, C., Bai, F., Wang, W., An, S., Zhao, K., Li, Z., Li, J., & Sun, H. (2024a). A comprehensive review of the promising clean energy carrier: Hydrogen production, transportation, storage, and utilization (HPTSU) technologies. *Fuel, 355,* 129455. doi:10.1016/j.fuel.2023.129455.

Zhang, Q., Jiao, Y., He, C., Ruan, R., Hu, J., Ren, J., Toniolo, S., Jiang, D., Lu, C., Li, Y., Man, Y., Zhang, H., Zhang, Z., Xia, C., Wang, Y., Jing, Y., Zhang, X., Lin, R., Li, G., … Tahir, N. (2024b). Biological fermentation pilot-scale systems and evaluation for commercial viability towards sustainable biohydrogen production. *Nature Communications, 15*(1). doi:10.1038/s41467-024-48790-4.

Zhang, Q., Zhu, S., Zhang, Z., Zhang, H., & Xia, C. (2021a). Enhancement strategies for photo-fermentative biohydrogen production: A review. *Bioresource Technology, 340,* 125601.

Zhang, T., Jiang, D., Zhang, H., Jing, Y., Tahir, N., Zhang, Y., & Zhang, Q. (2020b). Comparative study on bio-hydrogen production from corn stover: Photo-fermentation, dark-fermentation and dark-photo co-fermentation. *International Journal of Hydrogen Energy, 45*(6), 3807–3814. doi:10.1016/j.ijhydene.2019.04.170.

Zhang, X., Jiang, D., Zhang, H., Wang, Y., Zhang, Z., Lu, C., & Zhang, Q. (2021e). Enhancement of the biohydrogen production performance from mixed substrate by photo-fermentation: Effects of initial pH and inoculation volume ratio. *Bioresource Technology, 319,* 124153. doi:10.1016/j.biortech.2020.124153.

Zhang, Y., Li, Q., Wang, X., Yang, H., & Guo, L. (2017). Enhanced biohydrogen production from cornstalk through a two-step fermentation: Dark fermentation and photofermentation. *International Journal of Energy Research, 41*(15), 2491–2501. doi:10.1002/er.3810.

Zhao, Z. T., Ding, J., Wang, B. Y., Bao, M. Y., Liu, B. F., Pang, J. W., Ren, N. Q., & Yang, S. S. (2024). Advances in the biomass valorization in dark fermentation systems: A sustainable approach for biohydrogen production. *Chemical Engineering Journal, 481.* doi:10.1016/j.cej.2023.148444.

Zheng, Y., Zhang, Q., Zhang, Z., Jing, Y., Hu, J., He, C., & Lu, C. (2022). A review on biological recycling in agricultural waste-based biohydrogen production: Recent developments. *Bioresource Technology, 347,* 126595. doi:10.1016/j.biortech.2021.126595.

Zhong, C., Hu, W. B., & Cheng, Y. F. (2013). Recent advances in electrocatalysts for electro-oxidation of ammonia. *Journal of Materials Chemistry A, 1*(10), 3216–3238. doi:10.1039/c2ta00607c.

CHAPTER

2

Biohydrogen production from biophotolysis

Yaping Zheng[a], *Yun Huang*[b,*], *and Shan Liu*[c]

[a]Henan Agricultural University, Zhengzhou, China, [b]Chongqing University, Chongqing, China, [c]China Agricultural University, Beijing, China

* Corresponding author. E-mail address: yunhuang@cqu.edu.cn

2.1 Introduction

Over a century ago, during investigations into the fundamental metabolic activities of microorganisms, the phenomenon of transient hydrogen evolution by bacteria and microalgae under dark, anaerobic conditions was observed. By the 1990s, driven by the increasing global emphasis on renewable and clean energy sources, microalgae hydrogen production emerged as a prominent research focus.

As early as in 1939 the renowned German plant physiologist Hans Gaffron, while investigating the metabolic characteristics of the green alga *Scenedesmus obliquus*, serendipitously observed that algae cells subjected to dark anaerobic incubation exhibited transient hydrogen release upon reillumination. Subsequently, in 1942, Gaffron provided a detailed characterization of this photochemical hydrogen production process, linking it to the photosynthetic electron transport chain. This discovery challenged the prevailing paradigm that photosynthesis resulted exclusively in oxygen evolution, revealing an ancient hydrogen metabolic pathway preserved in microalgae under specific evolutionary pressures (Gaffron & Rubin, 1942).

2.2 Microalgae species and key enzyme systems

Microalgae capable of photosynthetic hydrogen production primarily include cyanobacteria and green algae.

Biohydrogen: Advances in Theory and Production
DOI: https://doi.org/10.1016/B978-0-443-27686-6.00007-4

As the earliest photoautotrophs on Earth, numerous cyanobacterial species have been confirmed to exhibit hydrogen evolution activity. In nitrogen-fixing cyanobacteria, such as *Anabaena cylindrica, Anabaena variabilis, Anabaena azollae,* and *Nostoc muscorum*, hydrogen production is catalyzed by nitrogenase as a byproduct of the nitrogen fixation reaction (Shah et al., 2003). During the catalysis of atmospheric nitrogen (N_2) reduction to ammonia (NH_3), nitrogenase inevitably reduces a portion of electrons and protons to hydrogen gas; this constitutes an inherent side reaction. Because of the oxygen sensitivity of nitrogenase, nitrogen-fixing cyanobacteria have evolved specialized cellular structures known as "heterocysts" to provide a microaerophilic environment for the enzyme (Liu et al., 2006). However, nitrogen fixation is a highly energy-intensive process, requiring the consumption of 16 adenosine triphosphate (ATP) molecules per reduced N_2 molecule. Consequently the energy conversion efficiency of hydrogen production mediated by nitrogenase is significantly lower than that of dedicated hydrogenases (Ghirardi et al., 2007). In nonnitrogen-fixing cyanobacteria, such as *Synechocystis* sp., *Spirulina platensis*, and *Synechococcus* sp., hydrogen production is catalyzed by reversible hydrogenases.

Green algae are eukaryotic microorganisms. To date, multiple species capable of biophotolytic hydrogen production have been identified, including *S. obliquus, Chlamydomonas reinhardtii,* and *Platymonas subcordiformis* (Baebprasert et al., 2010). Green algae photolysis utilizes the reversible hydrogenase system located within the chloroplast to split water. Characterized by high catalytic efficiency, low energy consumption, and a clean production process, this mechanism enables the self-organization of light-harvesting systems, spontaneous energy accumulation, and rapid directional conversion. Consequently, it is internationally recognized as one of the most promising approaches for solar-driven hydrogen production.

Hydrogenases are the primary enzymes responsible for catalyzing the reduction of protons to molecular hydrogen ($2H^+ + 2e^- \rightarrow H_2$). Based on the metal composition of their active centers, they are categorized primarily into [FeFe]-hydrogenases and [NiFe]-hydrogenases. Green algae (e.g., *C. reinhardtii*) predominantly possess [FeFe]-hydrogenases (such as HydA1 and HydA2). These enzymes are characterized by a high turnover frequency, making them among the most efficient biological catalysts for hydrogen evolution found in nature. However, they are extremely sensitive to oxygen; even trace amounts can lead to irreversible inactivation of the active site (Németh et al., 2020). In contrast, many cyanobacteria contain [NiFe]-hydrogenases, which typically exhibit superior oxygen tolerance but possess significantly lower catalytic activity for hydrogen evolution compared to [FeFe]-hydrogenases (Shepard et al., 2013) (Table 2.1).

TABLE 2.1 Comparison of the characteristics of key hydrogen-producing enzyme systems in microalgae.

	[FeFe]-Hydrogenase	**[NiFe]-Hydrogenase**	**Nitrogenase**
Metal center	Fe–Fe (H-cluster)	Ni–Fe	Mo–Fe
Typical source	Green algae (*Chlamydomonas reinhardtii*)	Cyanobacteria (*Synechocystis*)	N_2-fixing cyanobacteria (*Anabaena*)
Catalytic bias	Biased toward H_2 evolution ($2H^+ \rightarrow H_2$)	Biased toward H_2 uptake ($H_2 \rightarrow 2H^+$)	Byproduct of N_2 fixation ($N_2 \rightarrow NH_3 + H_2$)
TOF	Extremely high (6000–9000 s^{-1})	Moderate (~100 s^{-1})	Extremely low (~6 s^{-1})
ATP requirement	ATP-independent	ATP-independent	High energy consumption (16 ATP/N_2)
Oxygen sensitivity	Extremely sensitive (irreversible inactivation by O_2)	Relatively tolerant (reversible inactivation)	Sensitive (requires protection by heterocysts)
Maturation proteins	HydE, HydF, HydG	HypA, B, C, D, E, F	*Nif* gene cluster
References	Britt et al. (2015), Lampret et al. (2019), Caserta and Papini (2018), Betz et al. (2015), Filmon et al. (2025), Leone et al. (2023)		

TOF, *turnover frequency*.

2.3 Pathways of microalgae photolytic hydrogen production

Microalgae generally possess a relatively complete photosynthetic system. Microalgae photolytic hydrogen production can be divided into two steps. In the first step, microalgae split water via photosystem II (PSII), generating protons and electrons while releasing oxygen. The reaction can be expressed as follows (Grechanik & Tsygankov, 2022):

$$2H_2O \rightarrow 4H^+ + 4e^- + O_2 \tag{2.1}$$

In the second step, cyanobacteria reduce protons to hydrogen via the nitrogenase system, whereas green algae utilize the reversible hydrogenase system. The pathway for cyanobacterial hydrogen production is illustrated in the figure. Under anaerobic conditions, cyanobacteria rely on their intracellular nitrogenase system to convert nitrogen gas into ammonia, simultaneously releasing hydrogen gas. Only 25% of the electrons passing through the nitrogenase are used to reduce protons for hydrogen release (Fig. 2.1). The reaction equation is as follows:

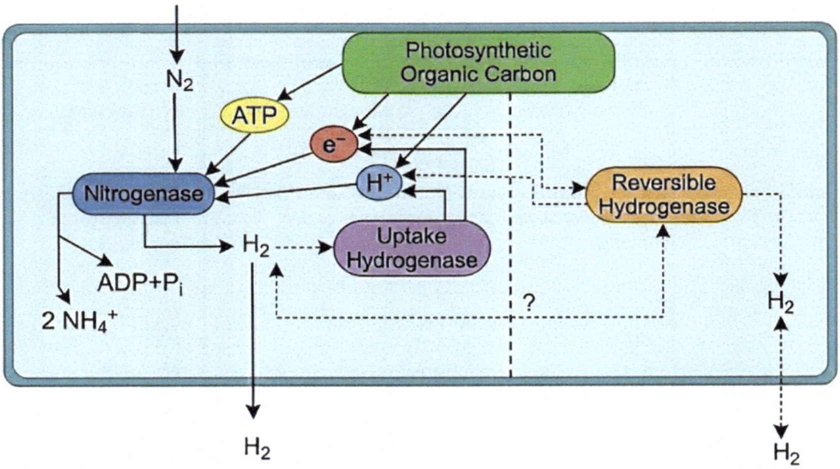

FIGURE 2.1 Schematic diagram of the hydrogen production pathway via cyanobacterial nitrogenase. *ADP*, Adenosine diphosphate; P_i, phosphate group.

$$N_2 + 8H^+ + 8e^- + 16ATP \rightarrow 2NH_3 + H_2 + 16ADP + 16Pi \quad (2.2)$$

This process not only consumes high-energy substances in the form of ATP but also has a proton utilization rate for hydrogen production of only one-fourth of that for the nitrogen fixation reaction, resulting in low hydrogen production efficiency.

Green algae rely on intracellular reversible hydrogenases to directly reduce protons to hydrogen under anaerobic illumination. The reaction equation is as follows:

$$2H^+ + 2e^- \overset{\text{Hydrogenases}}{\rightarrow} H_2 \quad (2.3)$$

Although the reaction catalyzed by reversible hydrogenase is simple, the pathway of hydrogen production in green algae is quite complex.

In the metabolic pathway of direct water photolysis for hydrogen production in green algae (Fig. 2.2), ferredoxin (Fd) serves as the primary electron donor for the hydrogenase. This process, which directly utilizes hydrogenase for gas production without involving dark reactions, possesses a high light energy conversion efficiency (Burgess et al., 2011). Theoretical calculations suggest it can reach 12%–14%, which is the maximum theoretical efficiency for converting solar energy into hydrogen energy. Direct biophotolysis is a highly promising approach because it utilizes solar energy to easily convert the substrate water into hydrogen and oxygen. However, although PSII photolysis of water generates electrons

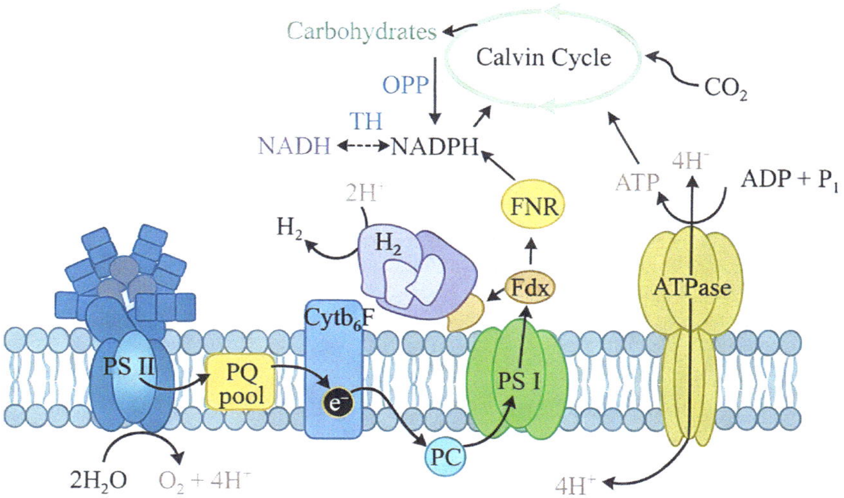

FIGURE 2.2 Direct biophotolysis pathway of microalgae hydrogen production. *ADP*, Adenosine diphosphate; *ATPase*, adenosine triphosphatase; *Cyt b6f*, cytochrome b6f; *Fdx*, ferredoxin; *FNR*, ferredoxin-NADP+ reductase; *NADH*, reduced nicotinamide adenine dinucleotide; *NADPH*, reduced nicotinamide adenine dinucleotide phosphate; *OPP*, oxidative pentose phosphate pathway; *PC*, plastocyanin; *Pi*, phosphate group; *PQ*, plastoquinone; *PSI*, photosystem I; *PSII*, photosystem II; *TH*, transhydrogenase.

and protons, it simultaneously produces a large amount of oxygen, and hydrogenases are extremely sensitive to oxygen. Researchers have indicated that reducing PSII activity can lower photosynthetic oxygen content, but this also limits hydrogen production. Therefore the key issue in utilizing microalgae for direct hydrogen production lies in resolving how to isolate the hydrogenase from the oxygen produced during photosynthesis or how to reduce oxygen content while ensuring sufficient proton transfer to the hydrogenase to improve hydrogen production efficiency (Hippler & Khosravitabar, 2024).

The process of indirect photosynthetic hydrogen production in green algae proceeds in two distinct stages. In the first stage, microalgae store energy in the form of carbohydrates, such as starch and glycogen, via photosynthesis. In the second stage the decomposition of these carbohydrates catalyzes the reduction of protons to generate hydrogen gas, as illustrated in Fig. 2.3. During indirect photolysis, PSII is involved in water splitting to produce oxygen and reduced ferredoxin. The primary advantage of this process is the avoidance of oxygen inhibition of the hydrogenase. Starch accumulated during photosynthesis is degraded during the dark reaction phase, generating a substantial flux of electrons that are transferred to the plastoquinone (PQ) pool; hydrogenases then utilize this electron transport

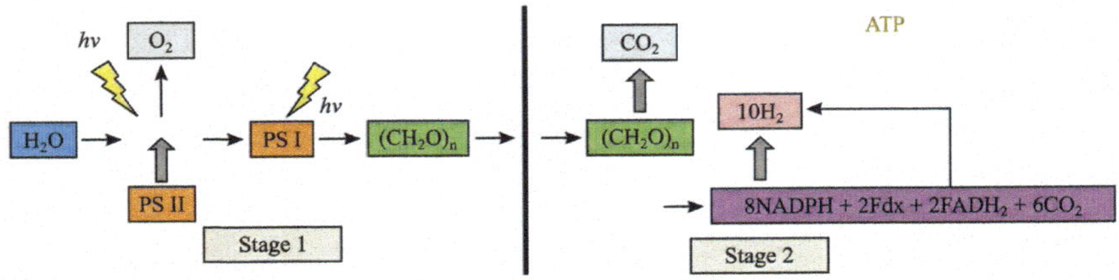

FIGURE 2.3 Process of indirect photosynthetic hydrogen production in green algae. *Fdx*, Ferredoxin; *NADPH*, reduced nicotinamide adenine dinucleotide phosphate; *PSI*, photosystem I; *PSII*, photosystem II.

chain to produce hydrogen. The disadvantages include high energy consumption and low light energy conversion efficiency.

In an alternative pathway for indirect photosynthetic hydrogen production, organic matter accumulated via the Calvin cycle in microalgae undergoes the tricarboxylic acid (TCA) cycle and glycolysis to produce reduced nicotinamide adenine dinucleotide phosphate (NADPH). NADPH oxidoreductase converts NADPH to NADP+, releasing electrons. These electrons enter the electron transport chain indirectly via the PQ pool and are transferred by ferredoxin to the hydrogenase for hydrogen production (Zhang et al., 2023).

From a thermodynamic perspective, biophotolytic hydrogen production is an endergonic process ($\Delta G° = +237$ kJ/mol) that elevates electrons from low-energy water molecules to high-energy hydrogen gas. This process relies entirely on the light reaction system of photosynthesis. Based on the electron transfer path from water to hydrogen and the involvement of intermediate metabolites, direct biophotolysis—where electrons originate directly from water splitting—represents the most ideal "one-step" process with the highest theoretical efficiency. However, it faces the direct challenge of hydrogenase inhibition by the product O_2. Indirect biophotolysis separates photosynthesis (oxygen production and carbon fixation for biomass accumulation) and the hydrogen production process (anaerobic consumption of biomass for hydrogen generation) either temporally or spatially. Although this mechanism circumvents oxygen inhibition, the overall thermodynamic efficiency is reduced because of the intermediate conversion of energy into chemical forms (sugars). Table 2.2 systematically compares the key characteristics of these two pathways.

Based on the preceding analysis the biological factors constraining the development of microalgae hydrogen production and the critical challenges requiring resolution can be summarized. These biological constraints include the sensitivity of hydrogenases to oxygen (Batyrova & Hallenbeck, 2017), regulatory issues arising from excessive ATP production during metabolic processes; competition for photosynthetic

TABLE 2.2 Characteristics of direct and indirect biophotolysis.

	Direct biophotolysis	Indirect biophotolysis
Core principle	Single-stage, light-driven H_2 evolution; water serves as the direct electron donor	Two-stage process: light-driven biomass accumulation → dark/photo-fermentation for H_2 production
Electron source	Water (H_2O), via PSII-mediated splitting	Stored endogenous carbohydrates (starch/glycogen)
Key enzyme systems	[FeFe]-Hydrogenase (green algae) [NiFe]-hydrogenase (cyanobacteria)	Hydrogenase Nitrogenase (cyanobacteria)
Main substrates	Water, photons	Water, photons, CO_2 (Stage 1) Endogenous starch (Stage 2)
Main products	H_2 and O_2 (simultaneously evolved)	O_2, biomass (Stage 1) H_2, CO_2, organic acids (Stage 2)
Representative organisms	Green algae (*Chlamydomonas reinhardtii*)	Cyanobacteria (*Anabaena, Synechocystis*) Green algae
Theoretical efficiency	High (direct solar-to-hydrogen conversion)	Lower (thermodynamic losses associated with intermediate metabolic conversion)
Technical bottlenecks	Extreme sensitivity of hydrogenase to coevolved O_2 (the "oxygen paradox")	Complex process control, limitations in substrate conversion rates, large areal footprint
References	Rather and Srivastav (2021), Faraloni et al. (2025), Jaramillo et al. (2025)	

PSII, *photosystem II*.

electrons by other metabolic pathways, low solar energy conversion efficiency; the competitive inhibition of photosynthetic hydrogen production by carbon dioxide; and the restriction of hydrogen production caused by the accumulation of proton gradients during photosynthesis.

2.4 Enhancement of hydrogen production via genetic and metabolic engineering

With the maturation of CRISPR/Cas9 genome editing technologies and synthetic biology toolkits for microalgae, the research focus has shifted from traditional physiological regulation to the rational genetic modification of algae cells. Current modification strategies can be summarized within the "push–pull–protect" framework (Table 2.3).

Protect. Construction of oxygen-tolerant hydrogenases. Tunnel engineering: Based on crystal structural data, site-directed mutagenesis is employed to narrow the internal gas diffusion channels of the hydrogenase. This strategy aims to hinder the access of O_2 to the catalytic active center while minimizing the obstruction of H_2 efflux. For instance, mutating channel residues in the *Clostridium* hydrogenase CpI (e.g., V268I) has been shown to significantly enhance oxygen tolerance. Kinetic control, screening, or designing hydrogenases capable of rapid reactivation: Biomimetic oxygen tolerance, certain bacteria, such as *Clostridium beijerinckii*, possess a CbA5H hydrogenase with a unique "safety cap" mechanism that protects the H-cluster through conformational changes in the presence of oxygen. Grafting such protective domains onto microalgae hydrogenases represents a current research hotspot .

Push. Truncated light-harvesting antenna (TLA). This strategy involves optimizing the light-harvesting antenna and electron supply to improve photon capture efficiency and ensure a sufficient flow of electrons to the reaction center. Wild-type microalgae possess large light-harvesting complexes (LHCs). Under high light intensities the absorbed light energy far exceeds photosynthetic requirements, resulting in substantial energy dissipation as heat (nonphotochemical quenching, NPQ) or fluorescence. This also causes insufficient light access for cells deep within the reactor because of the self-shading effect. By suppressing chlorophyll *b* synthesis genes (e.g., CAO) or LHC protein genes (TLA series) via RNA interference (RNAi) or gene knockout, mutants with reduced antenna sizes (e.g., *tla*1, *stm*6) have been constructed. This not only reduces energy waste but also improves light penetration depth within the photobioreactor (PBR). Experimental data indicate that the solar-to-biomass conversion efficiency of tla mutants can be two to three times higher than that of the wild type, with particularly pronounced advantages in high-density cultures (Kumar et al., 2021).

Pull. Construction of photosystem–hydrogenase fusion proteins. This represents the most direct "pull" strategy, aiming to direct electron flow specifically toward the hydrogenase, thereby reducing electron diversion to competing pathways and forcing electrons into hydrogen production. Using gene fusion technology the hydrogenase protein is directly tethered to the terminal electron acceptor subunit of PSI. This constructs a direct electron transfer channel, allowing high-efficiency electron transfer from PSI to the hydrogenase, thereby bypassing competition with natural electron acceptors such as ferredoxin–NADP + reductase (FNR) and preventing electron flow into the Calvin cycle (Kanygin et al., 2020). Experiments have demonstrated that such fusion proteins significantly enhance electron transfer efficiency, drastically increase hydrogen

TABLE 2.3 Enhancement of microalgae hydrogen production performance via genetic engineering strategies.

Classification of strategy	Target gene/protein	Mechanism of action	Performance enhancement/specific data	References
Light-harvesting optimization (push)	*TLA1, TLA2, TLA3* (truncated antenna)	Reduces LHC size, decreases NPQ, and mitigates self-shading effects	Light energy conversion efficiency increased by two to three times compared to wild type; significant increase in H_2 yield under high-density culture	Kumar et al. (2021)
Light-harvesting optimization (push)	*stm6* (*MOC1* mutation)	Regulates electron partitioning between mitochondrial respiration and photosynthesis; increases dark respiratory oxygen consumption	H_2 yield approximately five to nine times that of wild type (under specific illumination conditions)	Kanygin and Yacoby (2017)
Electron redirection (pull)	PSI–HydA fusion protein	Physically tethers PSI to hydrogenase, bypassing competition from FNR and enabling direct electron transfer	H_2 production rate reached 84.9 ± 3.1 μmol/H_2 mg/Chl h	Rumbaugh et al. (2024)

(*continued on next page*)

TABLE 2.3 Enhancement of microalgae hydrogen production performance via genetic engineering strategies—cont'd

Classification of strategy	Target gene/protein	Mechanism of action	Performance enhancement/specific data	References
Electron redirection (pull)	*pgr5* knockout	Blocks PGR5-dependent CEF, diverting more electrons toward hydrogenase	H_2 yield increased by fivefold (when combined with optimized reactor design)	Elman et al. (2025)
Metabolic flux regulation (indirect)	*sta7* overexpression (starch synthesis)	Enhances starch accumulation during the light phase, providing increased substrate availability for dark fermentation	Total H_2 yield during dark fermentation increased by three to five times	Quiroz et al. (2023)
Oxygen protection (protect)	*lba* (leghemoglobin)	Expression of leghemoglobin in the chloroplast to facilitate intracellular oxygen scavenging	H_2 yield increased by four times	Jiao et al. (2024)

CEF, *cyclic electron flow;* LHC, *light-harvesting complex;* NPQ, *nonphotochemical quenching;* PSI, *photosystem I.*

production rates, and exhibit improved utilization of high light intensities (Rumbaugh et al., 2024). Furthermore, strategies involving the knockout of competing pathways can be employed. Downregulating FNR expression reduces electron flow to NADPH. Alternatively, knocking out key proteins involved in cyclic electron flow (CEF) (such as PGRL1 or PGR5) forces electrons toward linear electron flow or hydrogen production, thereby providing more electrons for the latter. Studies show that *pgrl*1 mutants exhibit significantly higher hydrogen production rates than the wild type during the initial phase of anaerobic induction, although this is accompanied by the side effect of impaired growth (Tolleter et al., 2011).

2.5 Optimization of culture conditions and process parameters

In addition to the genetic modification of the microalgae itself, the optimization of external culture conditions is an indispensable component for enhancing hydrogen production efficiency. Methods for optimizing culture conditions include the synchronization of cell growth, the addition of low-concentration sulfate during the hydrogen production phase, and the optimization of culture medium pH, light intensity, light regimes, cell immobilization, and growth conditions.

2.5.1 Impact of Light

The core of microbial biophotolytic hydrogen production lies in utilizing the Photosystem II (PSII) of microalgae or cyanobacteria to split water molecules and transfer electrons to hydrogenase for proton reduction to generate hydrogen. As the energy substrate for this reaction system, light—specifically its intensity, spectrum, and light/dark cycles—plays a decisive regulatory role in hydrogen production performance.

2.5.1.1 Light intensity

Light serves as the energy source driving the entire process; however, its intensity acts as a "double-edged sword." Under low light intensity the hydrogen production rate is linearly and positively correlated with light intensity. At this stage, light conversion efficiency (LCE) is maximal, as nearly all captured photons are utilized for electron transport (Melitos et al., 2021). As light intensity increases, enzymatic reactions in the electron transport chain (particularly PQ reoxidation) become rate-limiting steps. The hydrogen production rate no longer increases significantly with increasing intensity, and excess light energy is dissipated as heat or fluorescence. When light intensity exceeds the saturation point (supra-optimal intensity), the rate of D1 protein damage in the PSII reaction center exceeds the repair rate, leading to a decline in photosynthetic activity and the cessation of hydrogen production (Takahashi et al., 2009).

Furthermore, higher light intensity can accelerate PSII photodamage under sulfur-deprived conditions, thereby shortening the induction time required to reach an anaerobic state and initiating the hydrogen production process earlier (Saifuddin & Priatharsini, 2016). However, excessively high light intensity causes severe photoinhibition, inflicting irreversible damage to the entire photosynthetic apparatus, including PSI, which ultimately reduces hydrogen production efficiency. Therefore an optimal light intensity range exists. This optimal value is not static but correlates with multiple factors, including cell density, initial sulfur concentration, and the pigment content of the algal strain. Identifying and maintaining optimal illumination conditions is a critical technical point in PBR design and process control (Williams & Bees, 2014). For *C. reinhardtii*, multiple studies indicate that the optimal light intensity range is typically between 50 and 200 $\mu E/m^2/s$. Intensities exceeding 200 $\mu E/m^2/s$ often lead to inhibition of hydrogen production. This is primarily because strong light accelerates PSII-mediated water splitting, causing intracellular oxygen concentrations to rise rapidly and exceed the oxygen consumption capacity of respiration, thereby shifting the environment to an aerobic state and inactivating the hydrogenase (Bonente et al., 2012; Laurinavichene et al., 2004).

Controlling the light-dark cycles during the cell culture process can induce synchronization of cell growth. Kosourov et al. (2002) found that synchronized cells required less time to establish an anaerobic environment and entered the hydrogen production phase more rapidly. Tsygankov et al. discovered that significant differences in hydrogen yield and the duration of the five stages of hydrogen production occurred depending on the time of cell harvest relative to the onset of illumination. Algal cells harvested 4 hours after the onset of illumination exhibited significantly higher hydrogen yields compared to nonsynchronized cells (Tsygankov et al., 2002).

2.5.1.2 Light quality

Light quality, or the wavelength distribution of light, significantly influences the biohydrogen production process by affecting the absorption efficiency of photosynthetic pigments and cellular metabolic regulatory signals. Microalgal photosynthesis is primarily driven by chlorophyll *a* (absorption peaks at ~440 and ~680 nm), chlorophyll *b* (~470 and ~650 nm), and carotenoids (400–500 nm). The "action spectrum" for hydrogen production typically aligns closely with the absorption spectra of PSII and PSI (Lubner et al., 2011).

Multiple studies indicate that red light (especially 660–690 nm) exhibits the highest hydrogen production efficiency per unit of photon energy input. A study by Li et al. further confirmed that 660 nm pulsed light significantly enhanced photosynthetic H_2 yield in *C. reinhardtii* in a system using sodium sulfite as an oxygen scavenger (Li et al., 2024). This is because red light avoids the absorption bands of carotenoids (the efficiency of energy transfer from carotenoids to the reaction center is typically lower than that of chlorophyll) and possesses photon energy that precisely meets the requirements for charge separation, reducing thermal effects caused by excess energy (Saifuddin & Priatharsini, 2016).

Although blue light can also drive photosynthesis, research has found that it tends to favor cell growth, protein synthesis, and the accumulation of secondary metabolites (such as astaxanthin) (Ooms et al., 2016). In certain photosynthetic bacteria (e.g., *Rhodopseudomonas palustris*), hydrogen production activity is higher under infrared or red light compared to the full visible spectrum or blue light, which is related to the properties of their light-harvesting pigments (bacteriochlorophylls). Furthermore, although blue light penetrates water better than red light (which is easily absorbed by water molecules), the strong absorption of blue light by chlorophyll in high-concentration algal cultures leads to severe "self-shading" effects (Teo et al., 2017).

Far-red light plays a key role in regulating the state of the photosynthetic apparatus. It preferentially excites PSI, promoting the transition to State 1, where the LHCII antenna moves toward PSII. Although pure far-red light cannot sustain hydrogen production (because of the lack of electrons from PSII), an appropriate proportion of far-red light can prevent the overreduction of the PQ pool, reducing "electron traffic jams." This facilitates ATP generation via the CEF pathway, which is crucial for maintaining intracellular homeostasis (Strand et al., 2015). However, for direct hydrogen production, the lack of PSII excitation (i.e., lack of red/blue light) is fatal.

2.5.1.3 *Pulsed light and the flashing light effects*

In natural environments or large-scale reactors, turbulence causes cells to cycle rapidly between illuminated zones (surface) and dark zones (deep layers), experiencing light/dark cycles. Simulating and optimizing this cycle using artificial light sources—known as "Pulsed Light" technology—to exploit the "Flashing Light Effect" has become an important strategy for enhancing LCE.

Under continuous high-intensity light the reaction center may receive a subsequent photon before completing the electron transfer initiated by the previous one, leading to the dissipation of the unusable photon. When high-intensity short pulses are provided, followed by a dark period, the enzyme system processes the accumulated electrons (PQ reoxidation) during the dark phase, readying the reaction center for the next flash. This utilizes the high excitation capacity of intense light while avoiding saturation-induced waste. Studies consistently show that to observe a significant "flashing light effect" (i.e., a yield higher than continuous light of the same average intensity), the frequency typically needs to exceed 100 Hz. For instance, in *C. reinhardtii* cultures, 100 Hz flashes resulted in a 35% increase in biomass yield compared to continuous light (Vejrazka et al., 2012). Conversely, low frequencies (<10 Hz) often lead to reduced yields because the dark period is too long, causing metabolic stagnation. A lower duty cycle (e.g., 5%–10%) combined with extremely high peak intensity allows light to penetrate deeper (because of high instantaneous intensity) while providing cells sufficient "rest" time to process electron flow, thereby reducing photoinhibition (Zarmi et al., 2020).

Research in 2024 demonstrated that employing optimized 660 nm pulsed light combined with chemical oxygen scavengers could significantly enhance microalgal photosynthetic H_2 yield (Kanygin et al., 2020). Pulsed light not only improves the quantum efficiency of the electron transport chain but may also mitigate photo-oxidative stress by reducing the continuous accumulation of reactive oxygen species (ROS). In addition, using light-emitting diodes (LEDs) for pulsed illumination can substantially reduce energy consumption compared to continuous lighting, which is of significant importance for the economic assessment of hydrogen production (Eilertsen et al., 2023).

2.5.2 Impact of nutrient limitation on hydrogen production

Micronutrients include trace elements such as iron, copper, calcium, zinc, molybdenum, manganese, and cobalt. A vitamin solution with the basic essential vitamins, such as thiamine, cyanocobalamine, and biotin, is usually added (Nagarajan et al., 2017). A nitrogen source in the form of ammonium or any nitrate salt is used. Except for photoautotrophic conditions a carbon source is also needed for growth. The presence or absence of any of these nutrients may induce a stress response in algal cells. Nutritional stress conditions accomplished by the deprivation of a particular macronutrient are one of the major turning points in microalgal technology, and the stress response brings about the accumulation of various energy reserves such as carbohydrates and lipids.

2.5.2.1 Impact of sulfur deprivation

As the successful induction of substantial hydrogen production in *C. reinhardtii* by Melis et al. (2000) using the "two-stage sulfur deprivation method," a new chapter in green algae hydrogen production research has been opened. This method provided a novel approach for utilizing green algae for hydrogen production, leading to rapid developments in condition optimization, physiological and biochemical studies, and metabolic pathway investigations based on this two-stage protocol. Utilizing this method to induce hydrogen production in wild-type *C. reinhardtii* yields 120–270 mL/L, with an average hydrogen production rate of up to 2 mL/L/h and a duration of up to 120 hours (Melis, 2007; Melis et al., 2000). Sulfur is an essential element for amino acids (cysteine and methionine).

The core mechanism of the sulfur-deprived hydrogen production system lies in the disruption of the PSII repair cycle. Under sulfur-deprived conditions, cells are unable to synthesize sufficient D1 protein (a core component of the PSII reaction center) to replace D1 proteins damaged under illumination, thereby blocking D1 protein turnover. As damaged D1 proteins accumulate the water-splitting activity of PSII gradually declines, leading to a decrease in the rate of photosynthetic oxygen evolution. When the rate of photosynthetic oxygen evolution falls below the rate of mitochondrial respiratory oxygen consumption, the dissolved oxygen within the closed system is depleted, and the cells spontaneously enter an anaerobic state. The establishment of this anaerobic environment induces the transcriptional expression of the hydrogenase gene (*HydA*) and its maturation protein genes (*HydEFG*), thereby initiating hydrogen production (Volgusheva et al., 2013).

The hydrogen production process induced by sulfur deprivation is a dynamic physiological progression, typically categorized into five consecutive phases (Kosourov et al., 2002):

Aerobic phase: During the initial 24 hours of sulfur deprivation, cells continue to perform photosynthesis and aerobic respiration, utilizing intracellularly stored sulfur to sustain limited growth. However, cell division subsequently ceases, and cells begin to accumulate substantial amounts of starch as a carbon reserve.

Oxygen consumption phase: As the repair of the D1 protein is blocked, the oxygen-evolving activity of PSII declines significantly. Concurrently, cells maintain a high rate of mitochondrial respiration (R_{O2}) through the degradation of accumulated starch. When R_{O2} exceeds the rate of photosynthetic oxygen evolution P_{O2}, the net oxygen concentration within the system rapidly drops to zero (Chochois et al., 2009).

Anaerobic phase: The establishment of an anaerobic environment triggers metabolic reprogramming, including the activation of fermentation pathways and the transcription and expression of the hydrogenase gene (*HydA*) (Cox, et al., 2013).

Hydrogen production phase: This phase can persist for 2–5 days or longer. Electrons primarily originate from residual PSII water-splitting activity (direct pathway) and from NADH/FADH2 generated via the degradation of endogenous starch, which enters the photosynthetic electron transport chain via the PQ pool (indirect pathway)

(Jaramillo et al., 2025). Studies indicate that even during the late stages of sulfur deprivation, PSII contributes approximately 80% of the electron flow for hydrogen production, demonstrating that water splitting remains the predominant electron source (Volgusheva et al., 2015).
Termination phase: As nutrient stress intensifies, severe disintegration of cellular structures occurs, key enzymatic activities are lost, and hydrogen production ceases, ultimately resulting in cell death (Kosea & Oncel, 2014).

In addition to *C. reinhardtii*, certain other green algae can also produce hydrogen under sulfur deprivation stress. For instance, Samira et al. isolated a freshwater *Chlorella* strain from acidic soil that achieved a hydrogen yield of 147 mL/L under sulfur deprivation (Chader et al., 2009). Ji et al. (2011) measured the hydrogen production of the marine green alga *Tetraselmis subcordiformis* under sulfur deprivation with the addition of CCCP, reaching a yield of 16.4 mL/L.

2.5.2.2 Impact of phosphorus deprivation

Phosphorus (P) is an integral constituent of ATP, NADPH, nucleic acids, and phospholipids. Phosphorus deficiency directly compromises the cellular energy charge and the regeneration of ATP. Given that the photosynthetic dark reaction (Calvin cycle) necessitates substantial ATP consumption for CO_2 fixation, the insufficient ATP supply resulting from phosphorus deprivation inhibits the dark reaction. Consequently, this leads to the overreduction of the photosynthetic electron transport chain (Batyrova et al., 2012a). This elevation in reduction pressure drives electron flow toward the hydrogenase, utilizing hydrogen production as an alternative pathway to dissipate excess electrons.

Wykoff et al. measured the photosynthetic net oxygen evolution rate and the photochemical activity of the photosynthetic systems in *C. reinhardtii* under phosphorus-deprived conditions. They found that the saturated net oxygen evolution rate of *C. reinhardtii* in phosphorus-free medium declined, dropping by 75% after 4 days. Although PSI activity was largely unaffected by phosphorus deficiency, both the photochemical activity of PSII and linear electron transport were downregulated (Wykoff et al., 1998). Batyrova et al. discovered that when *C. reinhardtii* was inoculated at a low cell density (<2 mg chl/L) into phosphorus-free medium, an anaerobic environment could be established in the closed culture system, as oxygen was gradually consumed, resulting in a hydrogen yield of ~70 mL/L. However, at inoculation densities higher than 2 mg chl/L, the system failed to establish the anaerobic conditions requisite for inducing hydrogenase activity. Because *C. reinhardtii* can accumulate substantial amounts of phosphorus intracellularly, a relatively strict phosphorus-free condition can only be achieved at low inoculation densities where cell growth naturally depletes intracellular phosphorus reserves. Compared to sulfur deprivation treatment, the decline in PSII net oxygen evolution activity during phosphorus deprivation is slower, requiring a longer time (~100 hours) to establish an anaerobic environment; hydrogen production ceases after 300 hours of treatment. The trends in starch and acetate content changes are similar to those under sulfur-deprived conditions: cells accumulate starch during the aerobic phase and consume it during the hydrogen production phase, whereas acetate content shows a declining trend throughout the cultivation process (Batyrova et al., 2012b).

Although phosphorus deprivation offers an alternative strategy, the kinetics of inducing anaerobiosis are typically slower compared to sulfur deprivation. Comparative transcriptomic and physiological studies indicate that phosphorus deprivation causes less severe damage to the photosynthetic membrane systems. However, in certain algal strains (e.g., *Chlamydomonas*), phosphorus deprivation alone often fails to achieve hydrogen production rates comparable to those obtained under sulfur deprivation, typically necessitating combination with other conditions (e.g., nitrogen limitation) (Pongpadung et al., 2018).

2.5.2.3 *Impact of nitrogen deprivation*

Nitrogen is an essential element indispensable for cell growth. In plant cells, it serves not only as a macroelement required for protein synthesis but also as a critical constituent of chlorophyll. Nitrogen atoms constitute the nitrogenous backbone of the magnesium-porphyrin ring in chlorophyll, rendering chlorophyll *a* substantial intracellular "nitrogen pool." Under conditions of nitrogen deficiency, chlorophyll undergoes degradation to release nitrogen, thereby supporting the synthesis of other essential biomolecules, such as vital proteins and nucleic acids.

Unlike sulfur deprivation, which specifically targets PSII, nitrogen deprivation results in a global inhibition of intracellular protein synthesis and the extensive degradation of photosynthetic pigments (chlorosis) to recycle nitrogen sources for the maintenance of basal survival. In *C. reinhardtii*, although nitrogen deprivation is also capable of inducing hydrogen production, its mechanism and efficacy differ significantly from those of sulfur deprivation. Under nitrogen-deprived conditions, the establishment of an anaerobic state is typically slower, as the decline in photosynthetic oxygen evolution activity is not as precipitous as observed under sulfur deprivation (because of the absence of a specific mechanism blocking D1 repair).

Philipps et al. investigated the hydrogen production process of *C. reinhardtii* under nitrogen-deprived conditions. They found that under nitrogen deficiency, PSII maintained relatively high activity for a short period; the system required 48–72 hours of treatment to establish an anaerobic environment, and hydrogenase activity was only detected after 72 hours. The hydrogen yield of *C. reinhardtii* under nitrogen-deprived conditions was approximately 58% of that under sulfur-deprived conditions. Although *C. reinhardtii* accumulates large amounts of starch under nitrogen deficiency, this accumulated starch is not rapidly degraded and utilized, preventing the rapid establishment of an intracellular anaerobic environment to induce hydrogenase-mediated hydrogen production. They speculated that one of the reasons for the low hydrogen yield under nitrogen deprivation might be the degradation of Cyt *b6f* and insufficient ferredoxin (Fd) supply, which hinders the electron transfer from carbohydrates such as starch to the hydrogenase (Philipps et al., 2012).

The most distinct metabolic feature of nitrogen deprivation is the redirection of carbon flux toward energy storage compounds. Although sulfur deprivation also results in starch accumulation, nitrogen deprivation preferentially induces the synthesis of lipids, such as triacylglycerols (TAGs), rendering it the preferred strategy for microalgal biodiesel production (Cakmak et al., 2012). This creates a competitive relationship between hydrogen production and oil accumulation. Studies indicate that within the metabolic network induced by nitrogen deprivation, a substantial amount of acetyl-CoA is channeled into fatty acid synthesis, which to a certain extent competes for the reducing power and carbon skeletons required for hydrogen production. Comparative studies have revealed that hydrogen yields under nitrogen-deprived conditions are typically lower than those under sulfur-deprived conditions. In *S. obliquus*, nitrogen deprivation primarily leads to lipid accumulation and is not conducive to hydrogen production (Limongi et al., 2021).

2.5.2.4 Synergistic effects of multielement deprivation

Research has indicated that combining sulfur deprivation with limitations of other nutrient elements may generate synergistic effects (Table 2.4). For example, in certain *Chlorella* strains, the simultaneous limitation of sulfur and nitrogen supply is more effective in promoting hydrogen production than sulfur deprivation alone. This combined stress typically further shortens the time required for anaerobic induction and enhances the final cumulative hydrogen yield. This suggests that the cellular hydrogen production response involves a complex regulatory network encompassing multiple nutrient signaling pathways. Through the combination of multiple stresses, it may be possible to more precisely modulate cellular metabolism, directing it toward a state most favorable for hydrogen production (Pongpadung et al., 2015).

2.6 Synergistic effects of algae-bacterial consortia

The fragility of single-species (axenic) microalgae systems and their stringent requirements for sterile conditions have constrained their

TABLE 2.4 Impact of different nutrient stress strategies on microalgae hydrogen production performance.

Stress type	Key metabolic characteristics	Advantages	Limitations	H_2 yield (mL/L)	References
Sulfur deprivation (–S)	Blocked PSII-D1 repair turnover; decline in PSII activity	Classic method; well-elucidated mechanism; prolonged H_2 production phase	Causes damage to the photosynthetic apparatus; cessation of cell growth	60–140 (*Chlamydomonas reinhardtii*)	Ghirardi et al. (2000)
Nitrogen deprivation (–N)	Inhibition of protein synthesis; massive accumulation of starch/lipids	Extremely high carbon reserves	Limited hydrogenase synthesis; generally lower H_2 production efficiency	~20–50	Németh et al. (2020)
Low-N preculture + sulfur deprivation	High starch accumulation + rapid anaerobic induction	Significant synergistic effects; overcomes limitations of single-stress approaches	Complex operational protocol	125.1 (Chlorella *sorokiniana*)	Melis et al. (2000)
Phosphorus deprivation (–P)	Restricted ATP synthesis; impairs the Calvin cycle	Milder stress; high cell viability	Weak capacity to induce anaerobiosis	~70	Wykoff et al. (1998)
Magnesium deprivation (–Mg)	Blocked chlorophyll synthesis; degradation of antenna complexes	Reduces photoinhibition	Unstable induction of H_2 production	~30	Pongpadung et al. (2015)

PSII, *photosystem II*.

TABLE 2.5 Enhancement of hydrogen production in algae-bacterial coculture systems.

Microalgae strain	Bacterial partner	Synergistic mechanism	H_2 yield enhancement	References
Chlamydomonas reinhardtii	*Pseudomonas* spp.	Oxygen scavenging via respiration maintains an anaerobic environment	22-fold increase	Fakhimi et al. (2019)
C. reinhardtii	*Bradyrhizobium japonicum*	Nitrogen-fixing bacteria consume oxygen and provide nitrogen sources/growth factors	14-fold increase	Xu et al. (2016)
C. reinhardtii	*Escherichia coli*	Bacteria utilize algae exudates and consume oxygen	24%–46% increase	Lakatos et al. (2017)
Chlorella sp.	*Bacillus* sp.	Bacteria degrade complex substrates for algae utilization and consume oxygen	Synergistic increase in biomass and H_2 yield	Ban et al. (2018)

industrial application. Constructing algae-bacterial consortia based on ecological principles to enhance system robustness and hydrogen production efficiency through interspecific mutualism has emerged as a prominent research hotspot. Utilizing aerobic bacteria as "biological oxygen scavengers" represents an ingenious strategy for maintaining the anaerobic environment required for microalgae hydrogen production (Table 2.5).

In coculture systems, heterotrophic bacteria (such as *Pseudomonas, Bacillus*, and *Rhizobium*) utilize respiration to rapidly consume the O_2 generated by microalgae photosynthesis. This maintains dissolved oxygen at extremely low levels, creating and sustaining an ideal anaerobic environment for the hydrogenase, thereby protecting its catalytic activity. Simultaneously the CO_2 produced by bacterial respiration serves as a carbon source for microalgae photosynthesis and helps maintain pH stability. This symbiotic relationship significantly improves the overall stability and hydrogen production efficiency of the system. Research indicates that in a coculture system of *C. reinhardtii* and aerobic bacteria, hydrogen yield increased by more than 22-fold compared to pure algae cultures, eliminating the need for expensive chemical oxygen scavengers or inert gas purging (Fakhimi et al., 2020).

In addition to oxygen scavenging, complex metabolic complementarity exists between algae and bacteria. Small-molecule organic carbon secreted by microalgae (such as glycolate, glycerol, and extracellular polysaccharides) can support bacterial growth. Conversely, many bacteria are capable of synthesizing vitamin B12 (cobalamin) or plant growth hormones (such as indole-3-acetic acid, IAA), which promote microalgae growth and metabolic activity. Furthermore, signaling molecules secreted by bacteria (such as *N*-acyl homoserine lactones, AHLs) may regulate microalgae biofilm formation and metabolic states (Giri et al., 2022).

2.7 Photobioreactor design optimization

Although genetic engineering constructs superior "cell factories," the PBR serves as the physical vessel for realizing their production potential. The design of PBRs must address multifaceted engineering challenges, including light attenuation, gas–liquid mass transfer—specifically the in situ removal of H_2 and O_2—and shear sensitivity. A summary of the characteristics of common reactors is presented in Table 2.6.

Targeting indirect photolysis designs, researchers have developed two-stage systems to physically separate growth from hydrogen production. The first stage (light/aerobic) optimizes cell growth and biomass/starch accumulation, whereas the second stage (dark/anaerobic or stressed) induces hydrogen production. This strategy effectively circumvents oxygen inhibition, allowing for the independent optimization of parameters (e.g., light intensity, pH, and nutrients) for each stage (Ghirardi et al., 2000).

Furthermore, to prolong the duration of hydrogen production, Laurinavichene et al. immobilized *C. reinhardtii* cells within an alginate membrane, extending hydrogen production up to 28 days. The hydrogen yield reached 2375 mL/L, approximately 10 times that of the wild-type suspension culture (210–250 mL/L), with an average production rate of 4.3 mL/L/h, nearly double the rate of the suspension culture (2 mL/L/h) (Laurinavichene et al., 2006). Upon immobilization in alginate, cells are concentrated within a confined space, drastically increasing cell density per unit volume. After entrapment in the alginate matrix, oxygen from the air or medium is consumed by cells at the surface layer, hindering diffusion to cells embedded within the matrix interior. This facilitates the establishment of an anaerobic environment and hydrogenase activation. Immobilized cells not only exhibited improved hydrogen production rates (12.5 vs. 9.6 mmol/mg Chl/h in suspension) but also effectively enhanced light energy conversion efficiency (1% vs. 0.24%) (Kosourov & Seibert, 2009).

TABLE 2.6 Comparison of performance parameters of different photobioreactor types for microalgae hydrogen production.

Reactor type	Key design features	Advantages	Limitations	Typical H_2 yield range	References
STR	Mechanical agitation; artificial light source	Precise control; homogeneous mixing	Extremely high energy consumption; high shear stress; severe light attenuation	Lower (primarily used for laboratory mechanistic studies)	Gaffron and Rubin (1942)
Tubular PBR	Solar/artificial light; airlift or pump circulation	Easy scale-up; high surface-to-volume ratio	Oxygen accumulation leads to photoinhibition, difficult temperature control, prone to biofouling	50–150 mL/L/culture	Elman et al. (2025)
Flat panel PBR	Short light path (1–5 cm); airlift mixing	Uniform illumination; high gas–liquid mass transfer coefficient	Scale-up requires modular arraying, large areal footprint, high capital cost	100–200 mL/L/culture	Skjånes et al. (2016)
Biofilm PBR	Cells attached to a carrier; direct contact with gas phase	Water-saving; no centrifugation required for harvesting; in situ formation of anaerobic microenvironments	Limited internal mass transfer; shallow light penetration; biofilm aging and sloughing	0.5–2.5 $mL/h/m^2$ (areal yield)	Delavar and Wang (2024)
Optical fiber/internally lit PBR	External light collection; internal transmission via optical fibers	Resolves light attenuation in deep layers; extremely uniform light distribution	Extremely expensive equipment; complex maintenance	High (volumetric yield), but poor economic viability	Jiao et al. (2024)

PBR, *photobioreactor;* STR, *stirred tank reactor.*

Another immobilized system is the biofilm PBR. Microalgae grow attached to porous hydrophilic substrates (e.g., filter membranes or fabrics). Certain designs, such as dual-chamber porous membrane reactors, allow H_2 to be released directly from the backside of the biofilm into the gas phase. This eliminates the need to overcome liquid film resistance and prevents thermodynamic inhibition caused by excessive partial pressure of H_2 (product inhibition). A natural oxygen gradient exists within the biofilm: surface cells perform photosynthetic oxygen evolution, whereas basal cells remain naturally anaerobic because of limited oxygen diffusion, providing an ideal "microreactor" structure for sustained hydrogen production under illumination.

To address complex hydrodynamics and mass transfer issues within hydrogen production reactors, advanced computational fluid dynamics (CFD) models have been introduced to aid design. In particular the coupled model of the lattice Boltzmann method (LBM) and cellular automata (CA) (LBM–CA) has gained widespread application in recent years. LBM excels at simulating complex boundaries (e.g., porous biofilms) and multiphase fluid flows, whereas CA simulates discrete biological behaviors such as individual microbial growth, division, death, and detachment. The coupling of these methods enables precise prediction of how fluid shear stress affects biofilm morphology and hydrogen production efficiency. Simulation studies indicate that the geometry of internal reactor obstacles is critical for hydrogen production. Compared to traditional smooth channels, installing rectangular or semicircular obstacles alters the shear stress distribution of the flow field, controlling the biofilm at an optimal thickness. This prevents basal cell death caused by excessive thickening while enhancing substrate mass transfer and the timely desorption of hydrogen bubbles. Research found that rectangular obstacles of appropriate height (e.g., 1.0 hour) can significantly improve hydrogen extraction efficiency (Delavar & Wang, 2024). The latest LBM–CA models also integrate pH field simulations, quantifying the inhibition of biofilm activity caused by local pH drops because of organic acid accumulation during dark fermentation. This guides precise injection strategies for acidic additives or buffers to maintain microenvironmental homeostasis (Delavar & Wang, 2021b).

With the development of Industry 4.0, artificial intelligence (AI) and machine learning (ML) are progressively permeating microalgae hydrogen production research. By integrating multiparameter sensors (light intensity, pH, DO, OD, flow rate, etc.) into reactors, data-driven predictive models are established using ML algorithms (e.g., XGBoost, ANN, Random Forest). These models can predict microalgae growth states and hydrogen production rates in real time based on current sensor data (Ahmad Sobri et al., 2023). Based on these predictions, AI control systems can dynamically adjust light/dark cycles, nutrient feed rates, and temperature, consistently maintaining the reactor under optimal operating conditions.

Case studies demonstrate that AI-assisted intelligent control systems can increase biomass productivity by 30%–50% and significantly reduce operational costs and batch failure rates caused by manual operation lags (Wu et al., 2025).

2.8 Techno-economic analysis (TEA) and life cycle assessment (LCA)

The commercialization potential of any technology ultimately hinges upon its economic feasibility and environmental benefits. Table 2.7 show the comparison of economic and environmental indicators for different hydrogen production pathways.

The primary barrier to the commercialization of microalgae hydrogen production remains the cost. Current production costs are significantly higher than those of hydrogen production via fossil fuels [steam methane reforming (SMR)]. According to multiple TEA studies conducted between 2024 and 2025, cost estimates for microalgae/biomass-based hydrogen production vary widely, being significantly influenced by specific technical pathways and operational scale. Current estimates range from approximately \$2.24 to \$8.97/kg H_2 (Ahmed et al., 2021). In contrast the cost of hydrogen production via natural gas reforming (gray hydrogen) is approximately \$1.00–\$2.10/kg. However, the cost of microalgae hydrogen production has approached or become more competitive than certain scenarios for green hydrogen production via water electrolysis (\$3.40–\$7.50/kg) (Saputro & Juangsa, 2025).

The major cost driver is the capital expenditure (CAPEX) for the construction and installation of PBRs (constituting over 50% of the total cost), particularly for closed systems requiring precise control and advanced materials. As the product is a gas mixture of H_2 and O_2, downstream gas separation and purification incur high costs. Operational expenditures (OPEXs) include energy consumption for mixing and pumping, as well as nutrient consumption. Recent predictive models suggest that by integrating high-efficiency genetically engineered strains (e.g., PSI–HydA fusion strains), optimized biofilm reactors, and economies of scale, the future cost of microalgae hydrogen production could be substantially reduced. Under idealized optimization scenarios, the cost could decrease to approximately \$1.48/kg H_2, thereby reaching the break-even point for competitiveness with fossil fuels (Elman et al., 2025).

LCA quantifies the environmental impact throughout the technology's entire life cycle, confirming the significant potential of microalgae hydrogen production in achieving carbon neutrality.

Greenhouse gas emissions [global warming potential (GWP)]: The GWP of microalgae photolytic hydrogen production is extremely low. Some studies

TABLE 2.7 Comparison of economic and environmental indicators for different hydrogen production pathways.

Technology pathway	Production cost ($/kg H_2)	GWP (g CO_2 eq/MJ)	Key advantages	Key challenges	References
Microalgae direct biophotolysis	2.70–7.24 (current) ~1.48 (future optimized)	31–55	Extremely high theoretical photosynthetic efficiency; zero carbon emissions; does not compete for arable land	Severe oxygen inhibition; high PBR capital costs; difficult scale-up	Ahmed et al. (2021)
Water electrolysis (green hydrogen)	3.40–7.50	~0–20	Mature technology; high gas purity; easy grid integration	Heavily dependent on electricity prices; expensive electrolyzer equipment	Saputro and Juangsa (2025)
Dark fermentation (biomass)	1.20–3.41	<30	Can utilize organic waste; relatively mature technology; light-independent	Lower yield; requires pretreatment; product purification required	Khalil (2025)
SMR (gray hydrogen)	1.00–2.10	>90	Lowest cost; most established supply chain; large scale	High carbon emissions; unsustainable and faces carbon tax risks	Akpasi et al. (2025)

SMR, *steam methane reforming*.

indicate that with optimal site selection (e.g., high-irradiance equatorial regions) and low-carbon electricity support, the life cycle carbon footprint can be as low as 31–45 g CO_2eq/MJ, representing a reduction of approximately 70% compared to traditional fossil diesel (Quiroz et al., 2023). Considering the active fixation of CO_2 during microalgae growth, combined with carbon capture and utilization (CCU) technologies, it may even be possible to achieve negative carbon emissions.

Water footprint and land use. Although direct biophotolysis does not consume substantial freshwater (if seawater or wastewater is utilized), evaporative water loss in open raceway ponds remains a concern. While closed PBRs are more water-efficient, they are associated with higher energy consumption.

2.9 Conclusion and perspectives

Microalgae biophotolytic hydrogen production technology has witnessed substantial advancements over the past few decades. Research focus has shifted from the exploration of phenomenological biology to multidisciplinary, systems-level research aimed at resolving core engineering challenges. Facilitated by a profound understanding of cellular metabolism, enzymatic catalytic mechanisms, reactor engineering, and process integration, the field is progressively transitioning from basic research toward application development.

Currently, investigations into the structure, catalytic mechanism, biosynthesis, and oxygen inactivation mechanisms of hydrogenases have reached the molecular level. The fundamental nature of nutrient deprivation-induced hydrogen production has been elucidated as a complex metabolic reprogramming process governed by cellular redox signaling networks. On the engineering front a suite of cultivation strategies—including nutrient stress, chemical regulation, genetic engineering, synthetic biology, advanced reactor design, and multispecies coculture—has been developed to overcome oxygen inhibition, enhance electron transfer efficiency, and intensify the process. Despite these advancements the commercialization of biophotolytic hydrogen production continues to face significant challenges:

1. Enzyme stability: Although progress has been made in engineering oxygen-tolerant hydrogenases, maintaining their stable and efficient catalytic activity in practical, large-scale, and long-term outdoor reactor operations remains a formidable challenge.
2. Solar energy conversion efficiency: The comprehensive solar-to-hydrogen (STH) conversion efficiency achieved under laboratory conditions currently remains below the 10% threshold requisite for economic

viability. Effectively overcoming light limitation and photoinhibition in high-density cultures is critical.

3. Scale-up: Scaling up PBRs from laboratory to industrial dimensions (hectare-level) presents a series of complex engineering issues regarding cost, stability, contamination control, and mass transfer efficiency.
4. Economic feasibility: High CAPEXs and OPEXs remain the primary factors constraining the commercialization of this technology.

To promote the translation of biophotolytic hydrogen production technology from laboratory research to industrial application, future research should focus on the following highly integrated directions:

Integrated synthetic biology: Guided by systems biology and genome-scale metabolic models, and combined with high-throughput screening and gene-editing tools like CRISPR, research should focus on the design and construction of microalgae chassis cells. These cells should integrate multiple optimized traits, such as oxygen-tolerant hydrogenases, TLAs, rearranged electron transport chains, and optimized carbon metabolic fluxes.

Intelligent PBRs: Development of PBRs integrating multiparameter online sensors (e.g., dissolved oxygen, pH, oxidation-reduction potential, biomass) and AI-driven advanced control systems. Such reactors will be capable of real-time monitoring of environmental changes (e.g., diurnal light fluctuations) and dynamic regulation of operating parameters (e.g., nutrient supply, gas flow rates) to maintain cells in an optimal state for hydrogen production.

Process integration within a circular bioeconomy framework: Shifting the research focus from independent hydrogen production processes to the construction of closed-loop, resource-oriented biorefinery systems. Future studies should systematically design and evaluate processes that couple hydrogen production with the treatment of municipal sewage, industrial wastewater, and CO_2-containing flue gas, thereby maximizing resource efficiency and the overall economic and environmental performance of the system.

In summary, microalgae biophotolytic hydrogen production is a promising clean energy technology. Through continuous synergistic innovation across molecular, cellular, and reactor scales, and by systematically designing within the macroframework of the circular economy, it is expected to overcome existing bottlenecks and contribute significantly to the global sustainable energy transition.

References

Ahmad Sobri, Redhwan, A., Ameen, F., Lim, J. W., Liew, C. S., … Rao, P. V. (2023). A review unveiling various machine learning algorithms adopted for biohydrogen productions from microalgae. *Fermentation, 9*, 243.

Ahmed, S. F., Rafa, N., Mofijur, M., Badruddin, I. A., Inayat, A., Ali, M. S., Farrok, O., & Yunus Khan, T. M. (2021). Biohydrogen production from biomass sources: Metabolic pathways and economic analysis. *Frontiers in Energy Research, 9*. http://www.frontiersin.org/Energy_Research/. doi:10.3389/fenrg.2021.753878.

Akpasi, S. O., Smarte Anekwe, I. M., Tetteh, E. K., Amune, U. O., Mustapha, S. I., & Kiambi, S. L. (2025). Hydrogen as a clean energy carrier: Advancements, challenges, and its role in a sustainable energy future. *Clean Energy, 9*(1), 52–88. https://academic.oup.com/ce/pages/About. doi:10.1093/ce/zkae112.

Baebprasert, W., Lindblad, P., & Incharoensakdi, A. (2010). Response of H_2 production and Hox-hydrogenase activity to external factors in the unicellular cyanobacterium *Synechocystis* sp. strain PCC 6803. *International Journal of Hydrogen Energy, 35*(13), 6611–6616. doi:10.1016/j.ijhydene.2010.04.047.

Ban, S., Lin, W., Wu, F., & Luo, J. (2018). Algal-bacterial cooperation improves algal photolysis-mediated hydrogen production. *Bioresource Technology, 251*, 350–357. doi:10.1016/j.biortech.2017.12.072.

Batyrova, K. A., & Hallenbeck, P. C. (2017). Hydrogen production by a *Chlamydomonas reinhardtii* strain with inducible expression of photosystem II. *International Journal of Molecular Sciences, 18*, 647.

Batyrova, K., Tsygankov, A. A., & Kosourov, S. (2012a). Sustainable hydrogen photoproduction by phosphorus-deprived marine green microalgae *Chlorella* sp. *International Journal of Hydrogen Energy, 37*, 8834–8839.

Batyrova, K. A., Tsygankov, A. A., & Kosourov, S. N. (2012b). Sustained hydrogen photoproduction by phosphorus-deprived *Chlamydomonas reinhardtii* cultures. *International Journal of Hydrogen Energy, 37*(10), 8834–8839. doi:10.1016/j.ijhydene.2012.01.068.

Betz, J. N., Boswell, N. W., Fugate, C. J., Holliday, G. L., Akiva, E., Scott, A. G., Babbitt, P. C., Peters, J. W., Shepard, E. M., & Broderick, J. B. (2015). [FeFe]-Hydrogenase maturation: Insights into the role HydE plays in dithiomethylamine biosynthesis. *Biochemistry., 54*(9), 1807–1818. doi:10.1021/bi501205e.

Bonente, G., Pippa, A., Castellano, A., Bassi, R., & Ballottari, L. (2012). Acclimation of *Chlamydomonas reinhardtii* to different growth light strengths. *Journal of Biological Chemistry, 287*, 5833–5847.

Britt, R. D., Rauchfuss, T. B., & Rao, G. (2015). The H-cluster of [FeFe] hydrogenases: Its enzymatic synthesis and parallel inorganic semisynthesis. *Accounts of Chemical Research, 48*, 2319–2326.

Burgess, S. J., Tamburic, B., Zemichael, F., Hellgardt, K., & Nixon, P. J. (2011). Solar-driven hydrogen production in green algae. *Advances in Applied Microbiology, 75*, 71–110. http://www.sciencedirect.com/science/journal/00652164. doi:10.1016/B978-0-12-387046-9.00004-9.

Cakmak, T., Angun, P., Demiray, Y. E., Ozkan, A. D., Elibol, Z., & Tekinay, T. (2012). Differential effects of nitrogen and sulfur deprivation on growth and biodiesel feedstock production of *Chlamydomonas reinhardtii*. *Biotechnology and Bioengineering, 109*(8), 1947–1957. doi:10.1002/bit.24474.

Caserta, G., & Papini, C. (2018). Engineering an [FeFe]-hydrogenase: Do accessory clusters influence O_2 resistance and catalytic bias? *Journal of the American Chemical Society, 140*, 5516–5526.

Chader, S., Hacene, H., & Agathos, S. N. (2009). Study of hydrogen production by three strains of *Chlorella* isolated from the soil in the Algerian Sahara. *International Journal of Hydrogen Energy, 34*(11), 4941–4946. doi:10.1016/j.ijhydene.2008.10.058.

Chochois, V., Dauvillée, D., Beyly, A., Tolleter, D., Cuiné, S., Timpano, H., Ball, S., Cournac, L., & Peltier, G. (2009). Hydrogen production in *Chlamydomonas*: Photosystem II-dependent and -independent pathways differ in their requirement for starch metabolism. *Plant Physiology, 151*(2), 631–640. http://www.plantphysiol.org/cgi/reprint/151/2/631. doi:10.1104/pp.109.144576.

Cox, G. M., McCrimmon, B. M., Nemati, T. A., & Olson, C. M. (2013). The production of hydrogen gas by *Chlamydomonas reinhardtii* in sulfur-deprived conditions under red light and white light. *The Expedition, 3.*

Delavar, M. A., & Wang, J. (2021b). Numerical investigation of pH control on dark fermentation and hydrogen production in a microbioreactor. *Fuel, 292*(5686), 120355.

Delavar, M. A., & Wang, J. (2024). Advancing biohydrogen production through indirect photolysis: Insights from numerical simulation using the lattice boltzmann method. *Fuel, 376*, 132707.

Eilertsen, H. C., Strømholt, J., Bergum, J. S., Eriksen, G. K., & Ingebrigtsen, R. (2023). Mass cultivation of microalgae: II. A large-scale pulse light concept. *BioTech, 12*, 40.

Elman, T., Isaac, S., & Yacoby, I. (2025). Photobiological hydrogen production at scale: Integrating bioprocess optimization and techno-economic modeling. *bioRxiv*. Israel bioRxiv. 26928205 https://www.biorxiv.org . doi:10.1101/2025.04.27.650233.

Fakhimi, N., Dubini, A., Tavakoli, O., & González-Ballester, D. (2019). Acetic acid is key for synergetic hydrogen production in *Chlamydomonas*-bacteria co-cultures. *Bioresource Technology, 289*, 121648. doi:10.1016/j.biortech.2019.121648.

Fakhimi, N., Gonzalez-Ballester, D., Fernández, E., Galván, A., & Dubini, A. (2020). Algae-bacteria consortia as a strategy to enhance H_2 production. *Cells, 9*(6), 1353. doi:10.3390/cells9061353.

Faraloni, C., Torzillo, G., Balestra, F., Moia, I. C., Zampieri, R. M., Jiménez-Conejo, N., & Touloupakis, E. (2025). Advances and challenges in biohydrogen production by photosynthetic microorganisms. *Energies, 18*(9). http://www.mdpi.com/journal/energies/. doi:10.3390/en18092319.

Filmon, D. T., Jaenecke, J., Winkler, M., Fourmond, V., Léger, C., & Plumeré, N. (2025). Turning the FeFe hydrogenase from *Clostridium beijerinckii* into an efficient H_2 oxidation catalyst using a redox-active matrix. *Proceedings of the National Academy of Sciences of the United States of America, 122*(41). https://www.pnas.org/doi/10.1073/pnas.2514698122. doi:10.1073/pnas.2514698122.

Gaffron, H., & Rubin, J. (1942). Fermentative and photochemical production of hydrogen in algae. *Journal of General Physiology, 26*(2), 219–240. doi:10.1085/jgp.26.2.219.

Ghirardi, M. L., Posewitz, M. C., Maness, P. C., Dubini, A., Yu, J., & Seibert, M. (2007). Hydrogenases and hydrogen photoproduction in oxygenic photosynthetic organisms. *Annual Review of Plant Biology, 58*, 71–91. doi:10.1146/annurev.arplant.58.032806.103848.

Ghirardi, M. L., Zhang, L., Lee, J. W., Flynn, T., Seibert, M., Greenbaum, E., & Melis, A. (2000). Microalgae: A green source of renewable H_2. *Trends in Biotechnology, 18*(12), 506–511. doi:10.1016/S0167-7799(00)01511-0.

Giri, D. D., Dwivedi, H., Khalaf D. Alsukaibi, A., Pal, D. B., Otaibi, A. A., Areeshi, M. Y., Haque, S., & Gupta, V. K. (2022). Sustainable production of algae-bacteria granular consortia based biological hydrogen: New insights. *Bioresource Technology, 352*. Www.elsevier.com/locate/biortech. doi:10.1016/j.biortech.2022.127036.

Grechanik, V. I., & Tsygankov, A. A. (2022). The relationship between photosystem II regulation and light-dependent hydrogen production by microalgae. *Biophysical Reviews, 14*(4), 893–904. doi:10.1007/s12551-022-00977-z.

Hippler, M., & Khosravitabar, F. (2024). Light-driven H2 production in *Chlamydomonas reinhardtii*: Lessons from engineering of photosynthesis. *Plants, 13*(15), 2114. doi:10.3390/plants13152114.

Jaramillo, A., Satta, A., Pinto, F., Faraloni, C., Zittelli, G. C., Benavides, A. M. S., Torzillo, G., Schumann, C., Méndez, J. Fá, Berggren, G., Lindblad, P., Parente, M., Esposito, S., & Diano, M. (2025). Outlook on synthetic biology-driven hydrogen production: Lessons from algal photosynthesis applied to cyanobacteria. *Energy & Fuels, 39*(11), 4987–5006. doi:10.1021/acs.energyfuels.4c04772.

Ji, C. F., Yu, X. J., Chen, Z. A., Xue, S., Legrand, J., & Zhang, W. (2011). Effects of nutrient deprivation on biochemical compositions and photo-hydrogen production of *Tetraselmis subcordiformis*. *International Journal of Hydrogen Energy, 36*(10), 5817–5821. doi:10.1016/j.ijhydene.2010.12.138.

Jiao, H., Tsigkou, K., Elsamahy, T., Pispas, K., Sun, J., Manthos, G., Schagerl, M., Sventzouri, E., Al-Tohamy, R., Kornaros, M., & Ali, S. S. (2024). Recent advances in sustainable hydrogen production from microalgae: Mechanisms, challenges, and future perspectives. *Ecotoxicology and Environmental Safety, 270*. https://www.sciencedirect.com/science/journal/01476513. doi:10.1016/j.ecoenv.2023.115908.

Kanygin, A., Milrad, Y., Thummala, C., Reifschneider, K., Baker, P., Marco, P., Yacoby, I., & Redding, K. E. (2020). Rewiring photosynthesis: A photosystem I-hydrogenase chimera that makes H2: In vivo. *Energy and Environmental Science, 13*(9), 2903–2914. http://pubs.rsc.org/en/journals/journal/ee. doi:10.1039/c9ee03859k.

Kosea, A., & Oncel, S. S. (2014). Biohydrogen production from engineered microalgae *Chlamydomonas reinhardtii*. *Advances in energy researchEnergies, 2*, 1.

Kosourov, S., Tsygankov, A., Seibert, M., & Ghirardi, M. L. (2002). Sustained hydrogen photoproduction by *Chlamydomonas reinhardtii*: Effects of culture parameters. *Biotechnology and Bioengineering, 78*(7), 731–740. doi:10.1002/bit.10254.

Kosourov, S. N., & Seibert, M. (2009). Hydrogen photoproduction by nutrient-deprived *Chlamydomonas reinhardtii* cells immobilized within thin alginate films under aerobic and anaerobic conditions. *Biotechnology and Bioengineering, 102*(1), 50–58. http://www3.interscience.wiley.com/cgi-bin/fulltext/121356483/PDFSTART. doi:10.1002/bit.22050United States.

Kumar, V., Sharma, N., Jaiswal, K. K., Vlaskin, M. S., Nanda, M., Tripathi, M. K., & Kumar, S. (2021). Microalgae with a truncated light-harvesting antenna to maximize photosynthetic efficiency and biomass productivity. *Process Biochemistry, 104*, 83–91.

Lakatos, G., Balogh, D., Farkas, A., Ördög, V., Nagy, P. T., Bíró, T., & Maróti, G. (2017). Factors influencing algal photobiohydrogen production in algal-bacterial co-cultures. *Algal Research, 28*, 161–171. http://www.sciencedirect.com/science/journal/aip/22119264. doi:10.1016/j.algal.2017.10.024.

Lampret, O., Adamska-Venkatesh, A., Vöpel, H. P., Gee, L. B., Cramer, S. P., Reijerse, E., ... Happe, T. (2019). The final steps of [FeFe]-hydrogenase maturation. *Proceedings of the National Academy of Sciences of the United States of America, 116*, 15802–15810.

Laurinavichene, T., Fedorov, A., Ghirardi, M., Seibert, M., & Tsygankov, A. (2006). Demonstration of sustained hydrogen photoproduction by immobilized, sulfur-deprived *Chlamydomonas reinhardtii* cells. *International Journal of Hydrogen Energy, 31*(5), 659–667. doi:10.1016/j.ijhydene.2005.05.002.

Laurinavichene, T., Tolstygina, I., & Tsygankov, A. (2004). The effect of light intensity on hydrogen production by sulfur-deprived *Chlamydomonas reinhardtii*. *Journal of Biotechnology, 114*(1–2), 143–151. doi:10.1016/j.jbiotec.2004.05.012.

Leone, L., Sgueglia, G., Gatta, S. L., Chino, M., Nastri, F., & Lombardi, A. (2023). Enzymatic and bioinspired systems for hydrogen production. *International Journal of Molecular Sciences, 24*(10), 8605. doi:10.3390/ijms24108605.

Li, W., Yao, Y., Yang, J., Ma, W., & Liu, M. (2024). Enhanced H2 photoproduction in sodium sulfite-treated *Chlamydomonas reinhardtii* by optimizing wavelength of pulsed light. *International Journal of Hydrogen Energy, 67*, 216–224.

Limongi, A. R., Viviano, E., De Luca, M., Radice, R. P., Bianco, G., & Martelli, G. (2021). Biohydrogen from microalgae: Production and applications. *Applied Sciences, 11*, 1616.

Liu, J., Bukatin, V., & Tsygankov, A. (2006). Light energy conversion into H2 by *Anabaena variabilis* mutant PK84 dense cultures exposed to nitrogen limitations. *International Journal of Hydrogen Energy, 31*(11), 1591–1596. doi:10.1016/j.ijhydene.2006.06.025.

Lubner, C. E., Applegate, A. M., Knörzer, P., Ganago, A., Bryant, D. A., Happe, T., et al. (2011). Solar hydrogen-producing bionanodevice designed to integrate photosystem I and [NiFe]-hydrogenase. *Proceedings of the National Academy of Sciences of the United States of America, 108*, 20988–20991.

Kanygin, A. L., & Yacoby, I. (2017). Synthetic biology for improved hydrogen production in *Chlamydomonas reinhardtii*. *Methods in Molecular Biology, 1655*, 209–224.

Khalil, Y. F. (2025). A comparative science-based viability assessment among current and emerging hydrogen production technologies. *Applied Sciences, 15*, 497.

Melis, A. (2007). Photosynthetic H2 metabolism in *Chlamydomonas reinhardtii* (unicellular green algae). *Planta, 226*(5), 1075–1086. doi:10.1007/s00425-007-0609-9.

Melis, A., Zhang, L., Forestier, M., Ghirardi, M. L., & Seibert, M. (2000). Sustained photobiological hydrogen gas production upon reversible inactivation of oxygen evolution in the green alga *Chlamydomonas reinhardtii*. *Plant Physiology, 122*(1), 127–135. http://www.plantphysiol.org/. doi:10.1104/pp.122.1.127.

Melitos, G., Voulkopoulos, X., & Zabaniotou, A. (2021). Waste to sustainable biohydrogen production via photo-fermentation and biophotolysis – A systematic review. *Renew. Energy Environ. Sustain, 6*, 45.

Nagarajan, D., Lee, D. J., Kondo, A., & Chang, J. S. (2017). Recent insights into biohydrogen production by microalgae – From biophotolysis to dark fermentation. *Bioresource Technology, 227*, 373–387. www.elsevier.com/locate/biortech. doi:10.1016/j.biortech.2016.12.104.

Németh, B., Land, H., Magnuson, A., Hofer, A., & Berggren, G. (2020). The maturase HydF enables [FeFe] hydrogenase assembly via transient, cofactor-dependent interactions. *Journal of Biological Chemistry, 295*, 11891–11901.

Ooms, M. D., Dinh, C. T., Sargent, E. H., & Sinton, D. (2016). Photon management for augmented photosynthesis. *Nature CommunicationsScientific Reports, 7*, 12699.

Philipps, G., Happe, T., & Hemschemeier, A. (2012). Nitrogen deprivation results in photosynthetic hydrogen production in *Chlamydomonas reinhardtii*. *Planta, 235*(4), 729–745. doi:10.1007/s00425-011-1537-2.

Pongpadung, P., Liu, J., Yokthongwattana, K., Techapinyawat, S., & Juntawong, N. (2015). Screening for hydrogen-producing strains of green microalgae in phosphorus or sulphur deprived medium under nitrogen limitation. *ScienceAsia, 41*, 97–107.

Pongpadung, P., Zhang, L., Sathasivam, R., Yokthongwattana, K., Juntawong, N., & Liu, J. (2018). Stimulation of hydrogen photoproduction in *Chlorella sorokiniana* subjected to simultaneous nitrogen limitation and sulfur- and/or phosphorus-deprivation. *International Journal of Pure & Applied Microbiology, 12*, 1719–1727.

Quiroz, D., Greene, J. M., Limb, B. J., & Quinn, J. C. (2023). Global life cycle and techno-economic assessment of algal-based biofuels. *Environmental Science and Technology, 57*(31), 11541–11551. http://pubs.acs.org/journal/esthag. doi:10.1021/acs.est.3c02892.

Rather, A. H., & Srivastav, A. K. (2021). A study on biohydrogen production based on biophotolysis from cyanobacteria. *Annals of the Romanian Society for Cell Biology, 25*, 12500–12509.

Rumbaugh, T. D., Gorka, M. J., Baker, C. S., Golbeck, J. H., & Silakov, A. (2024). Light-induced h2 generation in a photosystem i-o2-tolerant [fefe] hydrogenase nanoconstruct. *Proceedings of the National Academy of Sciences, 121*(34), 8.

Saifuddin, N., & Priatharsini, P. (2016). Developments in bio-hydrogen production from algae: A review. *Research Journal of Applied Sciences, Engineering and Technology, 12*(9), 968–982. doi:10.19026/rjaset.12.2815.

Saputro, E. D., & Juangsa, F. B. (2025). Techno-economic assessment of green hydrogen production via PEM electrolysis and integrated water treatment using hydropower energy. *Clean Energy, 9*(6), 81–95. doi:10.1093/ce/zkaf032.

Shah, V., Garg, N., & Madamwar, D. (2003). Ultrastructure of the cyanobacterium *Nostoc muscorum* and exploitation of the culture for hydrogen production. *Folia Microbiologica, 48*(1), 65–70. www.biomed.cas.cz/mbu/folia/index.html. doi:10.1007/BF02931278.

Shepard, E. M., Byer, A. S., Betz, J. M., & Peters, J. W. (2013). Synthesis of the 2Fe subcluster of the [FeFe]-hydrogenase H cluster on the HydF scaffold. *Proceedings of the National Academy of Sciences of the United States of America, 110*, 4699–4704.

Skjånes, K., Andersen, U., Heidorn, T., & Borgvang, S. A. (2016). Design and construction of a photobioreactor for hydrogen production, including status in the field. *Journal of Applied Phycology, 28*(4), 2205–2223. doi:10.1007/s10811-016-0789-4.

Strand, D. D., Livingston, A. K., Satoh-Cruz, M., Froehlich, J. E., Maurino, V. G., & Kramer, D. M. (2015). Activation of cyclic electron flow by hydrogen peroxide in vivo. *Proceedings of the National Academy of Sciences of the United States of America, 112*(17), 5539–5544. http://www.pnas.org/content/112/17/5539.full.pdf. doi:10.1073/pnas.1418223112.

Takahashi, S., Milward, S. E., Fan, D. Y., Chow, W. S., & Badger, M. R. (2009). How does cyclic electron flow alleviate photoinhibition in *Arabidopsis*? *Plant Physiology, 149*, 1560–1567.

Teo, C. L., Idris, A., & Sukiran, A. N. (2017). Synergistic effects of pulse light emitting diode on growth rate, lipid content and its carbon chain of *Nannochloropsis* sp. in Fatty Acid Methyl Ester. *Production.Chemical Engineering Transactions, 56*, 283–288.

Tolleter, D., Ghysels, B., Alric, J., Petroutsos, D., Tolstygina, I., Krawietz, D., Happe, T., Auroy, P., Adriano, J. M., Beyly, A., Cuiné, S., Plet, J., Reiter, I. M., Genty, B., Cournac, L., Hippler, M., & Peltier, G. (2011). Control of hydrogen photoproduction by the proton gradient generated by cyclic electron flow in *Chlamydomonas reinhardtii*. *Plant Cell, 23*(7), 2619–2630. http://www.plantcell.org/content/23/7/2619.full.pdf+html. doi:10.1105/tpc.111.086876.

Tsygankov, A., Kosourov, S., Seibert, M., & Ghirardi, M. L. (2002). Hydrogen photoproduction under continuous illumination by sulfur-deprived, synchronous *Chlamydomonas reinhardtii* cultures. *International Journal of Hydrogen Energy, 27*(11–12), 1239–1244 03603199. doi:10.1016/S0360-3199(02)00108-8.

Vejrazka, C., Janssen, M., Streefland, M., & Wijffels, R. H. (2012). Photosynthetic efficiency of *Chlamydomonas reinhardtii* in attenuated, flashing light. *Biotechnology and bioengineering, 109*, 2567–2574.

Volgusheva, A., Kukarskikh, G., Krendeleva, T., Rubin, A., & Mamedov, F. (2015). Hydrogen photoproduction in green algae *Chlamydomonas reinhardtii* under magnesium deprivation. *RSC Advances, 5*(8), 5633–5637. http://pubs.rsc.org/en/journals/journalissues. doi:10.1039/c4ra12710b.

Volgusheva, A., Styring, S., & Mamedov, F. (2013). Increased photosystem II stability promotes H_2 production in sulfur-deprived *Chlamydomonas reinhardtii*. *Proceedings of the National Academy of Sciences of the United States of America, 110*(18), 7223–7228. doi:10.1073/pnas.1220645110.

Williams, C. R., & Bees, M. A. (2014). Mechanistic modeling of sulfur-deprived photosynthesis and hydrogen production in suspensions of *Chlamydomonas reinhardtii*. *Biotechnology and Bioengineering, 111*, 320–335.

Wu, Y., Shan, L., Zhao, W., & Lu, X. (2025). Harnessing artificial intelligence to revolutionize microalgae biotechnology: Unlocking sustainable solutions for high-value compounds and carbon neutrality. *Marine Drugs, 23*(5), 184. doi:10.3390/md23050184.

Wykoff, D. D., Davies, J. P., Melis, A., & Grossman, A. R. (1998). The regulation of photosynthetic electron transport during nutrient deprivation in *Chlamydomonas reinhardtii*. *Plant Physiology, 117*(1), 129–139. http://www.plantphysiol.org/. doi:10.1104/pp.117.1.129.

Xu, L., Li, D., Wang, Q., & Wu, S. (2016). Improved hydrogen production and biomass through the co-cultivation of *Chlamydomonas reinhardtii* and *Bradyrhizobium japonicum*. *International Journal of Hydrogen Energy, 41*(22), 9276–9283. https://doi.org/10.1016/j.ijhydene.2016.04.009.

Zarmi, Y., Gordon, J. M., Mahulkar, A., Khopkar, A. R., Patil, S. D., Banerjee, A., … Sapre, A. (2020). Enhanced algal photosynthetic photon efficiency by pulsed light. *Iscience, 23*.

Zhang, J., Xue, D., Wang, C., Fang, D., Cao, L., & Gong, C. (2023). Genetic engineering for biohydrogen production from microalgae. *Iscience, 26*(8), 107255. https://doi.org/10.1016/j.isci.2023.107255.

CHAPTER

3

Microbial electrolysis cell for biohydrogen production

Shengnan Zhu[a], *Chao Wang*[b], *Haixing Chang*[c], *Huan Zhang*[a,*], *and Xiuqiang Zhang*[d]

[a]Henan Agricultural University, Zhengzhou, China, [b]Henan University of Science and Technology, Luoyang, China, [c]Wuhan University, Wuhan, China, [d]Henan Academy of Sciences, Energy Research Institute Co. Ltd., Zhengzhou, China

[*]Corresponding author. E-mail address: zhanghuan5754@163.com

3.1 Introduction

Microbial electrolysis cells (MECs) combine the advantages of biological and electrochemical reactions, providing dual benefits in treating wastewater and producing biohydrogen. The mechanism of MEC is to utilize the catalytic action of electroactive microorganisms (EAM) to decompose organic substrates under mild conditions while driving efficient biohydrogen production through a small external voltage. Compared to other hydrogen production methods (dark fermentation, photofermentation, and water electrolysis), it has the characteristics of high conversion efficiency, high hydrogen purity, wide substrate utilization range, and low energy consumption (Kadier et al., 2016a).

The MEC system comprises an anode, cathode, membrane, EAM, and power source. In an MEC system, EAM oxidizes organic substrates (e.g., acetate) on the anode into electrons (e^-), protons (H^+), and CO_2 (Eq. 1). Then the produced electrons are transferred to the cathode via an external circuit, whereas the generated protons diffuse to the cathode via membrane. Finally, protons combine with electrons to form biohydrogen at the

Biohydrogen: Advances in Theory and Production
DOI: https://doi.org/10.1016/B978-0-443-27686-6.00005-0

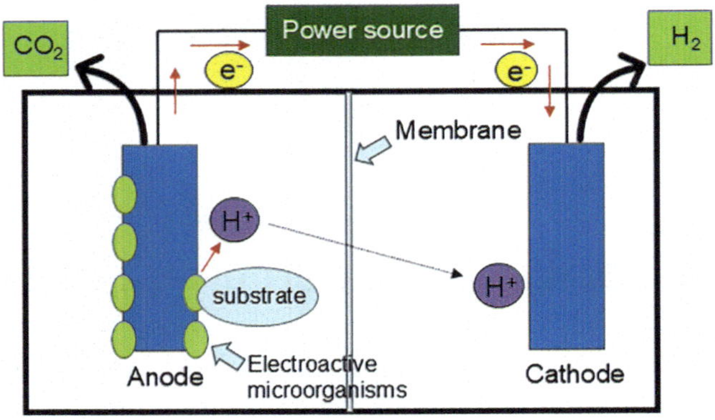

FIGURE 3.1 Schematic diagram of biohydrogen production in MEC (double-chambered). *MEC*, Microbial electrolysis cells.

cathode (Eq. 2). However, under standard biological conditions, hydrogen evolution reaction (HER) cannot occur spontaneously, thus requiring a cathodic potential of at least −0.414 V relative to the standard hydrogen electrode to generate hydrogen. Therefore an external voltage of at least 0.114 V must be applied to the MECs system to promote HER. However, in practical operations, a voltage higher than this calculated value is required to overcome the inherent ohmic loss, activation loss, and mass transfer limitations in MECs (Bahari et al., 2024; Yang et al., 2021) (Fig. 3.1).

Anodic chamber:

$$\text{Organic substrates} + H_2O \rightarrow CO_2 + H^+ + e^-$$

Cathodic chamber:

$$2H^+ + 2e^- \rightarrow H_2$$

Based on the principle of biohydrogen production in MECs, it can be concluded that the main parameters affecting biohydrogen production in MECs include strains, substrate type, reactor structure and operation, electrode materials (Kadier et al., 2016a).

3.2 Substrate

Substrate is one of the important factors affecting the biohydrogen production in MECs. The biohydrogen evolution reaction can occur in MEC with different substrates, including acetate, glucose, and various types of wastewater. Different substrates offer varying amounts and types of organic matter, as well as electron donors, which directly affect microbial metabolism and electrochemical reactions in MECs. In the selection of substrate types, availability, cost-effectiveness, and the desired final product are usually the main factors considered (Bahari et al., 2024; Sharma et al., 2024).

3.2.1 Substrate type

3.2.1.1 *Purely organic compounds*

Organic acids: In MEC, simple carbon sources such as acetate and propionate can be rapidly degraded by microorganisms with high electron transfer efficiency. Acetate, as the main end product of dark fermentation, is widely used as substrate for MEC. Chae et al. (2008) used acetate (2.0 mM) as sole substrate for biohydrogen production in a two-chambered MEC with total volume of 180 mL. The hydrogen production gradually increased with the applied voltage from 0.1 to 1.0 V, reaching a maximum efficiency of 52.5% at 0.8 V. According to a study by Liang et al. (2011), biohydrogen production was carried out using sodium acetate as substrate (1 g/L) in a single-chamber MEC, and the current and hydrogen production rate (HPR) were enhanced by 72% and 118% at 0.8 V applied voltage by modifying the anode into stacking mode. The maximum HPRs reached 1.58 L/L/day from sodium acetate in a cathode-on-top single-chamber MEC with applied voltages 1.0 V (Guo et al., 2010). Chen et al. (2016) studied the Cd(II) removal and hydrogen production in biocathode MECs using sodium acetate or $NaHCO_3$ as substrates. The results showed that both substrates could effectively promote the cathodic reduction reaction of the system. When sodium acetate was used as the substrate, the system exhibited better Cd(II) removal and hydrogen production, mainly attributed to the enhanced reduction ability of the biocathode by sodium acetate as an additional electron donor. $NaHCO_3$ mainly maintains microbial activity by adjusting the pH of the system and providing inorganic carbon sources.

Saccharides: Glucose is a simple carbohydrate. In single-chamber MEC, glucose was consumed rapidly with acetate and propionate as the main intermediates. Hydrogen yield of 7.2 mol H_2/mol was obtained from glucose (1 g/L), which was higher than that anaerobic fermentation (Selembo et al., 2009). Cui et al. (2023) studied biohydrogen production from glucose (1 g/L) in a single chamber MEC under hyperalkaline conditions. Results showed that high pH improved the acetate production from glucose fermentation and inhibited hydrogenotrophic methanogenesis. At pH 11.2 the maximum current density of 180 ± 9 A/m^3 and the average hydrogen production of 7.72 ± 0.23 mol H_2/mol glucose were obtained under 1.6 V. Lu et al. (2012b) proved that hydrogen can be obtained with glucose as electron donor at 4°C in a single-chamber MEC. The hydrogen yield was about 6 mol H_2/mol glucose, and rates were 0.25 ± 0.03 to 0.37 ± 0.04 m^3 H_2/m^3/day.

Glycerol: Glycerol is a byproduct in the production of biodiesel. P-glycerol and B-glycerol were used as substrates for hydrogen production in MEC system (Selembo et al., 2009). Results indicated that hydrogen production potential was influenced by the applied voltage. P-glycerol obtained better hydrogen yield (3.9 mol H_2/mol glycerol) at applied voltage 0.9 V, whereas B-glycerol showed higher hydrogen yield (2.3 mol H_2/mol glycerol) at applied voltage 0.5 V. The hydrogen production of MEC using B-glycerol as substrate was lower than that of P-glycerol, which may be because of the presence of methanol. Methanol can be toxic to microorganisms (Selembo et al., 2009). Badia-Fabregat et al. (2019) evaluated the performance of hydrogen production from crude glycerol in an alkaline MEC. Results showed that glycerol degradation was feasible, with a maximum hydrogen production of 0.46 L H_2/L/day and cathodic recovery of 85%.

3.2.1.2 *Wastewater*

Biohydrogen production from wastewater with a high organic matter content using bioelectrosynthesis reactions is an attractive option, which can produce biohydrogen from wastewater along with water remediation. Generally, wastewater contains complex organic compounds such as carbohydrates, proteins, and fats, which need to be decomposed into small molecules by microorganisms before participating in the reaction. In MEC, many types of wastewater have been used to produce biohydrogen, such as paper industry wastewater (Chaurasia et al., 2021), sugar industry effluent, landfill leachate (Hassan et al., 2018), distillery wastewater (Samsudeen et al., 2020).

Sugar industry wastewater: Sugar industry wastewater is a typical high-load organic wastewater, and different cathode materials are used in the MEC to improve its degradation efficiency and recover energy. In the report of Li et al. (Jayabalan et al., 2019), cane sugar wastewater and raw sugar wastewater were used as the substrates of dual-chamber MEC, and the effects of different cathode materials of nickel plate, nickel foam, stainless steel mesh on biohydrogen production were studied. Results showed that cane sugar wastewater had better hydrogen production potential than raw sugar wastewater in MEC, and when nickel foam was used as cathode material, hydrogen production in MEC at applied voltage of 1.0 V was 2 and 1.2 times higher than that in the stainless steel mesh and nickel plate, respectively. Chaurasia and Mondal (2022) applied Ni, Ni–Co, and Ni–Co–P electrodeposits as cathodes in MEC with 0.6 V at 30°C ± 2°C temperature and neutral pH, achieving better biohydrogen production rate (0.14–0.24 m^3/day) and chemical oxygen demand (COD) removal (50%) from sugar industrial wastewater (COD 4850 ± 50 mg/L, Biochemical oxygen demand (BOD) 1950 ± 20 mg/L). Similarly the use of foam nickel coated with nickel molybdate ($NiMoO_4$) as the cathode of the MEC enhanced the HPR (0.12 ± 0.01 L/L/day) of sugar industry wastewater (COD 4200 ± 20 mg/L) because the modified catalyst had higher catalytic activity and lower impedance, which was conducive to faster hydrogen evolution kinetics (Jayabalan et al., 2021). Besides the foam nickel modified by $NiCo_2O_4$–graphene nanocomposite as cathode in MEC also showed high catalytic performance in the treatment of sugar industry wastewater, and the HPR was 3.2 times higher than that of the control group (Jayabalan et al., 2020). However, the research on different cathode materials in MEC mentioned above is on a laboratory scale, and there are few applications on a pilot scale. Hence the hydrogen production from diluted sugar-based industrial wastewater was evaluated in a 150-L MEC pilot plant with nickel-foam and stainless-steel wool cathodes (Guerrero-Sodric et al., 2024). The reactor achieved continuous hydrogen production for over 80 days, with a maximum COD removal of 40%. Nickel-foam-enhanced energy efficiency and hydrogen production achieved 19.07 ± 0.46 L H_2/m^2/day (Guerrero-Sodric et al., 2024).

Food waste leachate: Food waste leachate is a high-concentration liquid phase that is released during the transportation and treatment of food waste through physical pressure filtration and microbial anaerobic decomposition. Food waste leachate has a high organic content and good biodegradability, making it a promising substrate for biohydrogen production in MEC. Ma et al. (2024) enhanced the hydrogen production from food waste leachate in a 1.2-L single-chamber MEC (carbon brushes as the anode, stainless steel brushes as the cathode) by adjusting the pH value of the leachate and applied voltage. Under the conditions of pH 11.2 and voltage 2.4 V the maximum current density of 121.9 ± 10.9 A/m^3 was obtained, and average hydrogen content reached 91.9% ± 3.2%. Zhang et al. (2022) explored the feasibility of hydrogen production in MEC of protein-rich food waste. Results showed that the anaerobic digestate liquid (meat digestate liquid) of protein-rich food waste was more suitable for hydrogen recovery in MEC than raw meat slurry because the main metabolites (propionic) of protein in raw meat slurry were not easily utilized by electroactive bacteria (EAB). Jwa et al. (2025) compared the performance of hydrogen production in a novel zero-gap bipolar membrane MEC from various wastewater sources, including brewery wastewater, food wastewater, and livestock wastewater. Results showed that the stable current density and high hydrogen yield was obtained from livestock wastewater in MEC because of the abundant fermentable substrates (such as formic, butyric, and valeric acids), which provided an optimal metabolic environment for EAM. The performance of brewery wastewater was lower than that of livestock wastewater because its high proportion of alcohol, resulting in a gradual decline in current density over 2 weeks. And food wastewater showed a lower performance because of the presence of nondegradable oil and lower pH (≤5.5).

Hydrothermal liquefaction (HTL) wastewater: HTL wastewater is the main byproduct generated during the production of biocrude oil from various biomass resources, containing a large amount of organic carbon, nitrogen, and toxic compounds (Gu et al., 2019). Such concentrated and toxic wastewater needs to be treated before being discharged into the environment. Shen et al. (2018) demonstrated that MECs system could effectively treat wastewater generated during the HTL of swine manure and simultaneously recover hydrogen. Besides the generated hydrogen could be used for onsite HTL biocrude upgrading. Under various organic loadings (958–6275 mg COD/L), applied voltages (0.4–1.2V), and flow rates (5.4–7.2 mL/min), excellent COD removal (90%–98%) and nitrogen removal (57%–93%) were achieved. A continuous upflow fixed-bed MEC was used to treat wastewater from HTL of cornstalk. The COD removal rate reached 80.2% under the conditions of organic load rate 2 g/L/day, hydraulic retention time (HRT) 16 h, and an applied voltage 1.2 V. Furfural and 5-hydroxymethylfurfural were completely degraded, and dimethyl phthalate and diethyl phthalate were significantly removed by 95.3% and 79.3%, respectively (Shen et al., 2016).

3.2.1.3 *Waste activated sludge*

Waste activated sludge (WAS) is composed of various complex organic compounds, including protein [accounting for 35%–61% of total COD (TCOD)], carbohydrates (7%–11% of TCOD), and other components (<1% of TCOD) (Wang et al., 2014). It can be used as a substrate for MEC hydrogen production and has the advantages of shortening the fermentation cycle and improving energy recovery rate. Hydrogen production of 5.67 ± 0.61 mg H_2/g volatile suspended solids (VSS) was obtained from raw WAS in a dual-chamber MEC, which was higher than that of fermentation. The alkaline-pretreated raw WAS doubled the hydrogen yield through cascade utilization of organic matter because alkaline pretreatment destroyed the structure of sludge flocs and accelerated hydrolysis and acidification, which provided more soluble substrates for MEC. At the end of reaction, TCOD removal rate ranged from 20% ± 3% to 28% ± 3%, total protein removal rate ranged from 20% ± 4% to 35% ± 7%, and total carbohydrate removal rate ranged from 26% ± 3% to 34% ± 12% (Lu et al., 2012). Liu et al. (2012) studied the hydrogen production in MECs with fermentation liquid of WAS as substrate. Before feeding into MEC system, WAS was pretreated through a combination of ultrasonic and fermentation. Accumulated short-chain fatty acids from WAS provided abundant substrate for MEC process, and the degradability of sludge fermentative liquid fractions in MECs for hydrogen production was acetate > propionate > *n*-butyrate > *n*-valerate > iso-valerate > iso-butyrate. Sun et al. (2014) studied the effect of WAS concentration of high-solid WAS on hydrogen production and sludge reduction in MEC, and highest hydrogen yield of 4.66 ± 1.90 mg H_2/g VSS was obtained at the concentration of 7.6 g VSS/L. Continuously increasing substrate concentration could lead to an excess supply of organic matter, causing other bacteria to proliferate and compete with hydrogen-producing bacteria. Besides, alkaline-pretreated WAS increased hydrogen yield in MEC by 2.75 times.

3.2.2 Substrate pretreatment methods

The organic matter in raw substrates (such as wastewater and WAS) typically exists in complex forms, making it difficult for electricity-generating bacteria in MEC to directly utilize. Hence selecting appropriate pretreatment methods for complex organic compounds to release more available organic matter is a crucial step to enhance waste degradation and hydrogen production in MEC.

3.2.2.1 *Prefermentation*

Prefermentation technology refers to the pretreatment process in which complex organic substrates are decomposed and converted into soluble small molecule organic compounds through anaerobic microorganisms prior to MEC. These compounds are more easily utilized by electrogenic bacteria in MEC systems. Hence this prefermentation has dual goals of dissolving suspended organic matter and providing more available electron sources for EAM in the anode of MEC.

Cha et al. (2024) applied prefermentation to enhance the exoelectrogenic utilization from food wastewater in dual-chamber MECs. Prefermentation increased the proportion of short-chain organic acids and ethanol in the TCOD from 27% to 48%–72%. Compared with MEC for treating raw food wastewater, prefermentation treatment of food wastewater could accelerate the start-up speed of MEC and increase the current density and HPR. And the highest hydrogen yield of 911 mL/g COD in MEC was obtained from prefermentation effluent treated with 2-day HRTs, which was 34.4% higher than MEC fed with raw food wastewater. Kim et al. (2025) demonstrated that prefermentation of food waste could improve its bioavailability in dual-chamber MECs by conditioning substrate composition, and greater abundance of exoelectrogenic bacteria (*Geobacter* and *Desulfovibrio*) was observed in bioanodes with prefermentation.The hydrogen yield of MECs treated by pre-fermentation and filtration increased by 39.1% compared to the untreated group. In addition, Mahmoud et al. (2014) evaluated the effect of prefermentation of landfill leachate on MEC performance. Results showed that the complex organic matter in the landfill leachate was transformed to simple volatile fatty acids (VFAs) during fermentation. And carbohydrates obtained the highest degree of fermentation, followed by protein and lipids. Compared to raw leachate, fermented leachate enhanced current density from 2.5 to 23 A/m^3, biochemical oxygen demand removal from 5.6% to 83% in MEC.

3.2.2.2 *Chemical pretreatment*

Chemical pretreatment method was applied to enhance VFAs accumulation from complex substrates with low cost and quick rate (Jiang et al., 2007). Most of organic matter in raw WAS was contained inside the cell, and organic matter exists in the form of macromolecule, which cannot be directly utilized by electrogenic bacteria, resulting in low biodegradation rates. Hence WAS needs to be pretreated before being used as a substrate for hydrogen production in MEC to release more available organic.

Four chemical methods, including acid, alkaline, lysis, and biosurfactant, were used to treat WAS for hydrogen production in MECs. The results showed that four kinds of pretreatment methods enhanced VFAs accumulation in WAS, and the highest yield of 4712.69 mg COD/L VFAs was obtained from β-cyclodextrin pretreatment. However, the highest hydrogen yield in MEC was obtained from sodium dodecyl sulfate pretreat because of the high proportion of acetic and propionic (Wang et al., 2014). Lu et al. (2012) fed alkaline pretreated WAS into a two-chamber MEC, and the hydrogen yield had almost doubled because of the abundance of VFAs, carbohydrates, protein, and COD in pretreated WAS solution. Wang et al. (2021b) used alkaline thermal pretreatment of WAS to improve hydrogen production in MECs. Results showed that low-temperature pretreatment (pH 11, 90°C, 3 h) could effectively break cell walls to release organic matter, causing efficient fermentation and further enhanced hydrogen production in MEC. Ahmadi et al. (2025) evaluated the hydrogen production of sewage sludge with acidic, alkaline, and a combined calcium carbonate-2 potassium hydroxide pretreatment (CCPH-P) by fermentation and MECs. Results showed that pretreatment significantly enhanced the accumulation of proteins, carbohydrates, and VFAs, and the dark fermentation stage further enhanced the solubilization of these key components. Compared to acidic pretreatment, alkaline pretreatment releases more intracellular organic matter through disrupting extracellular polymeric substances. The synergistic physicochemical interactions between $CaCO_3$ and KOH made CCPH-P show the best performance. Calcium carbonate promotes the overall stability of sludge by providing pH buffering. Meanwhile, potassium hydroxide, as a potent alkaline reagent, promotes lipid degradation and hydrolysis of complex organic polymers. The VFAs concentration in CCPH-P sludge achieved 1739 ± 51 mg/L, which was 10.6 times that of raw sludge. And the highest hydrogen yield of 132.14 mL/g VS was achieved in MEC with CCPH-P sludge.

3.3 Strains

Compared with traditional biological treatment processes, MEC introduces EAM, which can achieve the biodegradation function of electrodes. EAM are the core of MEC systems, capable of extracellular electron transfer. EAM use organic matter as electron donors to transfer electrons to the anode, which directly affects the hydrogen production in MEC (Cheng et al., 2022). There are three main mechanisms of electron transfer in electrogenic bacteria: (1) direct electron transfer using cytochrome attached to the outer membrane of the cell; (2) the nanowires of conductive pili facilitate electron transfer; (3) utilizing autocrine electron mediators to achieve electron transfer (Lovley & Nevin, 2011).

Both pure culture and mixed culture can be applied in MEC. The pure culture MEC system is well controlled and can avoid hydrogen losses to methanogens. Meantime, pure culture is beneficial for studying the electron transport mechanism between microorganisms and electrodes (Munoz et al., 2010). *Pseudomonas* spp. and *Shewanella* spp. are common pure electrogenic species. On the other hand the mixed-culture MEC system has shown advantages in adapting to various environmental conditions, utilizing complex substrates, and maintaining long-term stability of the system. Common mixed culture inoculants include sludge and various types of wastewater. Hu et al. (2008) investigated the hydrogen production between mixed culture (domestic wastewater as the inoculum) and pure culture (*Shewanella oneidensis* MR-1) in single-chamber membrane-free MECs. Results showed that there was no methane production when using pure culture (*S. oneidensis* MR-1) in the MEC system, but the HPR was much lower than that of the mixed culture system. In mixed culture MEC system at applied voltage of 0.6 V, higher HPR of 0.69 $m^3/day/m^3$ was achieved at pH 5.8. However, when using mixed cultures as inoculation sources in MECs, methanogens can utilize the hydrogen produced in MECs, leading to a decrease in hydrogen yield. To improve the electrochemical performance of the system, various pretreatment methods (physical, chemical, and biological) were used on the mixed culture before inoculation to inhibit the growth of methanogens (Jadhav et al., 2019). In addition, adding antibiotics [2-chloroethane sulfonate (20 mM), 2-bromoethane sulfonate (20 mM), 8-aza-hypoxanthine (3.6 mM)] has also been proven to be an alternative biotechnological method for inhibiting methane production and increasing hydrogen production in single chamber MECs with domestic wastewater as inoculum (Catal et al., 2015).

External voltage, internal resistance, pH value, substrate type and concentration, temperature, anode material, and system configuration are key factors, affecting EAM activity in the hydrogen production system of MECs. The effect of applied voltage on EAM and performance of single-chamber membrane-free MEC was studied (Ndayisenga et al., 2023).

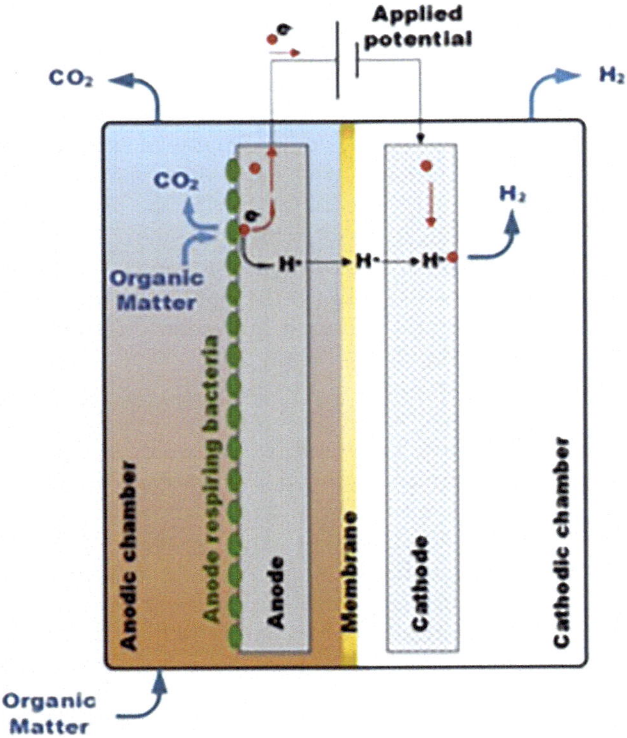

FIGURE 3.2 Schematic diagram of a dual-chamber MEC reactor. *MEC*, Microbial electrolysis cells. *Reproduced from Escapa et al. (2016), Copyright 2016 Elsevier.*

Results showed the anodic biofilm viability and bacterial cell shape were less affected in MEC at applied voltage 0.5 and 0.8 V, and highest hydrogen production of 6.017 mmol H_2/g COD was obtained at 0.8V. However, further increasing the voltage to 1.0 V would cause irreversible "electrical fatigue" of the anode biofilm, damaging cells, inhibiting microorganism activity, and ultimately leading to a decrease in hydrogen yield (Ndayisenga et al., 2023).

3.4 Reactors

MEC fundamentally consists of an anode, a cathode, and an ion exchange membrane (Fig. 3.2) (Jafary et al., 2019; Swaminathan et al., 2024; Varanasi et al., 2018). Electrochemically active bacteria attached to the anode oxidize organic substrates (e.g., wastewater), releasing CO_2, protons

(H^+), and electrons. These electrons transfer through an external circuit to the cathode, while protons migrate across the membrane. At the cathode, applied voltage drives hydrogen formation through proton–electron recombination. The design of MEC reactors is a critical determinant of overall system performance. It directly influences key physical factors such as internal resistance and current density, which, in turn, govern crucial outcomes including hydrogen yield, purity, and production cost (Gautam et al., 2023; Kadier et al., 2016a; Sharma et al., 2024).

3.4.1 Reactor structure

3.4.1.1 Reactor configurations

Over the past decade, research has explored various types of MEC configurations, primarily focusing on dual-chambered, single-chambered, or modified designs (Bora et al., 2022; Kadier et al., 2016b; Karthikeyan et al., 2017). To improve the performance and HPR as well as to decrease the cost of MECs, these reactors have utilized a wide range of materials under different operational conditions (Kim et al., 2022; Noori et al., 2024).

The dual-chamber MECs are the predominant reactor configuration. It consists of separate anode and cathode chambers divided by an ion-exchange membrane (Fig. 3.3A) (Abd-Elrahman et al., 2022). The chambers can be made of different shapes such as bottle-type, dual-chambered disc-shaped, rectangular-shaped, H-type, cube, and cylindrical type (Fig. 3.3C, E, F, K, and L) (Chae et al., 2009; Kadier et al., 2016a; Nayak et al., 2018; Varanasi et al., 2018). The membrane is crucial for maintaining high hydrogen purity by preventing microbial consumption of the produced hydrogen and minimizing crossover between chambers (Rozendal et al., 2006). Common membrane types include proton exchange membranes (PEMs), anion-exchange membranes (AEMs), and cation exchange membranes (CEMs) (Rozendal et al., 2007). Although the membrane increases system complexity and cost, it enables efficient substrate degradation (~90%) in the anode and hydrogen production efficiencies typically ranging from 60% to 75% (Park et al., 2017).

Single-chamber MECs (SCMECs) are developed to enhance biohydrogen production and streamline reactor design. In this configuration the anode and cathode coexist within a shared electrolyte compartment within a single chamber, typically constructed from glass or plastic materials [e.g., bottles, tubes, cubes, or cylinder (Fig. 3.3B and G–J) (Call & Logan, 2008; Guo et al., 2013; Hu et al., 2008, 2009; Varanasi et al., 2018)]. Crucially, MECs could operate under anaerobic conditions. This inherent characteristic permits the removal of the ion-exchange membrane without introducing oxygen to the anode, thereby avoiding the loss on operational efficiency

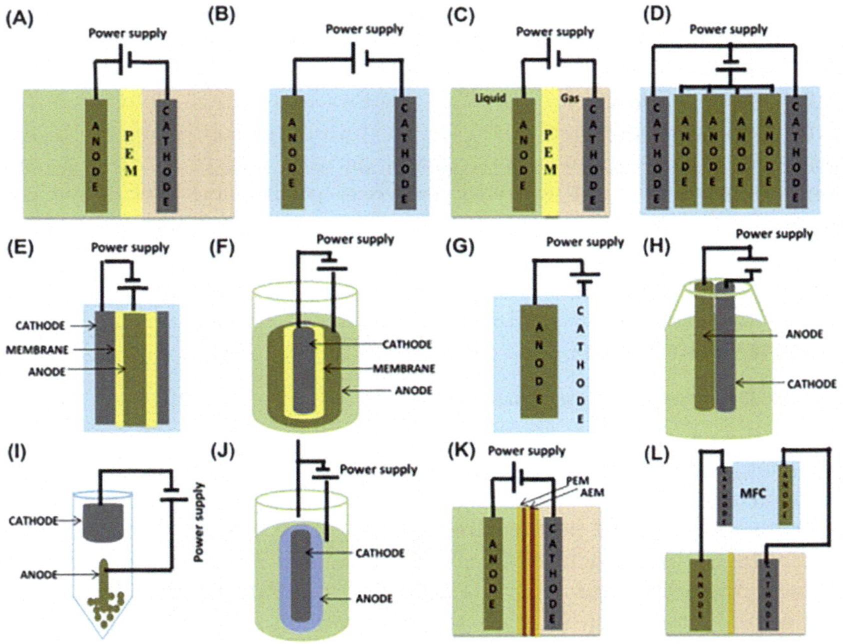

FIGURE 3.3 (A) Ion-exchange membrane separated dual chambers; (B) membraneless single chamber; (C) dual-chamber with gaseous cathode; (D) single chamber with multiple anodes; (E and F) membrane-separated tubular/cylindrical single chambers; (G) SS (stainless steel) anaerobic digester modified MEC; (H–J) membraneless bottle/tubular/cylindrical single chambers; (K) dual-chamber using stacked proton/anion-exchange membranes; and (L) an MFC–MEC coupled system powered by the MFC. *AEM*, Anion-exchange membranes; *MEC*, microbial electrolysis cells; *MFC*, microbial fuel cells; *PEMs*, proton exchange membranes. *Reproduced from Varanasi et al. (2018), Copyright 2019 Elsevier.*

(Call & Logan, 2008; Hu et al., 2008; Lee & Rittmann, 2010). The membrane-free architecture also reduces reactor complexity and cost, reduces internal resistance, and minimizes potential losses associated with membrane use. Furthermore the compact design enables stacked electrode arrangements to further reduce resistance (Fig. 3.3D) (Guo & Kim, 2018, 2019; Wang et al., 2023). However, a major challenge in SCMECs is the interference of methanogens, which consume substrate to produce methane, thereby reducing hydrogen purity (Chae et al., 2010; Cui et al., 2023). Comparative studies under equivalent conditions consistently demonstrate that while SCMECs achieve superior HPRs relative to dual-chamber systems, the resultant hydrogen exhibits lower purity owing to methane coproduction (Karthikeyan et al., 2017). Thus effective suppression of methanogenesis is paramount for upgrading biohydrogen purity in SCMEC applications.

To broaden the applicability of microbial electrolysis and overcome inherent limitations associated with conventional MEC designs, researchers have developed various advanced reactor configurations. These include multichamber, upflow, packed bed, fluidized bed, and trickle-bed reactors, as well as some integrated systems with other bioelectrochemical processes (Bora et al., 2022; Cheng et al., 2006; Nwanebu et al., 2024; Zeng et al., 2016). Among them, upflow reactors facilitate continuous operation and efficient wastewater treatment, demonstrating significant potential for scalability (Cui et al., 2020). Across all designs, sustained electrochemical activity hinges on mitigating scale formation, membrane fouling and biofilm overgrowth, while leveraging renewable feedstocks such as lignocellulosic biomass to secure economically viable hydrogen generation.

3.4.1.2 Anode

The anode in MEC serves as the terminal electron acceptor for electrogenic bacteria during anaerobic respiration, enabling extracellular electron transfer to the cathode (Almatouq et al., 2020; Varanasi et al., 2018). This critical function necessitates anode materials that combine biocompatibility for robust biofilm colonization, high electrical conductivity with low resistance, corrosion resistance, and optimized porosity to maximize surface area and retain sufficient void space to avoid mass-transfer bottlenecks (Lim et al., 2017).

Carbon-based materials naturally satisfy most of these criteria and have therefore become the default choice (Roubaud et al., 2019; Xu et al., 2023). Conventional forms such as graphite granules, graphite felt, carbon cloth, carbon paper, and carbon mesh, as well as engineered structures such as ammonia- or heat-treated graphite brushes and carbon-nanotube coatings, combine robust chemical stability with tunable porosity and low cost (Gautam et al., 2023; Sharma et al., 2024). Porous electrodes routinely outperform planar ones in current density because the enlarged interfacial area supports denser biofilms, but the mass transfer limitations within deep pore networks may impede nutrient diffusion, presenting a key scalability challenge (Baş & Kaya, 2022; Srivastava et al., 2023). Consequently, current works focuses on hierarchical porosity and surface functionalization to enhance extracellular electron transfer kinetics while preserving structural integrity over months of operation.

3.4.1.3 Cathode

The cathode in MEC serves as the electron conductor and microbial habitat where electrons harvested from organic matter at the anode finally meet protons to generate hydrogen and other value-added chemicals (Kumar et al., 2025; Parkhey & Gupta, 2017). The HER mechanism proceeds through three consecutive pathways, including the initial proton adsorption via Volmer reaction, followed by either chemical recombination (Tafel step) or electrochemical desorption (Heyrovsky step) (Liao et al., 2023; Weng et al., 2020). As HER kinetics depend on interfacial mass transfer across the solid–liquid–gas interfaces, each steps could be the rate limiting during hydrogen production in MEC. Tafel slope analysis provides essential mechanistic diagnostics for identifying dominant reaction pathways, whereas electrode microstructure governs proton diffusion kinetics and gas bubble evolution dynamics (Anantharaj & Noda, 2022; Tian et al., 2019). Consequently, cathode design critically determines system energy efficiency, hydrogen purity thresholds, and long-term operational stability.

Cathode materials can be grouped into four broad groups that balance performance, stability, and cost. Platinum-based catalysts remain to be the most promising catalyst with exceptional activity, but its susceptibility to sulfide poisoning and exorbitant cost preclude scalable development (Jwa et al., 2021; Sánchez-Peña et al., 2022). Transition metal alternatives offer moderate activity through optimized d-band electron configurations: The nickel-based alloys present satisfactory overpotential with significantly reduced cost and stainless steel meshes offer corrosion resistance with the demands of higher voltage (Chaurasia et al., 2020; Selembo et al., 2010). In addition, carbon-based materials serve as viable cathode alternatives when functionalized with transition metal sulfides, nitrides, or single-atom catalysts (SACs), achieving conductivity-enhanced performance while eliminating the demand of precious metal (Choi et al., 2019; Wang et al., 2017a; Xiao et al., 2012; Zhang et al., 2014). These systems operate under neutral pH conditions and reduce capital expenditures compared to precious metal cathodes. However, they require extended biofilm maturation periods (several weeks) and remain susceptible to methanogen contamination.

3.4.1.4 Membrane

Ion-exchange membranes function as essential separators in MECs, physically isolating anode and cathode chambers to preserve hydrogen purity by preventing microbial consumption of electrochemically generated hydrogen while mitigating fuel crossover and short-circuit risks (Choi et al., 2023; Dizge et al., 2019; Yang et al., 2020). The common membranes used in MECs are PEM, AEM, CEM, bipolar membrane, etc. (Pärnamäe et al., 2021; Zhao et al., 2021, 2023). During operation the membrane allows targeted ions (protons in PEM or other cations in CEM) to migrate from the anodic chamber to the cathodic chamber, thereby completing the electrical circuit and sustaining hydrogen evolution at the cathode. This selectivity also suppresses fuel crossover and prevents anodic bacteria from consuming the electrochemically generated hydrogen. However, the membranes also introduce several performance-limiting factors: It raises internal resistance, causes pH drift when cations other than protons dominate transport, and is susceptible to fouling that further elevates ohmic losses and hinders the HER (Colantoni et al., 2025; Ye & Logan, 2018). Specifically, CEMs induce substantial pH imbalances: rapid transport of cations (Na^+, Ca^+, K^+, and NH_4^+) relative to protons accelerates anodic acidification and cathodic alkalization, which, combined with inevitable membrane fouling, elevates internal resistance, suppresses current densities, and impedes hydrogen evolution kinetics through cumulative ohmic losses (Kazemi et al., 2016; Rozendal et al., 2006). Consequently, although membranes enhance gas purity, their inherent limitations have stimulated the search for membrane-free or antifouling designs.

3.4.2 Reactor operational process parameters

The performance of MECs is governed by a complex interplay of physicochemical and biological factors, such as the applied voltage, electrolyte pH, and temperature. This synergistic optimization of interdependent parameters dictates the efficiency and viability of MEC system.

3.4.2.1 *Applied voltage*

The cathodic HER in MEC imposes a fundamental thermodynamic requirement of 0.2 V (Cho et al., 2019; Logan et al., 2008). However, practical operation necessitates significantly higher applied voltages (typically 0.6–0.8 V or more) because of substantial cathode overpotential and other system losses, including activation, mass transport, and bacterial metabolism losses (Baek et al., 2021a). This increased voltage requirement directly impacts the process efficiency. Although hydrogen production generally increases with higher applied potential, an optimal range must be identified to balance scalability and performance; reported optimal values vary between studies, often falling within 0.2–1.4 V, with some optimized systems operating between 0.6 and 0.8 V for effective anode oxidation and cathodic hydrogen production (Lim et al., 2020; Nam et al., 2011). Critically the applied voltage profoundly influences microbial viability and activity: Excessive potential can lead to decreased metabolic function, increased cell rupture, altered biofilm dynamics, and inhibited enzyme activity (e.g., dehydrogenases), negatively affecting substrate degradation and biohydrogen yield (Baek et al., 2021b; Liu et al., 2023). Furthermore, lower voltage ranges could prevent effective electron transfer to the cathode, halting hydrogen production, whereas higher voltage ranges risk oxygen evolution and electrode damage, shifting the process toward conventional water electrolysis (Dong et al., 2024; Lim et al., 2020). Consequently, optimizing the applied voltage is essential not only for maximizing hydrogen output and minimizing energy consumption but also for maintaining integrated electrogenic microbial communities and ensuring long-term system stability. Real-time voltage monitoring and low-loss power supplies are crucial for future industrial implementation.

3.4.2.2 *Electrolyte pH*

The electrolyte pH functions as a master switch on MEC performance by concurrently regulating HER thermodynamics, microbial metabolic activity, and ion transport kinetics (Ruiz et al., 2016; Wang et al., 2019). The HER at the cathode is intrinsically pH-dependent, achieving the optimal efficiency at pH 5.0–6.0 because of favorable proton reduction kinetics, whereas alkaline conditions (pH > 10.0) induce substantial overpotential because of proton deficiency and altered reaction pathways (Cristiani et al., 2024; Yossan et al., 2013). Conversely, anodic biofilms require circumneutral to alkaline environments (pH 7.5–9.0) for optimal extracellular electron transfer, with acidification below pH 6.0 suppressing EAB metabolism and substrate oxidation (Liu et al., 2014; Wang et al., 2017b). These conflicting demands establish a transmembrane pH gradient, which drifts toward a highly alkaline catholyte and a progressively acidic analyte, amplifying internal resistance and ionic imbalances. Periodic polarity reversal or the use of hybrid biotic/abiotic electrodes has been shown to flatten this gradient and boost hydrogen yields more than five-fold (Jiang et al., 2016; Venkata Mohan et al., 2007). Buffer strength further modulates the system: increasing electrolyte conductivity with NaCl or phosphate buffers lowers ohmic losses and facilitates proton shuttling, yet excessive buffering can mask local pH microenvironments that microbes experience (Munoz et al., 2010; Nam & Logan, 2011). Empirically, peak hydrogen recovery is often reported at an initial catholyte pH near 5.0–6.0, provided the anode is held close to neutrality, underscoring the need for compartment-specific control (Nam & Logan, 2012). Ultimately, sustained performance demands dynamic pH management that reconciles the cathodic requirement for protons with the anodic necessity for a hospitable microbial milieu.

3.4.2.3 *Temperature*

Temperature is a master variable that exerts multifaceted effects on MEC functionality through both biological and electrochemical pathways. Operating under strictly anaerobic conditions, MECs rely on the metabolic rate of EAB; this rate is tightly linked to the thermal environment. Most MEC studies operate under mesophilic conditions (30°C ~ 40°C), because cellular growth, extracellular electron transfer, and substrate hydrolysis proceed optimally within this thermal range (Omidi & Sathasivan, 2013; Varanasi et al., 2018). As temperature deviates downward, mass-transfer limitations become more pronounced: diffusion coefficients shrink, cell membranes stiffen, and the overall microbial community shifts toward less electrogenic species, which favors methanogenesis over hydrogen evolution (Heidrich et al., 2014; Lu et al., 2011). Conversely, excessive heat accelerates enzyme denaturation and may precipitate unwanted side reactions. Temperature also influences the thermodynamics and kinetics of the hydrogen-evolution reaction itself by altering activation overpotentials and electrode potentials; even minor pH drifts that accompany thermal excursions can shift the local redox environment and further suppress HER. Although conventional water electrolysis researches demonstrate strong temperature dependencies on HER kinetics, these correlations are inapplicable to MECs because of fundamental differences in biocatalyst behavior and reactor configurations (Bi et al., 2025; Stühmeier et al., 2021). The absence of dedicated MEC temperature-kinetic studies represents a critical knowledge gap. Future research requires establishing temperature-pH-biofilm interaction models to optimize thermal management strategies for scalable systems.

3.4.2.4 *Other factors*

Beyond the abovementioned factors the overall efficiency of MEC is also affected by electrochemical physical properties and biological variables. Internal resistance, arising from activation, ohmic, mass transport, and bacterial metabolic losses, can be mitigated by optimizing electrode physicochemical characteristics, including surface area, morphology, conductivity, interelectrode distance, and activation resistance (Cheng & Logan, 2011; Liang et al., 2011; Wang et al., 2010). Electrolyte ionic strength and pH significantly influence MEC performance. Increased ionic strength enhances solution conductivity, reducing ohmic losses by facilitating proton transfer between electrodes (Merrill & Logan, 2009; Yossan et al., 2013).

Microbial factors also govern MEC performance. The density and metabolic activity of exoelectrogens that catalyze substrate oxidation and mediate extracellular electron transfer to the anode set the maximum attainable electron flux. Inoculation strategy is decisive; stable, high-current biofilms can be established by transplanting mature anodes from operating microbial fuel cells (MFCs), introducing axenic electroactive cultures, or leveraging the indigenous consortium present in wastewater sludge. Suppression of methanogens is equally critical, because these organisms divert reducing equivalents toward methane and can consume the hydrogen product. Selective suppression is routinely achieved through the addition of specific inhibitors such as 2-bromoethanesulfonate or by adopting defined inocula that lack methanogenic representatives (Commault et al., 2015; Logan, 2009; Zhang et al., 2016).

3.5 Net zero effect

MECs enable multifaceted CO_2 mitigation through direct bioelectrocatalytic conversion to electrofuels and indirect fossil fuel displacement via renewable energy storage (Nelabhotla & Dinamarca, 2018; Soggia et al., 2025a; Zhang et al., 2019). This dual-path approach synergizes direct carbon capture [microbial electrosynthesis (MES) of CO_2 into biomethane] with indirect emission avoidance (storing surplus renewables as carbon-neutral methane, displacing fossil natural gas). The technology's global scalability is underpinned by compatibility with existing gas infrastructure, high energy-density storage capacity, and utilization of intermittent renewable electricity. Integrating MECs into net zero portfolios is justified by their unique advantages: simultaneous CO_2 sequestration and utilization under ambient conditions, catalyst longevity, and tunable product selectivity (Nevin et al., 2010; Rabaey & Rozendal, 2010). By converging renewable electrification with carbon circularity, MECs exemplify a cross-sectoral decarbonization strategy directly targeting atmospheric CO_2 drawdown and fossil carbon substitution. Nevertheless, critical research gaps persist in optimizing electrode materials, elucidating extracellular electron transfer mechanisms, and enhancing biocatalyst efficiency at scale (Bajracharya et al., 2017; Pérez et al., 2021). Addressing these barriers is essential to fully harness the dual-path CO_2 reduction strategy, thereby advancing MECs as a scalable nexus of carbon circularity and net-zero energy systems.

3.5.1 Sources of CO_2 emissions

Carbon dioxide utilized in processes like MES can originate from diverse anthropogenic and natural sources. Primary pathways include (1) capture from concentrated industrial off-gases (e.g., fossil-fuel power generation, cement production, steel manufacturing, oil refining) representing significant point-source emissions (~65% of total); (2) direct air capture (DAC); (3) biological anaerobic digestion (AD) processes; and (4) emissions from geothermal reservoirs or power plants (Javad Bardi et al., 2025; Katuri et al., 2019; Wang et al., 2021; Yerli et al., 2022). Although industrial point sources offer high volumes, their utilization presents challenges: flue gases often require substantial cooling (incurring energy penalties and cost) from high temperatures (200°C–400°C) and may contain detrimental impurities (e.g., O_2, SO_x, NO_x, and particulate matter) inhibitory to sensitive biological processes like strict anaerobic electromethanogenesis (Ahn et al., 2017; Dykstra & Pavlostathis, 2021; Zhang et al., 2018).

Beyond these primary sources, valorizing specific waste streams offers distinct advantages and carbon management benefits. Industrial wastewater: Utilizing organics in industrial wastewater as a carbon source for MES represents a strategic alternative to conventional aerobic treatment. This shift avoids the significant indirect CO_2 emissions associated with the high energy demand of aeration systems. Furthermore, anaerobic MES processes inherently avoid the production of potent greenhouse gases like nitrous oxide (N_2O), which can occur during biological nitrogen removal in aerobic systems (San-Martín et al., 2021). This pathway transforms a treatment burden into a resource, providing a potential carbon-negative route for wastewater management while generating valuable products. Facultative anaerobes or engineered consortia in MES can efficiently convert soluble organics and potentially even captured CO_2 within the wastewater matrix (Jiang et al., 2024; Liu et al., 2025). In addition, agricultural waste streams (e.g., crop residues and manure) and wastewater derived from biomass processing (e.g., food processing effluent and digestate liquor) are rich in organic carbon (Zou et al., 2021). Processing these via AD yields biogas (CH_4/CO_2) suitable for subsequent upgrading or direct CO_2 extraction for MES. Crucially the solid digestate produced (biosolids) holds significant value for soil carbon sequestration. When applied to agricultural land as an amendment, this stabilized organic matter contributes to building stable soil organic carbon pools, enhancing soil health and fertility. This creates a dual carbon benefit: (1) utilizing the gaseous CO_2 fraction for bioproduction and (2) sequestering carbon long-term in soils via digestate application, effectively generating soil carbon sink gains (Ndayisenga et al., 2022). Integrating MES with AD for these streams enables a circular bioeconomy approach with cascading carbon utilization and sequestration.

3.5.2 Extracellular electron uptake (EEU) mechanisms

Within a sustainable MEC-based CO_2-mitigation framework, methane production serves as a representative product. Three mechanistically distinct routes of methanogenesis can be distinguished: (1) acetoclastic methanogenesis, wherein a mixed anaerobic community at the anode cleaves acetate to CH_4 and CO_2 using internally derived electron donors; (2) hydrogenotrophic methanogenesis at the cathode, where CO_2 is reduced to CH_4 by hydrogenotrophic archaea using electro-generated H_2 as the reductant; and (3) direct electron uptake (DEU) at the cathode, in which electroactive methanogens accept electrons directly from the electrode surface, eliminating the need for an H_2 intermediate (Nelabhotla & Dinamarca, 2018; Salehizadeh et al., 2020; Soggia et al., 2025b). To align with CO_2 consumption objectives the cathodic CO_2 reduction processes constitute our primary focus, given that anodic substrate oxidation releases CO_2 during concomitant methane generation.

The efficiency of electromethanogenesis is principally affected by electron transfer mechanisms at the biocathode: DEU (Eq. 3) and indirect electron transfer [indirect electron uptake (IEU), (Eqs. 4 and 5)] (Dykstra & Pavlostathis, 2017; Hagos et al., 2018). The DEU constitutes a bioelectrochemical mechanism wherein EAM physically conducts electrons from cathodic surfaces through direct cell–electrode contact, circumventing soluble redox mediators. The efficacy of DEU exhibits significant interspecies

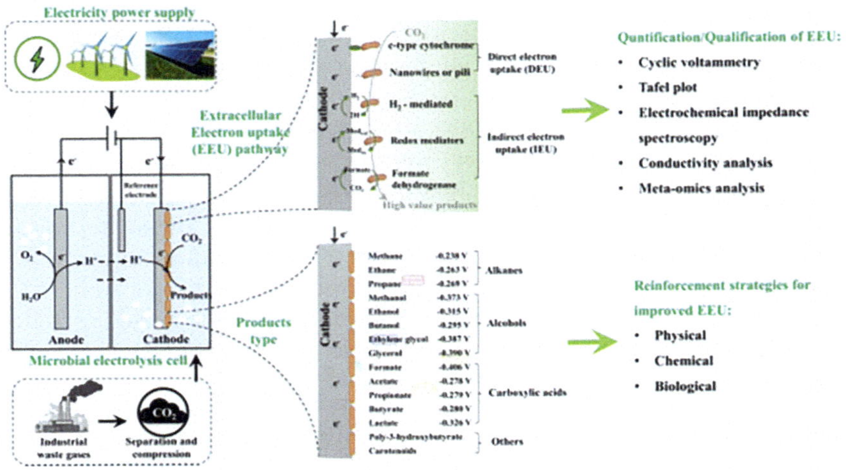

FIGURE 3.4 Schematic diagram of the cathodic electron uptake pathways in MEC and equilibrium potential (vs. standard hydrogen electrode, SHE) of various electrosynthesis products under biochemical standard conditions with carbon dioxide as the electron acceptor. *MEC*, Microbial electrolysis cells. *Reproduced from Li and Dong (2025), Copyright 2025 Elsevier.*

variation, governed by the density, spatial organization, and redox potentials of these biomolecular complexes. In electroactive methanogens such as *Methanosarcina* and *Methanothrix* species, DEU supports energy-efficient CO_2 reduction to methane by directly injecting low-potential electrons into microbial electron transport chains. This pathway eliminates thermodynamic penalties associated with intermediate hydrogen formation, thereby mitigating kinetic overpotentials and mass transfer limitations inherent in mediated electron transfer systems while maintaining faradaic efficiencies commensurate with biological energy conservation mechanisms (Fig. 3.4).

$$CO_2 + 8H^+ + 8e^- \rightarrow CH_4 + 2H_2O \ E = -0.244 \text{ vs. SHE}$$

$$2H + +8e^- \rightarrow H_2E = -0.244 \text{ vs. SHE}$$

$$CO_2 + 4H_2 \rightarrow CH_4 + 2H_2O\Delta G_0 = -131 \text{ kJ/mol}$$

The predominant IEU pathway is hydrogenotrophic electromethanogenesis, wherein cathodically evolved H_2 via HER drives CO_2 reduction to CH_4 by methanogenic archaea (e.g., *Methanobacterium* and *Methanosarcina*) (Kracke et al., 2019; Li et al., 2018; Martinez & Alvarez, 2018). This two-step process initiates with abiotic H_2 generation at potentials below −0.65 V (vs. SHE), followed by biological conversion via the Wolfe cycle (Enzmann et al., 2018; Zabranska & Pokorna, 2018). Despite incurring higher kinetic overpotentials (0.1–0.2 V) and lower coulombic efficiencies than DEU routes, IEU demonstrates superior operational robustness (Zhang et al., 2019). Crucially, both cathodic pathways necessitate precise management of electrode potential, biofilm thickness, and buffering capacity to simultaneously enhance CO_2 mass transfer and suppress carbonate precipitation and pH drift (Sugnaux et al., 2017). Consequently, these inherent constraints favor energy-efficient DEU for electro-methanogenesis where technically feasible.

3.5.3 Configuration integrations

Beyond conventional single-chamber designs, innovative MECs have emerged to address inherent limitations in energy efficiency, product spectrum, and feedstock flexibility. A prominent development involves MEC–MFC cascades, which strategically integrate MFCs with MECs (Hussain et al., 2016; Yu et al., 2017). This architecture harnesses the chemical energy of waste organics oxidized in the MFC anode to generate electricity, which subsequently drives cathodic CO_2 reduction in the coupled MEC. Such symbiosis transforms the MEC from an energy consumer to an energy-neutral or even energy-positive system, enabling surplus electricity export while converting CO_2 to acetate or methane at high coulombic efficiencies. The integration of MEC–AD enables in situ biogas upgrading (Ghaderikia & Yilmazel, 2024; Lu & Ren, 2016; Yu et al., 2018). Here, electroactive biofilms on the MEC cathode directly consume CO_2 from AD-derived biogas, reducing it to methane using electrons supplied from anodic organic oxidation or external renewables. This process elevates biogas methane content, simultaneously avoiding CO_2 sequestration costs and enhancing fuel quality. Critically, it achieves CO_2 utilization without biogas decompression, reducing energy inputs compared to ex situ upgrading. A particularly innovative frontier is solid-state MEC systems integrated with DAC (Sanz-Pérez et al., 2016; Zheng et al., 2022). These utilize moisture-swing adsorption or electro-swing adsorption on conductive, porous cathodes (e.g., carbon felt functionalized with ion-exchange resins) to concentrate atmospheric CO_2 directly onto electrode surfaces. Subsequent microbial reduction by cathodophilic electrotrophic communities converts sorbed CO_2 into organics without energy-intensive gas compression or liquid medium requirements. Although scalability challenges persist, particularly in maintaining biofilm viability in low-humidity solid-state systems and managing electron flux across cascade interfaces. These configurations collectively represent a strategic evolution toward decentralized, carbon-negative biomanufacturing platforms capable of utilizing dispersed CO_2 sources.

3.5.4 Technical analysis

MECs demonstrate distinct thermodynamic and operational advantages over biophotolysis or fermentative CO_2 emission approaches. By utilizing an external electrical bias, MECs overcome intrinsic endothermic barriers that constrain dark- and photofermentation, enabling efficient CO_2 reduction at ambient conditions. Critically, their light-independent operation permits continuous functionality irrespective of illumination, eliminating land-use conflicts and facilitating integration into industrial infrastructure. Furthermore, potentiostatic or galvanostatic control enables precise regulation of electron flux, allowing real-time tuning of product selectivity (e.g., CH_4 vs. multiple-carbonyl organic compounds), which is lacking in metabolic pathway-dependent biological systems. This signifies a superior electron recovery rate, whose the coulombic efficiency is much higher than that of biophotolysis or fermentation production. MECs concurrently deliver environmental cobenefits through anodic oxidation of organic waste substrates, enabling simultaneous wastewater treatment and carbon valorization within a unified reactor. Modular three-dimensional electrodes enhance scalability by decoupling kinetics from reactor volume, circumventing light-penetration limitations inherent in algal systems. Finally, MECs serve as a direct power-to-gas interface, converting intermittent renewable electricity into storable carbon-neutral fuels while mitigating intermittency challenges faced by biological platforms. Collectively, these attributes position MECs as a robust, tunable technology for deep decarbonization.

When compared with abiotic electrochemical CO_2 reduction (CO_2RR), MECs present distinct advantages in product spectrum and system tolerance. MECs leverage autotrophic biocatalysts to enable multicarbon (C_{2+}) product formation (e.g., acetate and butyrate) through inherent biological carbon chain elongation, a capability largely unattainable with conventional CO_2RR, which is predominantly limited to C_1–C_2 products and requires complex catalyst design for higher hydrocarbons (Ahmed et al., 2025; Lin et al., 2021; Xiong & Fu, 2025). Furthermore, MECs exhibit greater operational robustness, tolerating impurities (e.g., trace O_2 and sulfur compounds) common in industrial flue gases and operating effectively under ambient conditions without stringent purity requirements. However, MECs face significant challenges in electron transfer kinetics, resulting in lower absolute reaction rates compared to CO_2RR, prolonged start-up times for biocatalyst development, and potential bottlenecks in mass transfer to biofilms. Critically, in situ MEC configurations, especially those integrating three-dimensional electrodes like gas diffusion electrodes, can enhance CO_2 availability for both electrochemical and biological steps, partially mitigating mass transfer limitations while leveraging the high product diversity of microbial synthesis (Izadi & Harnisch, 2022;

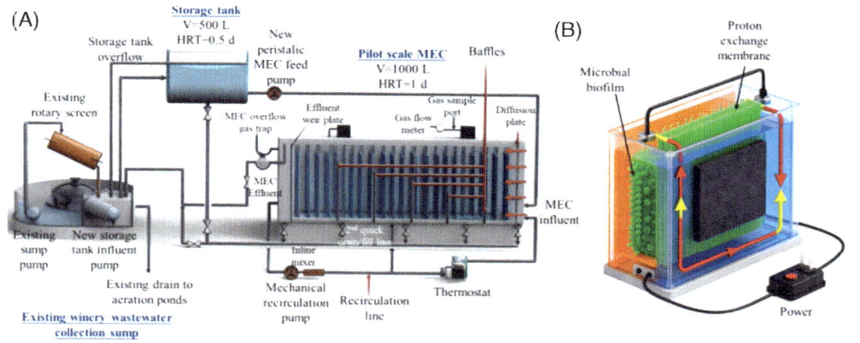

FIGURE 3.5 (A) System diagram of a 1000-L microbial electrolysis cell (MEC) hydrogen production demonstration project (B) structural diagram of MEC bioreactor. *HRT*, Hydraulic retention time. *Redraw after (A) Cusick et al. (2011).*

Wang et al., 2015). MECs uniquely coutilize organic waste streams at the anode, yielding a net-negative CO_2 balance and simultaneous wastewater treatment, which purely electrochemical systems cannot provide. Consequently, MECs excel in selectivity and carbon circularity but remain constrained by rate and volumetric productivity, positioning them as complementary rather than competing platforms for CO_2 valorization.

3.6 Analysis of engineering demonstration cases

Here, we introduce a practical project with a 1000-L pilot-scale MEC reactor by Penn State University as an example for engineering demonstration case analysis (Fig. 3.5) (Table 3.1). This project, led by the research group of Bruce Logan at Penn State University, is one of the most cited and significant demonstrations of MEC technology at a scale beyond the laboratory bench. It was a critical step in proving the feasibility of scaling up this technology.

This work aimed to validate the feasibility of scaling up MEC performance from laboratory to practical application, investigating its effectiveness in treating winery wastewater and producing hydrogen, along with identifying key operational conditions. As an emerging technology, MECs can convert soluble organic matter in wastewater into energy products such as hydrogen (H_2). The anode requires microbial catalysis to oxidize organics and release electrons, whereas the cathode necessitates an external voltage to drive hydrogen evolution, as the anode's open-circuit potential (approximately −0.3 V) cannot overcome the minimum potential required for hydrogen evolution (−0.4 V).

TABLE 3.1 Performance summary of the 1000-L MEC pilot plant.

Parameter	Value	Description/significance
Total reactor volume	1000 L	Demonstrates significant scale-up from mL/L lab systems. The working fluid volume is 910 L (90 L occupied by the electrode), and there is an additional 280 L space at the top
Reactor size	Height: 0.7 m; width: 0.64 m; length: 2.3 m	Rectangular structure, made of steel-reinforced high-density polyethylene
Applied voltage	0.9 V per module	There are 24 electrode modules (operating in parallel), each containing 6 anodes and 6 cathodes, totaling 144 pairs of electrodes. Anode: Graphite fiber brush (diameter 5.1 cm, length 66 cm, heat treated at 290°C for 1 h); Cathode: 304 stainless steel mesh (60 mesh, 7.6 cm width, 66 cm length), total cathode projected area 7.2 m^2, specific surface area 18.1 m^2/m^3
Flow field design	Continuous flow operation, HRT of 1 day, inflow rate of 1 m^3/day	A diffuser plate (with a pore size of 0.6 cm and an opening rate of 30%) ensures uniform water flow. Five baffles with increasing heights (20–56 cm) reduce short-circuit flow. A circulation pipeline (0.82 m^3/day) achieves mixing and uniform distribution of substrates
Operating condition	Temperature: 31°C ± 1°C, pH: ≥6	The temperature later dropped to 15°C because of a thermostat failure. Sodium hydrogen phosphate buffer was added in the early stage, and in the later stage, the wastewater was diluted with boiler water for adjustment
Feedstock	Winery wastewater	The Napa Wine Company in California, United States, generates approximately 5.1×10^4 m^3 of wastewater annually, mainly during the harvest season from August to November. The SCOD of the raw water is 4–7 g/L, and the VFA content is 0.05–0.21 g/L

(*continued on next page*)

TABLE 3.1 Performance summary of the 1000-L MEC pilot plant—cont'd

Parameter	Value	Description/significance
Inoculating method	Phased inoculation	On day 0, raw wastewater was added. On day 4300 L of sludge from the first aeration tank of the winery were added. On days 8 and 15, 500 L of activated sludge from the local sewage treatment plant was added in total. On day 15 an additional 210 L of enriched bacterial solution was added (winery wastewater + 8 g/L fumarate + 1.4 g/L acetate, cultured for 7 days to promote the growth of *Geobacter*)
Hydrogen production rate	0.09 ± 0.04 L/L/day during 22–43 days; 0.19 ± 0.04 L/L/day during 43–91 days; ~0.16 L/L/day after day 91	Mainly influenced by temperature
COD removal efficiency	<50% during start phase (0–60 days); ~62% ± 20% during stationary phase (60–100 days)	During stationary phase, the influent SCOD is controlled within 0.7–2.0 g/L, and the effluent SCOD decreases from 0.46 to 0.15 g/L. Moreover, the SCOD removal rate shows a positive correlation with the increase in current density (from 2.0 to 7.4 A/m^3), rising from 50% to 70%
Energy efficiency	Input energy: 6.0 W/m^3; output energy: 99.1 W/m^3 as highest value	The output energy is mainly from CH_4

COD, *chemical oxygen demand;* HRT, *hydraulic retention time;* MEC, *microbial electrolysis cells;* SCOD, *soluble chemical oxygen demand;* VFA, *volatile fatty acid.*

The pilot-scale system utilized a rectangular reactor constructed from steel-reinforced high-density polyethylene, with a total volume of 1000 L and a working liquid volume of 910 L. It contained 24 parallel electrode modules (totaling 144 electrode pairs). The anodes were graphite fiber brushes (heat-treated at 290°C for 1 h), and the cathodes were 304 stainless steel mesh, yielding a total cathodic specific surface area of 18.1 m^2/m^3. During operation, an external voltage of 0.9 V per module was applied. pH was maintained at ≥6 using buffer solution or wastewater dilution, and temperature was controlled at 31°C ± 1°C. The system operated in continuous flow mode (HRT of 1 day, inflow rate of 1 m^3/day) and incorporated diffuser plates, baffles, and recirculation piping to optimize flow distribution and mixing. The feedstock was wastewater from a winery in Napa Valley, California, United States (annual production approximately 5.1×10^4 m^3; characteristics during harvest season: soluble chemical oxygen demand (SCOD) 4–7 g/L, VFA 0.05–0.21 g/L). Inoculation was performed in stages using raw wastewater, aeration basin sludge, activated sludge, and an enriched culture containing *Geobacter*.

The study spanned 100 days of operation, with a reactor start-up period of 60 days. Key limiting factors were VFA content (requiring VFA/SCOD ≥ 0.5), temperature stability, and pH. Acetate addition effectively enriched exoelectrogenic bacteria (*Geobacter*), increasing current density from 0.3 to 2 A/m^3. Acetate addition was ceased after 65 days because of an increasing VFA proportion in the wastewater, and the current density peaked at 7.4 A/m^3 by day 91. During the stable period (days 60–100) the SCOD removal rate was 62% ± 20%, increasing from 50% to 70% as current density rose. Regarding gas production the average rate during the submesophilic phase (15°C–22°C) was 0.09 ± 0.04 L/L/day (primarily H_2 and CO_2), while it reached 0.19 ± 0.04 L/L/day during the mesophilic phase (31°C ± 1°C), with the CH_4 proportion rising to 86% ± 6%. Gas production decreased to 0.16 L/L/day after a temperature control failure. The actual H_2 production rate (0.07 ± 0.04 L/L/day) was close to the theoretical value, whereas the actual CH_4 production rate (0.24 L/L/day) significantly exceeded its theoretical value, indicating that methanogenesis also stemmed from direct anaerobic fermentation of organics. In terms of energy balance the energy input was only 6.0 W/m^3, the peak gas energy production rate was 99.1 W/m^3, and the electrical energy output after conversion by a microturbine (27.7 W/m^3) far exceeded the input, achieving net energy recovery.

Mechanistically the enrichment of *Geobacter* depended on acetate (preferred carbon source), mesophilic temperature, and suitable pH. Methane was primarily produced via hydrogenotrophic methanogenesis and direct electromethanogenesis. The inability to periodically expose the system to oxygen to inhibit methanogens in continuous single-chamber flow was key to H_2 conversion to CH_4. The 44% lower current density in the pilot-scale system compared to lab-scale results was attributed to the cathode material (stainless steel activity being lower than Pt-coated carbon cloth), system resistance (contact resistance and proton transfer resistance through separators), and insufficient start-up time.

In summary the pilot-scale MEC demonstrated successful SCOD removal from winery wastewater and net energy recovery. The critical operational conditions were VFA/SCOD ≥ 0.5, temperature of 31°C ± 1°C, and pH ≥ 6. However, the single-chamber continuous-flow design predisposed the system to H_2 conversion to CH_4.

3.7 Conclusion

In summary, MEC technology provides a highly promising approach for recovering energy from biodegradable waste and producing high-purity biohydrogen. This chapter reveals the complex bioelectrochemical processes within the system through in-depth analysis of the hydrogen production mechanism. Theoretical analysis shows that through substrate pretreatment, optimization of electrode materials, improvement of reactor configuration, inhibition of methane-producing bacteria activity, and precise control of operating conditions, the hydrogen production and coulombic efficiency can be significantly enhanced. However, to achieve large-scale application of this technology, breakthroughs are still needed in long-term operational stability, inhibition of biological side reactions, and reduction of construction costs.

References

Abd-Elrahman, N. K., Al-Harbi, N., Al-Hadeethi, Y., Alruqi, A. B., Mohammed, H., Umar, A., & Akbar, S. (2022). Influence of nanomaterials and other factors on biohydrogen production rates in microbial electrolysis cells—A review. *Molecules, 27*(23). http://www.mdpi.com/journal/molecules. doi:10.3390/molecules27238594.

Ahmed, S., Hussain, M. S., Khan, M. K., & Kim, J. (2025). Innovations in catalysis towards efficient electrochemical reduction of CO_2 to C1 chemicals. *Journal of Energy Chemistry, 107*, 622–649. https://www.sciencedirect.com/science/journal/20954956. doi:10.1016/j.jechem.2025.03.055.

Ahn, Y., Im, S., & Chung, J. W. (2017). Optimizing the operating temperature for microbial electrolysis cell treating sewage sludge. *International Journal of Hydrogen Energy, 42*(45), 27784–27791. http://www.journals.elsevier.com/international-journal-of-hydrogen-energy/. doi:10.1016/j.ijhydene.2017.05.139.

Almatouq, A., Babatunde, A. O., Khajah, M., Webster, G., & Alfodari, M. (2020). Microbial community structure of anode electrodes in microbial fuel cells and microbial electrolysis cells. *Journal of Water Process Engineering, 34*. http://www.journals.elsevier.com/journal-of-water-process-engineering/. doi:10.1016/j.jwpe.2020.101140.

Anantharaj, S., & Noda, S. (2022). How properly are we interpreting the Tafel lines in energy conversion electrocatalysis? *Materials Today Energy, 29*. https://www.journals.elsevier.com/materials-today-energy. doi:10.1016/j.mtener.2022.101123.

Badia-Fabregat, M., Rago, L., Baeza, J. A., & Guisasola, A. (2019). Hydrogen production from crude glycerol in an alkaline microbial electrolysis cell. *International Journal of Hydrogen Energy, 44*(32), 17204–17213. http://www.journals.elsevier.com/international-journal-of-hydrogen-energy/. doi:10.1016/j.ijhydene.2019.03.193.

Baek, G., Rossi, R., & Logan, B. E. (2021a). Changes in electrode resistances and limiting currents as a function of microbial electrolysis cell reactor configurations. *Electrochimica Acta, 388*. http://www.journals.elsevier.com/electrochimica-acta/. doi:10.1016/j.electacta.2021.138590.

Baek, G., Shi, L., Rossi, R., & Logan, B. E. (2021b). The effect of high applied voltages on bioanodes of microbial electrolysis cells in the presence of chlorides. *Chemical Engineering Journal, 405*. www.elsevier.com/inca/publications/store/6/0/1/2/7/3/index.htt. doi:10.1016/j.cej.2020.126742.

Bahari, M. B., Mamat, C. R., Jalil, A. A., Hassan, N. S., Sawal, M. H., Rajendran, S., & Alam, M. N. H. Z. (2024). Molybdenum as cathode materials: Paving the way for sustainable biohydrogen production in microbial electrolysis cells. *Process Safety and Environmental Protection, 191*, 1633–1647. doi:10.1016/j.psep.2024.09.032.

Bajracharya, S., Srikanth, S., Mohanakrishna, G., Zacharia, R., Strik, D. P., & Pant, D. (2017). Biotransformation of carbon dioxide in bioelectrochemical systems: State of the art and future prospects. *Journal of Power Sources, 356*, 256–273. https://www.journals.elsevier.com/journal-of-power-sources. doi:10.1016/j.jpowsour.2017.04.024.

Baş, F., & Kaya, M. F. (2022). 3D printed anode electrodes for microbial electrolysis cells. *Fuel, 317*. http://www.journals.elsevier.com/fuel/. doi:10.1016/j.fuel.2022.123560.

Bi, X., Wang, G., Cui, D., Qu, X., Shi, S., Yu, D., Cheng, M., & Ji, Y. (2025). Simulation study on the effect of temperature on hydrogen production performance of alkaline electrolytic water. *Fuel, 380*, 133209. doi:10.1016/j.fuel.2024.133209.

Bora, A., Mohanrasu, K., Angelin Swetha, T., Ananthi, V., Sindhu, R., Chi, N. T. L., Pugazhendhi, A., Arun, A., & Mathimani, T. (2022). Microbial electrolysis cell (MEC): Reactor configurations, recent advances and strategies in biohydrogen production. *Fuel, 328*. http://www.journals.elsevier.com/fuel/. doi:10.1016/j.fuel.2022.125269.

Call, D., & Logan, B. E. (2008). Hydrogen production in a single chamber microbial electrolysis cell lacking a membrane. *Environmental Science and Technology, 42*(9), 3401–3406. http://pubs.acs.org/doi/pdfplus/10.1021/es8001822. doi:10.1021/es8001822.

Catal, T., Lesnik, K. L., & Liu, H. (2015). Suppression of methanogenesis for hydrogen production in single-chamber microbial electrolysis cells using various antibiotics. *Bioresource Technology, 187*, 77–83. www.elsevier.com/locate/biortech. doi:10.1016/j.biortech.2015.03.099.

Cha, J., Choi, Y., Park, H., Kim, D., Baek, G., & Lee, C. (2024). Combining pre-fermentation and microbial electrolysis for efficient hydrogen production from food wastewater. *Process Safety and Environmental Protection, 187*, 1471–1480. https://www.sciencedirect.com/science/journal/09575820. doi:10.1016/j.psep.2024.05.062.

Chae, K. J., Choi, M. J., Kim, K. Y., Ajayi, F. F., Chang, I. S., & Kim, I. S. (2009). A solar-powered microbial electrolysis cell with a platinum catalyst-free cathode to produce hydrogen. *Environmental Science and Technology, 43*(24), 9525–9530. http://pubs.acs.org/doi/pdfplus/10.1021/es9022317. doi:10.1021/es9022317.

Chae, K. J., Choi, M. J., Kim, K. Y., Ajayi, F. F., Chang, I. S., & Kim, I. S. (2010). Selective inhibition of methanogens for the improvement of biohydrogen production in microbial electrolysis cells. *International Journal of Hydrogen Energy, 35*(24), 13379–13386. doi:10.1016/j.ijhydene.2009.11.114.

Chae, K. J., Choi, M. J., Lee, J., Ajayi, F. F., & Kim, I. S. (2008). Biohydrogen production via biocatalyzed electrolysis in acetate-fed bioelectrochemical cells and microbial community analysis. *International Journal of Hydrogen Energy, 33*(19), 5184–5192. doi:10.1016/j.ijhydene.2008.05.013.

Chaurasia, A. K., Goyal, H., & Mondal, P. (2020). Hydrogen gas production with Ni, Ni–Co and Ni–Co–P electrodeposits as potential cathode catalyst by microbial electrolysis cells. *International Journal of Hydrogen Energy, 45*(36), 18250–18265. http://www.journals.elsevier.com/international-journal-of-hydrogen-energy/. doi:10.1016/j.ijhydene.2019.07.175.

Chaurasia, A. K., & Mondal, P. (2022). Enhancing biohydrogen production from sugar industry wastewater using Ni, Ni–Co and Ni–Co–P electrodeposits as cathodes in microbial electrolysis cells. *Chemosphere, 286*. www.elsevier.com/locate/chemosphere. doi:10.1016/j.chemosphere.2021.131728.

Chaurasia, A. K., Shankar, R., & Mondal, P. (2021). Effects of nickle, nickle-cobalt and nickle-cobalt-phosphorus nanocatalysts for enhancing biohydrogen production in microbial electrolysis cells using paper industry wastewater. *Journal of Environmental Management,*

298. https://www.sciencedirect.com/journal/journal-of-environmental-management. doi:10.1016/j.jenvman.2021.113542.

Chen, Y., Shen, J., Huang, L., Pan, Y., & Quan, X. (2016). Enhanced Cd(II) removal with simultaneous hydrogen production in biocathode microbial electrolysis cells in the presence of acetate or $NaHCO_3$. *International Journal of Hydrogen Energy, 41*(31), 13368–13379. http://www.journals.elsevier.com/international-journal-of-hydrogen-energy/. doi:10.1016/j.ijhydene.2016.06.200.

Cheng, D., Ngo, H. H., Guo, W., Chang, S. W., Nguyen, D. D., Zhang, S., Deng, S., An, D., & Hoang, N. B. (2022). Impact factors and novel strategies for improving biohydrogen production in microbial electrolysis cells. *Bioresource Technology, 346*. www.elsevier.com/locate/biortech. doi:10.1016/j.biortech.2021.126588.

Cheng, S., Liu, H., & Logan, B. E. (2006). Increased power generation in a continuous flow MFC with advective flow through the porous anode and reduced electrode spacing. *Environmental Science and Technology, 40*(7), 2426–2432. doi:10.1021/es051652w.

Cheng, S., & Logan, B. E. (2011). High hydrogen production rate of microbial electrolysis cell (MEC) with reduced electrode spacing. *Bioresource Technology, 102*(3), 3571–3574. doi:10.1016/j.biortech.2010.10.025.

Cho, S. K., Lee, M. E., Lee, W., & Ahn, Y. (2019). Improved hydrogen recovery in microbial electrolysis cells using intermittent energy input. *International Journal of Hydrogen Energy*, 2253–2257. http://www.journals.elsevier.com/international-journal-of-hydrogen-energy/. doi:10.1016/j.ijhydene.2018.07.025.

Choi, M. J., Yang, E., Yu, H. W., Kim, I. S., Oh, S. E., & Chae, K. J. (2019). Transition metal/carbon nanoparticle composite catalysts as platinum substitutes for bioelectrochemical hydrogen production using microbial electrolysis cells. *International Journal of Hydrogen Energy*, 2258–2265. http://www.journals.elsevier.com/international-journal-of-hydrogen-energy/. doi:10.1016/j.ijhydene.2018.07.020.

Choi, Y., Kim, D., Choi, H., Cha, J., Baek, G., & Lee, C. (2023). A study of electron source preference and its impact on hydrogen production in microbial electrolysis cells fed with synthetic fermentation effluent. *Bioengineered, 14*(1), 2244759. doi:10.1080/21655979.2023.2244759.

Colantoni, S., Pillot, G., Cvoro, S., Kerzenmacher, S., & Santiago, Ó. (2025). Evaluation of separators for potential use in microbial electrolysis cells under anaerobic digester conditions. *Journal of Membrane Science, 722*, 123887. doi:10.1016/j.memsci.2025.123887.

Commault, A. S., Barrière, F., Lapinsonnière, L., Lear, G., Bouvier, S., & Weld, R. J. (2015). Influence of inoculum and anode surface properties on the selection of *Geobacter*-dominated biofilms. *Bioresource Technology, 195*, 265–272. http://www.journals.elsevier.com/bioresource-technology/. doi:10.1016/j.biortech.2015.06.141.

Cristiani, L., Zeppilli, M., Feriaud, D., Marandola, C., Petrangeli Papini, M., Da Silva, S., Erable, B., & Villano, M. (2024). Enhancing energy efficiency and H_2 production in lab-scale dual chamber microbial electrolysis cells: A focus on catholyte composition and voltage losses. *Journal of Environmental Chemical Engineering, 12*(1). https://www.sciencedirect.com/science/journal/22133437. doi:10.1016/j.jece.2023.111782.

Cui, M. H., Sangeetha, T., Gao, L., & Wang, A. J. (2020). Hydrodynamics of up-flow hybrid anaerobic digestion reactors with built-in bioelectrochemical system. *Journal of Hazardous Materials, 382*. www.elsevier.com/locate/jhazmat. doi:10.1016/j.jhazmat.2019.121046.

Cui, W., Luo, H., & Liu, G. (2023). Efficient hydrogen production in single-chamber microbial electrolysis cell with a fermentable substrate under hyperalkaline conditions. *Waste Management, 171*, 173–183. www.elsevier.com/locate/wasman. doi:10.1016/j.wasman.2023.08.017.

Dizge, N., Unal, B. O., Arikan, E. B., Karagunduz, A., & Keskinler, B. (2019). Recent progress and developments in membrane materials for microbial electrochemistry technologies: A review. *Bioresource Technology Reports, 8*. https://www.journals.elsevier.com/bioresource-technology-reports. doi:10.1016/j.biteb.2019.100308.

Dong, X., Pang, D., Luo, G., & Zhu, X. (2024). Microbial water electrolysis cells for efficient wastewater treatment and H_2 production. *ACS Sustainable Chemistry and Engineering, 12*(10), 4203–4212. http://pubs.acs.org/journal/ascecg. doi:10.1021/acssuschemeng.3c07953.

Dykstra, C. M., & Pavlostathis, S. G. (2017). Zero-valent iron enhances biocathodic carbon dioxide reduction to methane. *Environmental Science and Technology, 51*(21), 12956–12964. http://pubs.acs.org/journal/esthag. doi:10.1021/acs.est.7b02777.

Dykstra, C. M., & Pavlostathis, S. G. (2021). Hydrogen sulfide affects the performance of a methanogenic bioelectrochemical system used for biogas upgrading. *Water Research, 200.* www.elsevier.com/locate/watres. doi:10.1016/j.watres.2021.117268.

Enzmann, F., Mayer, F., Rother, M., & Holtmann, D. (2018). Methanogens: Biochemical background and biotechnological applications. *AMB Express, 8*(1). http://www.amb-express.com/. doi:10.1186/s13568-017-0531-x.

Gautam, R., Nayak, J. K., Ress, N. V., Steinberger-Wilckens, R., & Ghosh, U. K. (2023). Biohydrogen production through microbial electrolysis cell: Structural components and influencing factors. *Chemical Engineering Journal, 455.* www.elsevier.com/inca/publications/store/6/0/1/2/7/3/index.htt. doi:10.1016/j.cej.2022.140535.

Ghaderikia, A., & Yilmazel, Y. D. (2024). Methane recovery from cattle manure: Role of granular activated carbon amendment into anaerobic digestion-microbial electrolysis cells (AD-MECs) integrated systems. *ACS Sustainable Chemistry and Engineering, 12*(4), 1437–1445. http://pubs.acs.org/journal/ascecg. doi:10.1021/acssuschemeng.3c05370.

Gu, Y., Zhang, X., Deal, B., & Han, L. (2019). Biological systems for treatment and valorization of wastewater generated from hydrothermal liquefaction of biomass and systems thinking: A review. *Bioresource Technology, 278,* 329–345. www.elsevier.com/locate/biortech. doi:10.1016/j.biortech.2019.01.127.

Guerrero-Sodric, O., Baeza, J. A, & Guisasola, A. (2024). Enhancing bioelectrochemical hydrogen production from industrial wastewater using Ni-foam cathodes in a microbial electrolysis cell pilot plant. *Water Research, 256,* 121616. doi:10.1016/j.watres.2024.121616.

Guo, H., & Kim, Y. (2018). Scalable multi-electrode microbial electrolysis cells for high electric current and rapid organic removal. *Journal of Power Sources, 391,* 67–72. doi:10.1016/j.jpowsour.2018.04.075.

Guo, H., & Kim, Y. (2019). Stacked multi-electrode design of microbial electrolysis cells for rapid and low-sludge treatment of municipal wastewater. *Biotechnology for Biofuels, 12*(1). http://www.biotechnologyforbiofuels.com/. doi:10.1186/s13068-019-1368-0.

Guo, K., Tang, X., Du, Z., & Li, H. (2010). Hydrogen production from acetate in a cathode-on-top single-chamber microbial electrolysis cell with a mipor cathode. *Biochemical Engineering Journal, 51*(1–2), 48–52. doi:10.1016/j.bej.2010.05.001.

Guo, X., Liu, J., & Xiao, B. (2013). Bioelectrochemical enhancement of hydrogen and methane production from the anaerobic digestion of sewage sludge in single-chamber membrane-free microbial electrolysis cells. *International Journal of Hydrogen Energy, 38*(3), 1342–1347. doi:10.1016/j.ijhydene.2012.11.087.

Hagos, K., Liu, C., & Lu, X. (2018). Effect of endogenous hydrogen utilization on improved methane production in an integrated microbial electrolysis cell and anaerobic digestion: Employing catalyzed stainless steel mesh cathode. *Chinese Journal of Chemical Engineering, 26*(3), 574–582. doi:10.1016/j.cjche.2017.08.005.

Hassan, M., Fernandez, A. S., San Martin, I., Xie, B., & Moran, A. (2018). Hydrogen evolution in microbial electrolysis cells treating landfill leachate: Dynamics of anodic biofilm. *International Journal of Hydrogen Energy, 43*(29), 13051–13063. http://www.journals.elsevier.com/international-journal-of-hydrogen-energy/. doi:10.1016/j.ijhydene.2018.05.055.

Heidrich, E. S., Edwards, S. R., Dolfing, J., Cotterill, S. E., & Curtis, T. P. (2014). Performance of a pilot scale microbial electrolysis cell fed on domestic wastewater at ambient temperatures for a 12 month period. *Bioresource Technology, 173,* 87–95. www.elsevier.com/locate/biortech. doi:10.1016/j.biortech.2014.09.083.

Hu, H., Fan, Y., & Liu, H. (2008). Hydrogen production using single-chamber membrane-free microbial electrolysis cells. *Water Research, 42*(15), 4172–4178. https://www.sciencedirect.com/science/journal/00431354. doi:10.1016/j.watres.2008.06.015.

Hu, H., Fan, Y., & Liu, H. (2009). Hydrogen production in single-chamber tubular microbial electrolysis cells using non-precious-metal catalysts. *International Journal of Hydrogen Energy, 34*(20), 8535–8542. doi:10.1016/j.ijhydene.2009.08.011.

Hussain, A., Manuel, M., & Tartakovsky, B. (2016). A comparison of simultaneous organic carbon and nitrogen removal in microbial fuel cells and microbial electrolysis cells. *Journal of Environmental Management, 173*, 23–33. https://www.sciencedirect.com/journal/journal-of-environmental-management. doi:10.1016/j.jenvman.2016.02.025.

Izadi, P., & Harnisch, F. (2022). Microbial | electrochemical CO_2 reduction: To integrate or not to integrate? *Joule, 6*(5), 935–940. https://www.journals.elsevier.com/joule. doi:10.1016/j.joule.2022.04.005.

Jadhav, D. A., Chendake, A. D., Schievano, A., & Pant, D. (2019). Suppressing methanogens and enriching electrogens in bioelectrochemical systems. *Bioresource Technology, 277*, 148–156. www.elsevier.com/locate/biortech. doi:10.1016/j.biortech.2018.12.098.

Jafary, T., Wan Daud, W. R., Ghasemi, M., Abu Bakar, M. H., Sedighi, M., Kim, B. H., Carmona-Martínez, A. A., Jahim, J. M., & Ismail, M. (2019). Clean hydrogen production in a full biological microbial electrolysis cell. *International Journal of Hydrogen Energy, 44*(58), 30524–30531. http://www.journals.elsevier.com/international-journal-of-hydrogen-energy/. doi:10.1016/j.ijhydene.2018.01.010.

Javad Bardi, M., Müller, F., Polag, D., Habtu, N. G, & Koch, K. (2025). The intriguing effect of CO_2 enrichment in anaerobic digestion. *Bioresource Technology, 416*, 131743. doi:10.1016/j.biortech.2024.131743.

Jayabalan, T., Manickam, M., & Naina Mohamed, S. (2020). $NiCo_2O_4$-graphene nanocomposites in sugar industry wastewater fed microbial electrolysis cell for enhanced biohydrogen production. *Renewable Energy, 154*, 1144–1152. http://www.journals.elsevier.com/renewable-and-sustainable-energy-reviews/. doi:10.1016/j.renene.2020.03.071.

Jayabalan, T., Matheswaran, M., & Naina Mohammed, S. (2019). Biohydrogen production from sugar industry effluents using nickel based electrode materials in microbial electrolysis cell. *International Journal of Hydrogen Energy*, 17381–17388. http://www.journals.elsevier.com/international-journal-of-hydrogen-energy/. doi:10.1016/j.ijhydene.2018.09.219.

Jayabalan, T., Matheswaran, M., Radhakrishnan, T. K., & Naina Mohamed, S. (2021). Influence of nickel molybdate nanocatalyst for enhancing biohydrogen production in microbial electrolysis cell utilizing sugar industrial effluent. *Bioresource Technology, 320*. www.elsevier.com/locate/biortech. doi:10.1016/j.biortech.2020.124284.

Jiang, S., Chen, Y., & Zhou, Q. (2007). Effect of sodium dodecyl sulfate on waste activated sludge hydrolysis and acidification. *Chemical Engineering Journal, 132*(1–3), 311–317. doi:10.1016/j.cej.2007.01.017.

Jiang, Y., Liang, P., Zhang, C., Bian, Y., Sun, X., Zhang, H., Yang, X., Zhao, F., & Huang, X. (2016). Periodic polarity reversal for stabilizing the pH in two-chamber microbial electrolysis cells. *Applied Energy, 165*, 670–675. http://www.elsevier.com/inca/publications/store/4/0/5/8/9/1/index.htt. doi:10.1016/j.apenergy.2016.01.001.

Jiang, Z., Ding, Y., Zhao, Z., Yu, Q., Sun, C., & Zhang, Y. (2024). Strengthening the cathodic CO_2 adsorption on the MIL-88B(Fe) to enhance methane production. *Chemical Engineering Journal, 492*, 152203. doi:10.1016/j.cej.2024.152203.

Jwa, E., Kim, M., Han, J. H., Jeong, N., Kim, H. C., Yun, Y. M., & Nam, J. Y. (2021). Feasibility of using electrodes with ultralow Pt loading in two-chamber microbial electrolysis cells. *Energies, 14*(22). https://www.mdpi.com/1996-1073/14/22/7752/pdf. doi:10.3390/en14227752.

Jwa, E., Kim, S., Jeong, N., Han, S., Song, M. J., Kang, S., & Nam, J. Y. (2025). Enhanced hydrogen production in zero-gap bipolar membrane microbial electrolysis with binderless cathodes in real wastewater. *Chemical Engineering Journal, 509*. https://www.sciencedirect.com/science/journal/13858947. doi:10.1016/j.cej.2025.161416.

Kadier, A., Kalil, M. S., Abdeshahian, P., Chandrasekhar, K., Mohamed, A., Azman, N. F., Logroño, W., Simayi, Y., & Hamid, A. A. (2016a). Recent advances and emerging challenges in microbial electrolysis cells (MECs) for microbial production of hydrogen and value-added chemicals. *Renewable and Sustainable Energy Reviews, 61*, 501–525. https://www.journals.elsevier.com/renewable-and-sustainable-energy-reviews. doi:10.1016/j.rser.2016.04.017.

Kadier, A., Simayi, Y., Abdeshahian, P., Azman, N. F., Chandrasekhar, K., & Kalil, M. S. (2016b). A comprehensive review of microbial electrolysis cells (MEC) reactor designs and configurations for sustainable hydrogen gas production. *Alexandria Engineering Journal, 55*(1), 427–443. http://www.elsevier.com/wps/find/journaldescription.cws_home/724292/description#description. doi:10.1016/j.aej.2015.10.008.

Karthikeyan, R., Cheng, K. Y., Selvam, A., Bose, A., & Wong, J. W. C. (2017). Bioelectrohydrogenesis and inhibition of methanogenic activity in microbial electrolysis cells - A review. *Biotechnology Advances, 35*(6), 758–771. www.elsevier.com/inca/publications/store/5/2/5/4/5/5/index.htt. doi:10.1016/j.biotechadv.2017.07.004.

Katuri, K. P., Ali, M., & Saikaly, P. E. (2019). The role of microbial electrolysis cell in urban wastewater treatment: Integration options, challenges, and prospects. *Current Opinion in Biotechnology, 57*, 101–110. http://www.elsevier.com/locate/copbio. doi:10.1016/j.copbio.2019.03.007.

Kazemi, S., Mohseni, M., & Fatih, K. (2016). A systematic study of separators in air-breathing flat-plate microbial fuel cells—Part 1: Structure, properties, and performance correlations. *Energies, 9*(2). http://www.mdpi.com/1996-1073/9/2/78/pdf. doi:10.3390/en9020078.

Kim, B., Yang, E., Kim, B., Obaid, M., Jang, J. K., & Chae, K. J. (2022). Recent application of nanomaterials to overcome technological challenges of microbial electrolysis cells. *Nanomaterials, 12*(8). https://www.mdpi.com/2079-4991/12/8/1315/pdf. doi:10.3390/nano12081316.

Kim, D., Choi, Y., Cha, J., Park, H., Baek, G., & Lee, C. (2025). Pre-fermentation and filtration pretreatments enhance hydrogen production from food waste through microbial electrolysis. *Bioresource Technology, 424*, 132267. doi:10.1016/j.biortech.2025.132267.

Kracke, F., Wong, A. B., Maegaard, K., Deutzmann, J. S., Hubert, M. K. A., Hahn, C., Jaramillo, T. F., & Spormann, A. M. (2019). Robust and biocompatible catalysts for efficient hydrogen-driven microbial electrosynthesis. *Communications Chemistry, 2*(1). nature.com/commschem/. doi:10.1038/s42004-019-0145-0.

Kumar, P., Singh Rana, S. P., Sakshi, A. A. M. A., & Singh, L. (2025). Nanoparticle-driven cathode for hydrogen production in microbial electrolysis cell: Synergies and impact. *Fuel, 393*. https://www.sciencedirect.com/science/journal/00162361. doi:10.1016/j.fuel.2025.135020.

Lee, H. S., & Rittmann, B. E. (2010). Significance of biological hydrogen oxidation in a continuous single-chamber microbial electrolysis cell. *Environmental Science and Technology, 44*(3), 948–954. http://pubs.acs.org/doi/pdfplus/10.1021/es9025358. doi:10.1021/es9025358.

Li, Z., Li, G., Chen, X., Xia, Z., Yao, J., Yang, B., Lei, L., & Hou, Y. (2018). Water splitting–biosynthetic hybrid system for CO_2 conversion using nickel nanoparticles embedded in N-doped carbon nanotubes. *ChemSusChem, 11*(14), 2382–2387. http://onlinelibrary.wiley.com/journal/10.1002/(ISSN)1864-564X. doi:10.1002/cssc.201800878.

Liang, D. W., Peng, S. K., Lu, S. F., Liu, Y. Y., Lan, F., & Xiang, Y. (2011). Enhancement of hydrogen production in a single chamber microbial electrolysis cell through anode arrangement optimization. *Bioresource Technology, 102*(23), 10881–10885. doi:10.1016/j.biortech.2011.09.028.

Liao, P., Kang, J., Zhong, Y., Xiang, R., Wang, S., Li, S., Liu, X., & Li, G. (2023). Recent advances of two-dimensional metal-organic frameworks in alkaline electrolysis water for hydrogen production. *Science China Chemistry, 66*(7), 1924–1939. https://www.springer.com/journal/11426. doi:10.1007/s11426-022-1545-0.

Lim, S. S., Fontmorin, J. M., Izadi, P., Wan Daud, W. R., Scott, K., & Yu, E. H. (2020). Impact of applied cell voltage on the performance of a microbial electrolysis cell fully catalysed by microorganisms. *International Journal of Hydrogen Energy, 45*(4), 2557–2568. http://www.journals.elsevier.com/international-journal-of-hydrogen-energy/. doi:10.1016/j.ijhydene.2019.11.142.

Lim, S. S., Yu, E. H., Daud, W. R. W., Kim, B. H., & Scott, K. (2017). Bioanode as a limiting factor to biocathode performance in microbial electrolysis cells. *Bioresource Technology, 238*, 313–324. www.elsevier.com/locate/biortech. doi:10.1016/j.biortech.2017.03.127.

Lin, R., Deng, C., Zhang, W., Hollmann, F., & Murphy, J. D. (2021). Production of bio-alkanes from biomass and CO_2. *Trends in Biotechnology, 39*(4), 370–380. www.elsevier.com/locate/tibtech. doi:10.1016/j.tibtech.2020.12.004.

Liu, S., Ding, G., Gu, R., Hao, J., Liu, P., Qin, W., Yu, Y., Han, Y. u, Huang, J., & He, W. (2025). Co-capture and recovery of ammonia and CO_2 driven by microbial electrolysis system coupling with mineral carbon sequestration by industrial wastes. *Resources, Conservation and Recycling, 212*, 107931. doi:10.1016/j.resconrec.2024.107931.

Liu, S., Zhang, Z., Gu, P., Yang, K., Huang, X., Li, M., & Miao, H. (2023). Elucidating applied voltage on the fate of antibiotic resistance genes in microbial electrolysis cell: Focusing on its transmission between anolyte and biofilm microbes. *Science of the Total Environment, 904*. www.elsevier.com/locate/scitotenv. doi:10.1016/j.scitotenv.2023.166901.

Liu, W., Huang, S., Zhou, A., Zhou, G., Ren, N., Wang, A., & Zhuang, G. (2012). Hydrogen generation in microbial electrolysis cell feeding with fermentation liquid of waste activated sludge. *International Journal of Hydrogen Energy, 37*(18), 13859–13864. doi:10.1016/j.ijhydene.2012.04.090.

Liu, Y. P., Wang, Y. H., Wang, B. S., & Chen, Q. Y. (2014). Effect of anolyte pH and cathode Pt loading on electricity and hydrogen co-production performance of the bio-electrochemical system. *International Journal of Hydrogen Energy, 39*(26), 14191–14195. http://www.journals.elsevier.com/international-journal-of-hydrogen-energy/. doi:10.1016/j.ijhydene.2014.02.127.

Logan, B. E. (2009). Exoelectrogenic bacteria that power microbial fuel cells. *Nature Reviews Microbiology, 7*(5), 375–381. doi:10.1038/nrmicro2113.

Logan, B. E., Call, D., Cheng, S., Hamelers, H. V. M., Sleutels, T. H. J. A., Jeremiasse, A. W., & Rozendal, R. A. (2008). Microbial electrolysis cells for high yield hydrogen gas production from organic matter. *Environmental Science and Technology, 42*(23), 8630–8640. doi:10.1021/es801553z.

Lovley, D. R., & Nevin, K. P. (2011). A shift in the current: New applications and concepts for microbe-electrode electron exchange. *Current Opinion in Biotechnology, 22*(3), 441–448. doi:10.1016/j.copbio.2011.01.009.

Lu, L., Ren, N., Zhao, X., Wang, H., Wu, D., & Xing, D. (2011). Hydrogen production, methanogen inhibition and microbial community structures in psychrophilic single-chamber microbial electrolysis cells. *Energy and Environmental Science, 4*(4), 1329–1336. doi:10.1039/c0ee00588f.

Lu, L., & Ren, Z. J. (2016). Microbial electrolysis cells for waste biorefinery: A state of the art review. *Bioresource Technology, 215*, 254–264. www.elsevier.com/locate/biortech. doi:10.1016/j.biortech.2016.03.034.

Lu, L., Xing, D., Liu, B., & Ren, N. (2012a). Enhanced hydrogen production from waste activated sludge by cascade utilization of organic matter in microbial electrolysis cells. *Water Research, 46*(4), 1015–1026. https://www.sciencedirect.com/science/journal/00431354. doi:10.1016/j.watres.2011.11.073.

Lu, L., Xing, D., Ren, N., & Logan, B. E. (2012b). Syntrophic interactions drive the hydrogen production from glucose at low temperature in microbial electrolysis cells. *Bioresource Technology, 124*, 68–76. doi:10.1016/j.biortech.2012.08.040.

Ma, S., Zhang, Y., Tu, L., Li, X., Chen, X., Lin, S., Luo, H., Zhan, X., & Liu, G. (2024). Efficient hydrogen production from food waste leachate using single-chamber microbial electrolysis cell. *Environmental Research, 263*(Pt 2), 120159. doi:10.1016/j.envres.2024.120159.

Mahmoud, M., Parameswaran, P., Torres, C. I., & Rittmann, B. E. (2014). Fermentation pretreatment of landfill leachate for enhanced electron recovery in a microbial electrolysis cell. *Bioresource Technology, 151*, 151–158. www.elsevier.com/locate/biortech. doi:10.1016/j.biortech.2013.10.053.

Martinez, C. M., & Alvarez, L. H. (2018). Application of redox mediators in bioelectrochemical systems. *Biotechnology Advances, 36*(5), 1412–1423. www.elsevier.com/inca/publications/store/5/2/5/4/5/5/index.htt. doi:10.1016/j.biotechadv.2018.05.005.

Merrill, M. D., & Logan, B. E. (2009). Electrolyte effects on hydrogen evolution and solution resistance in microbial electrolysis cells. *Journal of Power Sources, 191*(2), 203–208. doi:10.1016/j.jpowsour.2009.02.077.

Munoz, L. D. S., Erable, B., Etcheverry, L., Riess, J., Basséguy, R., & Bergel, A. (2010). Combining phosphate species and stainless steel cathode to enhance hydrogen evolution in microbial electrolysis cell (MEC). *Electrochemistry Communications, 12*(2), 183–186. doi:10.1016/j.elecom.2009.11.017.

Nam, J. Y., & Logan, B. E. (2011). Enhanced hydrogen generation using a saline catholyte in a two chamber microbial electrolysis cell. *International Journal of Hydrogen Energy, 36*(23), 15105–15110. doi:10.1016/j.ijhydene.2011.08.106.

Nam, J. Y., & Logan, B. E. (2012). Optimization of catholyte concentration and anolyte pHs in two chamber microbial electrolysis cells. *International Journal of Hydrogen Energy, 37*(24), 18622–18628. doi:10.1016/j.ijhydene.2012.09.140.

Nam, J. Y., Tokash, J. C., & Logan, B. E. (2011). Comparison of microbial electrolysis cells operated with added voltage or by setting the anode potential. *International Journal of Hydrogen Energy, 36*(17), 10550–10556. doi:10.1016/j.ijhydene.2011.05.148.

Nayak, J. K., Amit, & Ghosh, U. K. (2018). An innovative mixotrophic approach of distillery spent wash with sewage wastewater for biodegradation and bioelectricity generation using microbial fuel cell. *Journal of Water Process Engineering, 23*, 306–313. http://www.journals.elsevier.com/journal-of-water-process-engineering/. doi:10.1016/j.jwpe.2018.04.003.

Ndayisenga, F., Yu, Z., Wang, B., Yang, J., Wu, G., & Zhang, H. (2022). Using bioelectrohydrogenesis left-over residues as a future potential fertilizer for soil amendment. *Scientific Reports, 12*(1). https://www.nature.com/srep/. doi:10.1038/s41598-022-22715-x.

Ndayisenga, F., Yu, Z., Wang, B., & Zhou, D. (2023). Effects of the applied voltage on electroactive microbial biofilm viability and hydrogen production in a recalcitrant organic waste-fed single-chamber membrane-free microbial electrolysis cell performance. *Chemical Engineering Journal, 469*. www.elsevier.com/inca/publications/store/6/0/1/2/7/3/index.htt. doi:10.1016/j.cej.2023.144002.

Nelabhotla, A. B. T., & Dinamarca, C. (2018). Electrochemically mediated CO_2 reduction for bio-methane production: A review. *Reviews in Environmental Science and Biotechnology, 17*(3), 531–551. http://www.kluweronline.com/issn/1569-1705. doi:10.1007/s11157-018-9470-5.

Nevin, K. P., Woodard, T. L., Franks, A. E., Summers, Z. M., & Lovley, D. R. (2010). Microbial electrosynthesis: Feeding microbes electricity to convert carbon dioxide and water to multicarbon extracellular organic compounds. *mBio, 1*(2). http://mbio.asm.org/content/1/2/e00119-10.full.pdf+html. doi:10.1128/mBio.00103-10.

Noori, M. T., Rossi, R., Logan, B. E., & Min, B. (2024). Hydrogen production in microbial electrolysis cells with biocathodes. *Trends in Biotechnology, 42*(7), 815–828. https://www.sciencedirect.com/science/journal/01677799. doi:10.1016/j.tibtech.2023.12.010.

Nwanebu, E., Tartakovsky, B., Wang, X., & Jugnia, L.-B. (2024). Electrochemical and bioelectrochemical treatment of steam-assisted gravity drainage water. *Electrochimica Acta, 502*, 144768. doi:10.1016/j.electacta.2024.144768.

Omidi, H., & Sathasivan, A. (2013). Optimal temperature for microbes in an acetate fed microbial electrolysis cell (MEC). *International Biodeterioration and Biodegradation, 85*, 688–692. doi:10.1016/j.ibiod.2013.05.026.

Park, S. G., Chae, K. J., & Lee, M. (2017). A sulfonated poly (arylene ether sulfone)/polyimide nanofiber composite proton exchange membrane for microbial electrolysis cell application under the coexistence of diverse competitive cations and protons. *Journal of Membrane Science, 540*, 165–173. www.elsevier.com/locate/memsci. doi:10.1016/j.memsci.2017.06.048.

Parkhey, P., & Gupta, P. (2017). Improvisations in structural features of microbial electrolytic cell and process parameters of electrohydrogenesis for efficient biohydrogen production: A review. *Renewable and Sustainable Energy Reviews, 69*, 1085–1099. https://www.journals.elsevier.com/renewable-and-sustainable-energy-reviews. doi:10.1016/j.rser.2016.09.101.

Pärnamäe, R., Mareev, S., Nikonenko, V., Melnikov, S., Sheldeshov, N., Zabolotskii, V., Hamelers, H. V. M., & Tedesco, M. (2021). Bipolar membranes: A review on principles, latest developments, and applications. *Journal of Membrane Science, 617*, 118538. doi:10.1016/j.memsci.2020.118538.

Pérez, D., Lie, T. T., & Weber, C. C. (2021). Operationalization of a microbial electrolysis cell: The interaction of the primary factors for energy storage efficiency. *Bioresource Technology, 326*. www.elsevier.com/locate/biortech. doi:10.1016/j.biortech.2021.124788.

Rabaey, K., & Rozendal, R. A. (2010). Microbial electrosynthesis—Revisiting the electrical route for microbial production. *Nature Reviews Microbiology, 8*(10), 706–716. http://www.nature.com/nrmicro/index.html. doi:10.1038/nrmicro2422.

Roubaud, E., Lacroix, R., Silva, S. D., Etcheverry, L., Bergel, A., Basséguy, R., & Erable, B. (2019). Benchmarking of industrial synthetic graphite grades, carbon felt, and carbon cloth as cost-efficient bioanode materials for domestic wastewater fed microbial electrolysis cells. *Frontiers in Energy Research, 7*. http://www.frontiersin.org/Energy_Research/. doi:10.3389/fenrg.2019.00106.

Rozendal, R. A., Hamelers, H. V. M., & Buisman, C. J. N. (2006). Effects of membrane cation transport on pH and microbial fuel cell performance. *Environmental Science and Technology, 40*(17), 5206–5211. doi:10.1021/es060387r.

Rozendal, R. A., Hamelers, H. V. M., Molenkamp, R. J., & Buisman, C. J. N. (2007). Performance of single chamber biocatalyzed electrolysis with different types of ion exchange membranes. *Water Research, 41*(9), 1984–1994. https://www.sciencedirect.com/science/journal/00431354. doi:10.1016/j.watres.2007.01.019.

Ruiz, Y., Baeza, J. A., & Guisasola, A. (2016). Microbial electrolysis cell performance using non-buffered and low conductivity wastewaters. *Chemical Engineering Journal, 289*, 341–348. www.elsevier.com/inca/publications/store/6/0/1/2/7/3/index.htt. doi:10.1016/j.cej.2015.12.098.

Salehizadeh, H., Yan, N., & Farnood, R. (2020). Recent advances in microbial CO_2 fixation and conversion to value-added products. *Chemical Engineering Journal, 390*. www.elsevier.com/inca/publications/store/6/0/1/2/7/3/index.htt. doi:10.1016/j.cej.2020.124584.

Samsudeen, N., Spurgeon, J., Matheswaran, M., & Satyavolu, J. (2020). Simultaneous biohydrogen production with distillery wastewater treatment using modified microbial electrolysis cell. *International Journal of Hydrogen Energy, 45*(36), 18266–18274. http://www.journals.elsevier.com/international-journal-of-hydrogen-energy/. doi:10.1016/j.ijhydene.2019.06.134.

Sánchez-Peña, P., Rodriguez, J., Gabriel, D., Baeza, J. A., Guisasola, A., & Baeza, M. (2022). Graphene functionalization with metallic Pt nanoparticles: A path to cost-efficient H_2 production in microbial electrolysis cells. *International Journal of Hydrogen Energy, 47*(34), 15397–15409. http://www.journals.elsevier.com/international-journal-of-hydrogen-energy/. doi:10.1016/j.ijhydene.2022.03.078.

San-Martín, M. I., Pelaz, G., Escapa, A., & Morán, A. (2021). Microbial electrolysis cells for return flow: Simultaneous nitrogen and carbon removal. *Journal of Environmental Management, 289*. https://www.sciencedirect.com/journal/journal-of-environmental-management. doi:10.1016/j.jenvman.2021.112499.

Sanz-Pérez, E. S., Murdock, C. R., Didas, S. A., & Jones, C. W. (2016). Direct capture of CO_2 from ambient air. *Chemical Reviews, 116*(19), 11840–11876. http://pubs.acs.org/journal/chreay. doi:10.1021/acs.chemrev.6b00173.

Selembo, P. A., Merrill, M. D., & Logan, B. E. (2010). Hydrogen production with nickel powder cathode catalysts in microbial electrolysis cells. *International Journal of Hydrogen Energy, 35*(2), 428–437. doi:10.1016/j.ijhydene.2009.11.014.

Selembo, P. A., Perez, J. M., Lloyd, W. A., & Logan, B. E. (2009). High hydrogen production from glycerol or glucose by electrohydrogenesis using microbial electrolysis cells. *International Journal of Hydrogen Energy, 34*(13), 5373–5381. doi:10.1016/j.ijhydene.2009.05.002.

Sharma, A., Hussain Mehdi, S. E., Pandit, S., Eun-Oh, S., & Natarajan, V. (2024). Factors affecting hydrogen production in microbial electrolysis cell (MEC): A review. *International Journal of Hydrogen Energy, 61*, 1473–1484. https://www.sciencedirect.com/science/journal/03603199. doi:10.1016/j.ijhydene.2024.02.193.

Shen, R., Jiang, Y., Ge, Z., Lu, J., Zhang, Y., Liu, Z., & Ren, Z. J. (2018). Microbial electrolysis treatment of post-hydrothermal liquefaction wastewater with hydrogen generation. *Applied Energy, 212*, 509–515. http://www.elsevier.com/inca/publications/store/4/0/5/8/9/1/index.htt. doi:10.1016/j.apenergy.2017.12.065.

Shen, R., Liu, Z., He, Y., Zhang, Y., Lu, J., Zhu, Z., Si, B., Zhang, C., & Xing, X. H. (2016). Microbial electrolysis cell to treat hydrothermal liquefied wastewater from cornstalk and recover hydrogen: Degradation of organic compounds and characterization of microbial community. *International Journal of Hydrogen Energy, 41*(7), 4132–4142. http://www.journals.elsevier.com/international-journal-of-hydrogen-energy/. doi:10.1016/j.ijhydene.2016.01.032.

Soggia, G., Goglio, A., Cristiani, P., & Adani, F. (2025). Bioelectrochemical conversion of CO_2 emissions as greenhouse gases capture and renewable electricity storage strategy: Advancing power-to-methane technologies—A critical review. *Sustainable Energy Technologies and Assessments, 80*, 104394. doi:10.1016/j.seta.2025.104394.

Srivastava, P., González, C., Palma, J., & Garcia-Quismondo, E. (2023). Nanoporous oxide coating on carbon paper electrodes to enable bio-hydrogen production in microbial electrolysis cells. *Catalysis Today, 422*. http://www.sciencedirect.com/science/journal/09205861. doi:10.1016/j.cattod.2023.114246.

Stühmeier, B. M., Pietsch, M. R., Schwämmlein, J. N., & Gasteiger, H. A. (2021). Pressure and temperature dependence of the hydrogen oxidation and evolution reaction kinetics on Pt electrocatalysts via PEMFC-based hydrogen-pump measurements. *Journal of the Electrochemical Society, 168*(6). https://iopscience.iop.org/article/10.1149/1945-7111/ac099c. doi:10.1149/1945-7111/ac099c.

Sugnaux, M., Happe, M., Cachelin, C. P., Gasperini, A., Blatter, M., & Fischer, F. (2017). Cathode deposits favor methane generation in microbial electrolysis cell. *Chemical Engineering Journal, 324*, 228–236. www.elsevier.com/inca/publications/store/6/0/1/2/7/3/index.htt. doi:10.1016/j.cej.2017.05.028.

Sun, R., Xing, D., Jia, J., Liu, Q., Zhou, A., Bai, S., & Ren, N. (2014). Optimization of high-solid waste activated sludge concentration for hydrogen production in microbial electrolysis cells and microbial community diversity analysis. *International Journal of Hydrogen Energy, 39*(35), 19912–19920. http://www.journals.elsevier.com/international-journal-of-hydrogen-energy/. doi:10.1016/j.ijhydene.2014.09.163.

Swaminathan, P., Ghosh, A., Sunantha, G., Sivagami, K., Mohanakrishna, G., Aishwarya, S., Shah, S., Sethumadhavan, A., Ranjan, P., & Prajapat, R. (2024). A comprehensive review of microbial electrolysis cells: Integrated for wastewater treatment and hydrogen generation.

Process Safety and Environmental Protection, 190, 458–474. https://www.sciencedirect.com/science/journal/09575820. doi:10.1016/j.psep.2024.08.032.

Tian, X., Zhao, P., & Sheng, W. (2019). Hydrogen evolution and oxidation: Mechanistic studies and material advances. *Advanced Materials, 31*(31). http://onlinelibrary.wiley.com/journal/10.1002/(ISSN)1521-4095. doi:10.1002/adma.201808066.

Varanasi, J. L., Veerubhotla, R., Pandit, S., & Das, D. (2018). Biohydrogen production using microbial electrolysis cell: Recent advances and future prospects. *Biomass, biofuels, biochemicals: Microbial electrochemical technology: Sustainable platform for fuels, chemicals and remediation* (pp. 843–869). Elsevier. http://www.sciencedirect.com/science/book/9780444640529. doi:10.1016/B978-0-444-64052-9.00035-2.

Venkata Mohan, S., Vijaya Bhaskar, Y., Murali Krishna, P., Chandrasekhara Rao, N., Lalit Babu, V., & Sarma, P. N. (2007). Biohydrogen production from chemical wastewater as substrate by selectively enriched anaerobic mixed consortia: Influence of fermentation pH and substrate composition. *International Journal of Hydrogen Energy, 32*(13), 2286–2295. doi:10.1016/j.ijhydene.2007.03.015.

Wang, A., Liu, W., Ren, N., Cheng, H., & Lee, D. J. (2010). Reduced internal resistance of microbial electrolysis cell (MEC) as factors of configuration and stuffing with granular activated carbon. *International Journal of Hydrogen Energy, 35*(24), 13488–13492. doi:10.1016/j.ijhydene.2009.11.124.

Wang, H., Fu, Z., Wang, S., & Zhang, W. (2021a). Analysis of CO_2 emissions in the whole production process of coal-fired power plant. *Sustainability (Switzerland), 13*(19). https://www.mdpi.com/2071-1050/13/19/11084/pdf. doi:10.3390/su131911084.

Wang, H., Liu, J., Zhang, Z., Li, J., Zhang, H., & Zhan, Y. (2021b). Alkaline thermal pretreatment of waste activated sludge for enhanced hydrogen production in microbial electrolysis cells. *Journal of Environmental Management, 294*. https://www.sciencedirect.com/journal/journal-of-environmental-management. doi:10.1016/j.jenvman.2021.113000.

Wang, J., Xu, F., Jin, H., Chen, Y., & Wang, Y. (2017a). Non-noble metal-based carbon composites in hydrogen evolution reaction: Fundamentals to applications. *Advanced Materials, 29*(14). http://onlinelibrary.wiley.com/journal/10.1002/(ISSN)1521-4095. doi:10.1002/adma.201605838.

Wang, L., Linowski, K., & Liu, H. (2023). Scalable membrane-less microbial electrolysis cell with multiple compact electrode assemblies for high performance hydrogen production. *Chemical Engineering Journal, 454*. www.elsevier.com/inca/publications/store/6/0/1/2/7/3/index.htt. doi:10.1016/j.cej.2022.140257.

Wang, L., Liu, W., Kang, L., Yang, C., Zhou, A., & Wang, A. (2014). Enhanced biohydrogen production from waste activated sludge in combined strategy of chemical pretreatment and microbial electrolysis. *International Journal of Hydrogen Energy, 39*(23), 11913–11919. http://www.journals.elsevier.com/international-journal-of-hydrogen-energy/. doi:10.1016/j.ijhydene.2014.06.006.

Wang, Q., Dong, H., Yu, H., Yu, H., & Liu, M. (2015). Enhanced electrochemical reduction of carbon dioxide to formic acid using a two-layer gas diffusion electrode in a microbial electrolysis cell. *RSC Advances., 5*(14), 10346–10351. http://pubs.rsc.org/en/journals/journalissues. doi:10.1039/c4ra14535f.

Wang, X., Rossi, R., Yan, Z., Yang, W., Hickner, M. A., Mallouk, T. E., & Logan, B. E. (2019). Balancing water dissociation and current densities to enable sustainable hydrogen production with bipolar membranes in microbial electrolysis cells. *Environmental Science and Technology, 53*(24), 14761–14768. http://pubs.acs.org/journal/esthag. doi:10.1021/acs.est.9b05024.

Wang, Y. Z., Zhang, L., Xu, T., & Ding, K. (2017b). Influence of initial anolyte pH and temperature on hydrogen production through simultaneous saccharification and fermentation of lignocellulose in microbial electrolysis cell. *International Journal of Hydrogen Energy, 42*(36), 22663–22670. http://www.journals.elsevier.com/international-journal-of-hydrogen-energy/. doi:10.1016/j.ijhydene.2017.07.214.

Weng, C. C., Ren, J. T., & Yuan, Z. Y. (2020). Transition metal phosphide-based materials for efficient electrochemical hydrogen evolution: A critical review. *ChemSusChem, 13*(13), 3357–3375. http://onlinelibrary.wiley.com/journal/10.1002/(ISSN)1864-564X. doi:10.1002/cssc.202000416.

Xiao, L., Wen, Z., Ci, S., Chen, J., & He, Z. (2012). Carbon/iron-based nanorod catalysts for hydrogen production in microbial electrolysis cells. *Nano Energy, 1*(5), 751–756. doi:10.1016/j.nanoen.2012.06.002.

Xiong, L., & Fu, X. (2025). Recent advances and challenges in electrochemical CO_2 reduction to CH_4. *Chinese Journal of Catalysis, 73*, 39–61. https://www.sciencedirect.com/science/journal/18722067. doi:10.1016/S1872-2067(25)64672-9.

Xu, L., Li, W., Luo, J., Chen, L., He, K., Ma, D., Lv, S., & Xing, D. (2023). Carbon-based materials as highly efficient catalysts for the hydrogen evolution reaction in microbial electrolysis cells: Mechanisms, methods, and perspectives. *Chemical Engineering Journal, 471*. www.elsevier.com/inca/publications/store/6/0/1/2/7/3/index.htt. doi:10.1016/j.cej.2023.144670.

Yang, E., Omar Mohamed, H., Park, S. G., Obaid, M., Al-Qaradawi, S. Y., Castaño, P., Chon, K., & Chae, K. J. (2021). A review on self-sustainable microbial electrolysis cells for electro-biohydrogen production via coupling with carbon-neutral renewable energy technologies. *Bioresource Technology, 320*. www.elsevier.com/locate/biortech. doi:10.1016/j.biortech.2020.124363.

Yang, K., Li, X., Guo, J., Zheng, J., Li, S., Zhang, S., Cao, X., Sherazi, T. A., & Liu, X. (2020). Preparation and properties of anion exchange membranes with exceptional alkaline stable polymer backbone and cation groups. *Journal of Membrane Science, 596*. www.elsevier.com/locate/memsci. doi:10.1016/j.memsci.2019.117720.

Ye, Y., & Logan, B. E. (2018). The importance of OH− transport through anion exchange membrane in microbial electrolysis cells. *International Journal of Hydrogen Energy, 43*(5), 2645–2653. http://www.journals.elsevier.com/international-journal-of-hydrogen-energy/. doi:10.1016/j.ijhydene.2017.12.074.

Yerli, C., Sahin, U., & Oztas, T. (2022). CO_2 emission from soil in silage maize irrigated with wastewater under deficit irrigation in direct sowing practice. *Agricultural Water Management, 271*. http://www.journals.elsevier.com/agricultural-water-management/. doi:10.1016/j.agwat.2022.107791.

Yossan, S., Xiao, L., Prasertsan, P., & He, Z. (2013). Hydrogen production in microbial electrolysis cells: Choice of catholyte. *International Journal of Hydrogen Energy, 38*(23), 9619–9624. doi:10.1016/j.ijhydene.2013.05.094.

Yu, N., Xing, D., Li, W., Yang, Y., Li, Z., Li, Y., & Ren, N. (2017). Electricity and methane production from soybean edible oil refinery wastewater using microbial electrochemical systems. *International Journal of Hydrogen Energy, 42*(1), 96–102. http://www.journals.elsevier.com/international-journal-of-hydrogen-energy/. doi:10.1016/j.ijhydene.2016.11.116.

Yu, Z., Leng, X., Zhao, S., Ji, J., Zhou, T., Khan, A., Kakde, A., Liu, P., & Li, X. (2018). A review on the applications of microbial electrolysis cells in anaerobic digestion. *Bioresource Technology, 255*, 340–348. www.elsevier.com/locate/biortech. doi:10.1016/j.biortech.2018.02.003.

Zabranska, J., & Pokorna, D. (2018). Bioconversion of carbon dioxide to methane using hydrogen and hydrogenotrophic methanogens. *Biotechnology Advances, 36*(3), 707–720. www.elsevier.com/inca/publications/store/5/2/5/4/5/5/index.htt. doi:10.1016/j.biotechadv.2017.12.003.

Zeng, X., Borole, A. P., & Pavlostathis, S. G. (2016). Performance evaluation of a continuous-flow bioanode microbial electrolysis cell fed with furanic and phenolic compounds. *RSC Advances, 6*(70), 65563–65571. http://pubs.rsc.org/en/journals/journalissues. doi:10.1039/c6ra13735k.

Zhang, B., Wen, Z., Ci, S., Chen, J., & He, Z. (2014). Nitrogen-doped activated carbon as a metal free catalyst for hydrogen production in microbial electrolysis cells. *RSC Advances, 4*(90), 49161–49164. doi:10.1039/c4ra08555h.

Zhang, J., Bai, Y., Fan, Y., & Hou, H. (2016). Improved bio-hydrogen production from glucose by adding a specific methane inhibitor to microbial electrolysis cells with a double anode arrangement. *Journal of Bioscience and Bioengineering, 122*(4), 488–493. http://www.elsevier.com/wps/find/journaldescription.cws_home/505516/description#description. doi:10.1016/j.jbiosc.2016.03.016.

Zhang, P., Ma, T., Li, W. D., Ma, G. Y., & Wang, Q. W. (2018). Design and optimization of a novel high temperature heat exchanger for waste heat cascade recovery from exhaust flue gases. *Energy, 160*, 3–18. www.elsevier.com/inca/publications/store/4/8/3/. doi:10.1016/j.energy.2018.06.216.

Zhang, T., Zhang, P., Liu, C., Zheng, Y., & Chen, X. (2022). Feasibility of hydrogen recovery and optimization of gas production from protein-rich food waste by bio-electrochemical system. *International Journal of Hydrogen Energy, 47*(73), 31241–31254. http://www.journals.elsevier.com/international-journal-of-hydrogen-energy/. doi:10.1016/j.ijhydene.2022.07.051.

Zhang, Z., Song, Y., Zheng, S., Zhen, G., Lu, X., Takuro, K., Xu, K., & Bakonyi, P. (2019). Electro-conversion of carbon dioxide (CO_2) to low-carbon methane by bioelectromethanogenesis process in microbial electrolysis cells: The current status and future perspective. *Bioresource Technology, 279*, 339–349. www.elsevier.com/locate/biortech. doi:10.1016/j.biortech.2019.01.145.

Zhao, N., Liang, D., Li, X., Meng, S., & Liu, H. (2021). Hydrophilic porous materials provide efficient gas-liquid separation to advance hydrogen production in microbial electrolysis cells. *Bioresource Technology, 337*. www.elsevier.com/locate/biortech. doi:10.1016/j.biortech.2021.125352.

Zhao, N., Liang, D., Liu, H., Meng, S., & Li, X. (2023). Efficient H_2 production in a novel separator electrode assembly (SEA) microbial electrolysis cell. *Chemical Engineering Journal, 451*. www.elsevier.com/inca/publications/store/6/0/1/2/7/3/index.htt. doi:10.1016/j.cej.2022.138561.

Zheng, T., Zhang, M., Wu, L., Guo, S., Liu, X., Zhao, J., Xue, W., Li, J., Liu, C., Li, X., Jiang, Q., Bao, J., Zeng, J., Yu, T., & Xia, C. (2022). Upcycling CO_2 into energy-rich long-chain compounds via electrochemical and metabolic engineering. *Nature Catalysis, 5*(5), 388–396. https://www.nature.com/natcatal/. doi:10.1038/s41929-022-00775-6.

Zou, L., Wang, C., Zhao, X., Wu, K., Liang, C., Yin, F., Yang, B., Liu, J., Yang, H., & Zhang, W. (2021). Enhanced anaerobic digestion of swine manure via a coupled microbial electrolysis cell. *Bioresource Technology, 340*. www.elsevier.com/locate/biortech. doi:10.1016/j.biortech.2021.125619.

CHAPTER

4

Dark fermentation biohydrogen production from organic matters

Yang Zhang[a], *Wachiranon Chuenchart*[b], *Shaoqing Cui*[c], *Yu Tang*[c], *and Jianzhi Yue*[a,*]

[a] Henan Agricultural University, Zhengzhou, China, [b] Quasar Energy Group, Independence, United States, [c] Zhejiang University, Hangzhou, China

* Corresponding author. E-mail address: jianzhiyue@126.com

4.1 Introduction

Some heterotrophic anaerobic bacteria can utilize organic matter for acid–hydrogen production, depending on the end products of the fermentation; the dark fermentation is mainly classified into butyl groups' acid fermentation, propionic acid fermentation, and ethanol fermentation. Compared with light fermentation for hydrogen production, advantages of the dark fermentation include that the dark fermentation process does not need light energy, which saves the maintenance cost of the light source and makes it easier to apply the design and operation of the reactor to the industrial scale; as heterotrophic bacteria, dark fermentation substrate sources are abundant and easy to obtain; industrial organic wastewater, agricultural waste materials, etc. are good raw materials for dark fermentation; in addition to this, the growth rate of dark fermentation bacteria is generally faster than that of photosynthesis bacteria in the process of cultivation, and the actual hydrogen production effect is better than that of photosynthesis bacteria because of more controllable conditions.

Biohydrogen: Advances in Theory and Production
DOI: https://doi.org/10.1016/B978-0-443-27686-6.00001-3

4.2 Substrate

The substrates for dark fermentation are primarily categorized as carbohydrate-rich feedstocks, including dedicated energy crops, lignocellulosic biomass [e.g., agricultural residues and organic wastes such as food waste and the organic fraction of municipal solid waste (OFMSW)]. The choice of substrate type is critical, as complex polymers, such as cellulose and hemicellulose, are not directly utilizable by most hydrogen-producing microorganisms. Consequently, substrate pretreatment is an indispensable step to enhance biodegradability and maximize hydrogen yield. Pretreatment methods—encompassing physical (e.g., milling and heating), chemical (e.g., acid and alkali), physicochemical (e.g., steam explosion and ammonia fiber expansion), and biological (e.g., enzymatic hydrolysis) approaches—aim to disrupt recalcitrant structures, solubilize carbohydrates, and break them down into readily fermentable sugars such as glucose and xylose, thereby making the organic matter more accessible for subsequent biohydrogen production.

4.2.1 Substrate type

The range of substrates that can be utilized for hydrogen production in dark fermentation is very broad and includes the following main categories (Sharma & Arya, 2017):

1. Easily degradable mono- and disaccharides (glucose, sucrose, fructose, xylose, and arabinose): The most commonly used and widely researched is glucose, which is usually efficient in hydrogen production, and sucrose from a wide range of sources is also commonly used as a highly efficient substrate; fructose, xylose, and arabinose are present in hydrolysates of fruits and plant hemicelluloses, with xylose being the predominant sugar in lignocellulosic hydrolysates other than glucose.
2. Polysaccharides requiring hydrolytic pretreatment (starch, cellulose, hemicellulose, and inulin): Starch from grains, such as corn, potatoes, and wheat, needs to be acidified or enzymatically cleaved into glucose or maltose before it can be efficiently utilized. Cellulose, the main cost of plant cell walls, is the most abundant renewable resource on earth, and when cellulose is used as a substrate for dark fermentation for hydrogen production (Ghaffar et al., 2018), it must be broken down into glucose monomers or oligosaccharides after intense physical, chemical, or biological pretreatment before it can be utilized by microorganisms. Hemicellulose is also an important component of lignocellulosic biomass, often accompanied by cellulose, which also has to be hydrolyzed and utilized, and hydrolysis mainly produces five-carbon sugars such as xylose and arabinose

(Glick & Pasternak, 1995). The inulin present in plants, such as inulin and chicory, can be hydrolyzed to produce fructose for fermentation and hydrogen production.

3. Lignocellulosic biomass (mainly agricultural waste, forestry waste, and energy crops): Agricultural wastes (e.g., corn stover, wheat straw, rice straw, bagasse, and rice hulls), forestry wastes (e.g., wood shavings, sawdust, and twigs), and energy crops (e.g., willow, jacaranda, and heu manchu) contain large amounts of cellulose and hemicellulose, which are highly promising sources of low-cost, renewable substrates but require effective pretreatment to disrupt the lignin structure and to release fermentable sugars.
4. Organic acids (including acetic, lactic, butyric, propionic, and pyruvic): Acetic, lactic, butyric, propionic, and pyruvic acids are often present in organic wastewater as intermediate or end products of certain fermentation processes. Some specific hydrogen-producing bacteria (HPB) (e.g., certain *Clostridia*) can utilize them for hydrogen metabolism (e.g., butyric acid–type fermentation), and the amount of hydrogen produced per unit mass is usually lower than that of sugars.
5. Glycerol: Glycerol is the main by-product of biodiesel production, and certain HPB can use glycerol as a carbon source for fermentation and hydrogen production, which is an important way to realize the resource utilization of biodiesel industry chain waste.
6. Proteins and their hydrolysis products (whey, algal biomass, and soybean meal): Proteins are first hydrolyzed to amino acids or peptides by proteases secreted by microorganisms, and then some of the amino acids can be used for hydrogen production by certain bacteria through specific pathways (e.g., Stickland reaction coupling) (Meyer et al., 2018). However, the process is relatively complex, and protein degradation is often accompanied by ammonia accumulation, which may inhibit hydrogen production.
7. Organic waste (food processing wastewater/waste, municipal sludge, organic domestic waste, livestock and poultry manure, and various industrial organic wastewater): Starch wastewater, sugar wastewater, dairy wastewater, and fruit and vegetable processing wastes are rich in carbohydrates, proteins, and lipids and can be utilized for dark fermentation to produce hydrogen. Municipal sludge and organic household waste contain biodegradable organic components but are complex in composition and have many impurities and usually require pretreatment. Livestock and poultry manure contains biodegradable organic components, but the composition is complex and has many impurities, and pretreatment is usually required for dark fermentation for hydrogen production. Industrial organic wastewater, such as brewing wastewater, citric acid wastewater, and black liquor from papermaking,

also contains components that can be fermented for hydrogen production.

In summary, it can be seen that the substrates that can be utilized for hydrogen production in dark fermentation are extremely diverse, ranging from pure compounds (e.g., glucose) to complex wastes (e.g., straw, food waste, and organic wastewater). Current research and development hotspots focus on the utilization of low-cost, renewable, and widely available substrates, especially various carbohydrate-rich agricultural/industrial wastes and by-products (e.g., lignocellulose, crude glycerol, and food-processing wastewater) to achieve the economic feasibility and environmental benefits of biohydrogen production.

4.2.2 Substrate pretreatment

Feedstock pretreatment is a crucial step in the dark fermentative hydrogen production process. Its core objectives are to improve the biodegradability of the feedstock (especially for complex substrates), to increase the release of fermentable sugars, to remove or inhibit inhibitory substances, to optimize the composition of microbial-available substrates, and ultimately to increase the hydrogen yield. Different feedstocks require different pretreatment techniques because of their vastly different chemical compositions and physical structures. The following is a description of the corresponding pretreatment technologies for the main types of feedstock.

1. Pretreatment technology for lignocellulosic raw materials: The goal of lignocellulosic pretreatment is to disrupt the lignin–hemicellulose network, reduce crystallinity, increase porosity, and expose cellulose so that enzymes or microorganisms can more easily access and degrade cellulose/hemicellulose into fermentable sugars. Pretreatment methods include physical, chemical, physicochemical, and biological methods.
   a. Physical method: Physical methods include mechanical pulverization, steam blasting, hydrothermal pretreatment, and microwave or ultrasonic pretreatment. Mechanical pulverization of lignocellulosic raw materials can reduce particle size, increase specific surface area, and reduce crystallinity. However, this method has high energy consumption and limited lignin destruction and is usually combined with other methods. Steam blasting is usually a high-temperature, high-pressure steam treatment (160°C–260°C, 0.7–4.8 MPa) for a few seconds to a few minutes, followed by an instantaneous release of pressure and the use of water vapor infiltration and bursting action to destroy the structure, dissolve part of the hemicellulose, and change the structure of lignin. Steam explosion method is highly efficient and widely used but will produce some inhibitors

[furfural and hydroxymethylfurfural (HMF)]. Microwave and ultrasonic pretreatment is the use of microwave and ultrasonic energy to destroy the cell wall structure and to promote the dissolution of components. The method is often used in conjunction with other methods, and the effect is greatly affected by parameters.

b. Chemical method: Chemical methods include acid treatment, alkali treatment, oxidation, and organic solvents. Acid treatment is subdivided into dilute acid pretreatment and concentrated acid pretreatment, and the commonly used acids for dilute acid pretreatment are H_2SO_4 and HCl, with a concentration of generally 0.5%–3% and a reaction temperature of 130°C–210°C. This method can efficiently hydrolyze hemicellulose into monosaccharides (xylose-based), partially dissolve cellulose, and destroy the structure. Dilute acid hydrolysis pretreatment method is less costly and effective but produces more inhibitors (organic acids and furans); corrosive equipment needs to be neutralized. Concentrated acid pretreatment commonly uses concentrated sulfuric acid, with a concentration of about 30%–70%. The method of dissolving cellulose and hemicellulose has a good effect, but the acid recovery is difficult, high cost, more inhibitors, and less application. Alkali treatment is the use of NaOH, $Ca(OH)_2$, KOH, or ammonia to dissolve lignin and part of the hemicellulose, destroying the lignin-carbohydrate connecting bond, dissolving cellulose, and increasing porosity. The effect is good for the raw materials with high lignin content (such as wheat straw), and the inhibitor produces less, but the alkali dosage is large, the cost is high, and it is difficult to dispose of the waste liquid. The oxidation method uses oxidizing agents to degrade lignin and some hemicelluloses. This method is more effective, but the oxidizer is costly and may also produce by-products. The organic solvent method utilizes organic solvents such as ethanol, methanol, and acetone, often with some acid or alkali catalysts, to selectively dissolve and remove lignin and retain cellulose and hemicellulose, but there are problems with solvent recovery and high costs.

c. Physical–chemical method: Physical and chemical methods include the use of steam blasting and acid–base impregnation combined method and sulfite pretreatment to overcome recalcitrance of lignocellulose (SPORL)method. Steam blasting and acid–alkali impregnation method is a combination of steam blasting before the use of dilute acid or alkali impregnation of raw materials to achieve the purpose of improving the efficiency of pretreatment. SPORL method with sulfite treatment under acidic conditions effectively removes lignin and reduces the degree of crystallization of cellulose; the effect of pretreatment for softwood is relatively good.

d. Biological method: Biological methods include fungal pretreatment and enzymatic pretreatment. The fungal pretreatment method utilizes white-rot fungi (e.g., *Phanerochaete chrysosporium*) and others to selectively degrade lignin (Kumar et al., 2025). This method has mild conditions, low energy consumption, and fewer inhibitors and is environmentally friendly, but the cycle time is long (weeks to months) and the efficiency is relatively low; in addition, the strains need to be strictly managed and controlled. The enzyme pretreatment method directly uses lignin-degrading enzymes (e.g., laccase and manganese peroxidase) or hemicellulases to degrade lignocellulosic biomass. The method has high specificity and mild conditions, but the enzyme cost is high and the efficiency is slow, so it is generally used as an auxiliary means.

2. Pretreatment technology for sugar-rich feedstock: Molasses (a by-product of the sugar industry), sweet sorghum juice, fruit processing waste liquid (containing fructose, glucose, and sucrose), certain food processing waste liquids, etc. themselves contain a high concentration of soluble sugars (sucrose, glucose, fructose, etc.), which microorganisms can directly utilize. However, they may contain inhibitors such as high salts, pigments, heavy metals, and residual chemicals. Therefore the pretreatment of such raw materials is mainly to remove or reduce the concentration of inhibitors and adjust the sugar concentration and pH to adapt to fermentation (Yunusa & Karapinar, 2025).

3. Pretreatment technology for starch-rich feedstock: Waste potatoes, grain processing waste, expired grain, starch-containing food waste, etc., whose main component is starch (a large molecule made from polymerized glucose). Starch granules are resistant to enzymatic degradation. The goal of starch pretreatment is to hydrolyze starch into fermentable sugars, mainly glucose, a process also called scarification. The pretreatment process starts with pasteurization (swelling and rupturing of starch granules at high temperature to absorb water and form a paste), a process that destroys their crystal structure and exposes the site of action to subsequent enzymatic hydrolysis, followed by the addition of high-temperature α-amylase, which randomly hydrolyzes long-chain starch into short-chain dextrins and oligosaccharides, reducing viscosity, and then subsequent lowering of the temperature and the addition of saccharolytic enzymes (mainly glucoamylase), which completely hydrolyze dextrins and oligosaccharides into glucose.

4. Pretreatment technology for organic wastewaters: Food processing wastewater, such as dairy, starch, brewing, slaughtering, fruit, and vegetable processing; domestic wastewater (primary sedimentation sludge and residual sludge after pretreatment); and livestock and poultry wastewater used for dark fermentation for hydrogen production are extremely complex and may contain sugars, proteins, fats, volatile fatty acid (VFA), dissolved organics, suspended solids, colloids, high ammonia and nitrogen concentrations, salts, oils and fats, sulfides, heavy metals, and residual disinfectants. High risk of inhibitors. Therefore the pretreatment objectives are to remove solid impurities, oil and grease, and toxic substances; adjust the concentration of organic matter [chemical oxygen demand (COD)/biochemical oxygen demand (BOD)], carbon-to-nitrogen (C/N) ratio, pH; reduce the concentration of ammonia nitrogen and sulfide; and improve biodegradability. The main pretreatment technologies are physical method, chemical method, physical–chemical method, and biological method. In physical method treatment, large particle solids are removed by grating or screening; suspended solids and some colloids and oil and grease are removed by sedimentation or flotation; water quality and quantity are made more stable by homogenization; and excessive organic matter concentration or inhibitor concentration is reduced by dilution. Chemical pretreatment is first adjusted to the range suitable for dark fermentation (usually 5.5–7.5), then coagulation or flocculation to remove colloids and fine suspended matter, and finally ozone or Fenton reagent is used to degrade difficult-to-degrade organic matter and some inhibitors. Biological methods use hydrolytic acidifying bacteria under anaerobic conditions to break down complex macromolecules (proteins, fats, and polysaccharides) into small molecules (mainly VFA) and, at the same time, reduce the rate of ammonia and nitrogen release (compared to complete anaerobic digestion).

5. Pretreatment technology for municipal solid organic waste and food waste: The composition of municipal solid organic waste and kitchen waste is often complex and unstable, and the components usually contain starch, protein, fat, cellulose, hemicellulose, lignin, inorganic salts, oils, and fats and may also contain impurities such as plastics and metals (Ali et al., 2024).

To address this situation the pretreatment goals for municipal solid organic waste and kitchen waste as feedstock for hydrogen production by dark fermentation are to remove impurities such as plastics and metals, homogenize, separate oils and fats, reduce particle size, and improve biodegradability. In the case of complex fractions, hydrolysis is also performed if necessary. The pretreatment process includes sorting, crushing, homogenization, oil and grease separation, desalination, hydrolysis acidification, and enzymatic treatment. Through sorting to use mechanical or manual removal of plastic, metal, glass, and other large impurities; through crushing to reduce the particle size to facilitate the subsequent treatment and microbial action; and then adding water and mixing to make a slurry to achieve homogenization, if the raw material has a high content of oil and grease to inhibit the production of hydrogen, it is also necessary to remove the floating oil through the static, centrifugal, or dissolved air flotation, etc., and if the raw material is kitchen waste, which may be salty, then it is necessary to wash it or treat it with a membrane, and if the raw material contains complex macromolecular organic matter, it needs to be converted into VFA through hydrolysis and acidification before utilization.

In short, there is no "one-size-fits-all" pretreatment method for all feedstock. For lignocelluloses the core is to crack the stubborn structure, and physicochemical methods (e.g., steam blasting, dilute acid/dilute alkali) are the mainstream. Pretreatment of sugar-containing raw materials focuses on detoxification and conditioning. Starch-based raw materials must undergo scarification. Organic wastewater and kitchen waste, on the other hand, emphasize impurity removal, homogenization, and the critical hydrolysis-acidification step, which converts complex substrates into small molecules, such as VFA, that can be easily utilized by HPB. When selecting pretreatment technologies, detailed feedstock characterization and process comparisons must be performed to achieve the optimal balance of cost-effectiveness and environmental sustainability.

4.3 Strains

The dark fermentation process is a process in which organic compounds are converted to hydrogen by the action of anaerobic microorganisms in the absence of light, whereas various acid/alcohol by-products are also produced. The microorganisms involved in this process may be either parthenogenetic or specialized anaerobic (Cabrol et al., 2017; Castelló et al., 2020; Goud & Mohan, 2012; Venkata Mohan et al., 2008). In nature, there are a variety of microorganisms that are capable of dark fermentation for hydrogen production, which mainly include *Clostridium, Enterobacter, Acetomicrobium, Trdiumbutyricum, Klebsiella*, and *Methanococcus* (Ndayisenga et al., 2021). These microorganisms achieve effective conversion of organic compounds and hydrogen production through their unique metabolic pathways in a specific anaerobic environment.

When the substrates are glucose, starch, and sucrose, the end products of fermentation of some bacteria are mainly butyric acid, acetic acid, hydrogen, and CO_2, so the fermentation is butyric acid type. The dominant species in butyric acid fermentation are mainly *Clostridium* spp. Among them, *Clostridium butyricum* and *Clostridium typhimurium* have good hydrogen production properties.

When the fermentation substrate is lactic acid, glucose, glycerol, nitrogenous organic matter, or fibrin, the end products of the fermentation of some bacteria are propionic acid, acetic acid, and CO_2, and the fermentation is propionic acid type. The dominant bacteria in propionic acid fermentation are mainly propionic acid bacilli, such as Propionibacterium acidipropionici and Aspergillus terreus. As there is no hydrogen production in the propionic acid pathway, it is not meaningful to study the hydrogen production from pure propionic acid. However, in the case of mixed-strain cultures, propionic acid–producing bacteria compete with other types of HPB for substrates and ecological niches, changing the direction of microbial community succession, which leads to a significant decrease in the hydrogen production capacity of the fermentation system. Therefore it is important to strictly control the type of fermentation and guide the direction of bacterial succession for biohydrogen production.

In the anaerobic fermentation of *Clostridium ethanolicum* (*C. ethanolicum* Harbin) with glucose as substrate the liquid-phase end products were mainly ethanol and acetic acid, which showed the characteristics of ethanol-type fermentation. The superiority of ethanol fermentation lies in the fact that ethanol is a neutral product, which reduces the interference of redox potential on bacterial hydrogen production and has higher metabolic rate and hydrogen production efficiency and has gradually become a hot spot of research on biological hydrogen production.

4.4 Parameters

The performance of dark fermentation is critically controlled by key operational parameters. An optimal slightly acidic pH (around 5.5–6.5) is essential to favor HPB. Temperature, typically mesophilic or thermophilic, governs the reaction rate and microbial consortium. Additives, such as iron or nickel nanoparticles, can significantly enhance hydrogen yield by activating hydrogenase enzymes. Conversely the process is inhibited by the accumulation of metabolic by-products (e.g., VFAs) and high levels of compounds like ammonia, which can suppress microbial activity.

4.4.1 pH

Hydrogen production in dark fermentation incorporates multiple stages and pathways, relying on the syntrophic relationship of fortitude microorganisms. pH affects microbial consortia, enzyme activity, substrate solubility, and degradability. In biohydrogen production, a wide range of optimal pH has been recommended to be 5.0–7.0 (Gull et al., 2025). This optimal pH range controls the pathways of biohydrogen production. The low pH range (4.5–6.0) is the favored condition for the acetate-butyrate pathway. On the other hand the lactate pathway may be favored at the higher pH range of 6.0–7.0, whereas ethanol and propionic acid tend to be produced when pH exceeds 7.0 (Jain et al., 2024; Palanivel et al., 2025). For example, in swine manure and glucose biohydrogen production, among 4.4–5.6, pH 5.0 was indicated as optimal at 180 NmL/g volatile solids (VS), with evidence of methane production at pH higher than 5.3 (Wu et al., 2010). For synthetic wastewater the optimal pH of 5 resulted in biohydrogen production at 71 NmL/g VS, whereas ethanol was predominant at pH 4.0 (Tang et al., 2022). Lower pH is typically observed to regulate the formation of acidic and alcoholic metabolites, which could penetrate through the cell membrane, interfering with metabolic activities (Elbeshbishy et al., 2017). Acetate, butyrate, lactate, and ethanol with half-maximum inhibition concentrations (IC_{50}) were reported at 209, 104, 273, and 150 mM, and maximum inhibition at concentrations (C_{max}) around 300–500, 200–300, 400, and 300 mM, respectively, depending on inoculums and substrates (Noguer et al., 2022; Tang et al., 2012; Wang et al., 2008).

Although a lower pH range favors acetate-butyrate pathways, a higher pH range (6.0–7.0) was favorable for lactate-driven dark fermentation (LD-DF), relying on lactic acid bacteria (LAB) and HPB to convert lactate to hydrogen, which emerges as an efficient pathway for easily biodegradable substrates, such as food waste or OFMSW. This is based on the fact that *Clostridium* species are generally favored by a pH range between 5.5 and 6.8 (Liu et al., 2011), whereas *Lactobacillus* has an optimal pH at 5.5–6.2 (Górska et al., 2016). Therefore this upper range of pH promotes HPB while controlling LAB. Several studies have shown the promising biohydrogen yield (BHY) at the upper range of optimal pH, including fruit and vegetable waste at pH 7.0 with 49.5 NmL/g VS_{fed} (Martínez-Mendoza et al., 2022); food waste at pH 6.5 with 81 NmL/g VS_{fed} (Regueira-Marcos et al., 2023); and OFMSW at pH 6.5 with 33.7 NmL/g VS_{fed} (Martínez-Fraile et al., 2024). However, pH more than 7.0 resulted in lower hydrogen production because of inhibition of metabolic activity and enzymes. In addition, higher pH resulted in the formation of free ammonia, the unionized form of ammonium, which may diffuse through and shuttle protons through the cell membrane, increasing energy consumption to maintain internal pH and resulting in cell inactivation. Methanogens thrive in natural to slightly alkaline conditions, outcompeting and diverting the primary pathways to methane production.

Even though most substrates are self-buffering substrates, such as sewage sludge and animal manure, pH adjustment methods can be used by using chemicals such as HCl, NaOH, H_2PO_4, $Ca(OH)_2$, $Mg(OH)_2$, or organic buffers acetic acid/acetate, butyric acid/butyrate, citric acid/citrate (Guo et al., 2022; Liu et al., 2011). In advance, CO_2 sparging, which dissolves CO_2 to carbonic acid form, can also be beneficial in both maintaining slightly acidic conditions and reducing hydrogen partial pressure, while resulting in higher hydrogen productivity (Kim et al., 2012). However, key factors, including ionic strength, salt buildup, and chemical cost, also need to be considered in practice or industrial level.

4.4.2 Temperature

The temperature range of biohydrogen production can be varied from psychrophilic (10°C–25°C), mesophilic (30°C–40°C), and thermophilic (45°C–60°C). Temperature plays an important role for the niche species as well as their enzyme activities. It is worth mentioning that at lower temperatures, psychrophilic microorganisms could produce hydrogen in the range of 0.32–1.57 mol H_2/mol glucose at 25°C. GA024 strain was observed to produce the highest 1.57 mol H_2/mol glucose at pH 6.5 (Alvarado-Cuevas et al., 2015). G088 strain (EU636029) was able to produce hydrogen of 1.81 mol H_2/mol glucose at the optimal conditions of 26.3°C and pH 6.24 (Alvarez-Guzmán et al., 2017). In addition, at the temperature close to mesophilic conditions, N92 could produce up to 1.7 mol H_2/mol glucose with an initial pH of 6.86, which is comparable to the mesophilic mixed culture of 1.9 at 40°C (Wang & Wan, 2008). Psychrophilic dark fermentation may allow us to produce hydrogen in limited conditions, including cold climates with low initial energy requirements.

Mesophilic conditions are the most common practice in biohydrogen production because of available microbial consortia, their optimal enzyme activities, and system controllability. Various organic wastes and inoculum sources yield hydrogen production around 99–360 mL H_2/g VS (Elbeshbishy et al., 2017; Ghimire et al., 2015). Common hydrogen producers (e.g., *Clostridium* and *Lactobacillus*) have their optimal temperature and predominance fall into mesophilic conditions (Elbeshbishy et al., 2017; Kamalaskar et al., 2010; Pérez-Rangel et al., 2026). Several mesophilic inocula also showed higher hydrogen productivity compared to the thermophilic condition, corresponding to their diverse majority of inoculum sources in the environments and engineered biological processes (Kim et al., 2024).

Thermophilic conditions could accelerate biological and chemical conversion, improve enzyme activities, promote thermodynamically favorable metabolic pathways, and ultimately improve hydrogen production. In thermophilic dark fermentation the main microbial structure relies on predomination of *Clostridium* (i.e., *C. thermocellum*) and *Thermoanaerobacterium* (i.e., *T. thermosaccharolyticum*) (Liu et al., 2008). A comprehensive qualitative review suggested 1.9-fold higher hydrogen production (2.30 mole H_2/mol substrate) by *Thermoanaerobacterales* compared to mesophiles (i.e., Clostridiaceae and Enterobacteriaceae) (Rittmann & Herwig, 2012). Thermophilic fermentation of cheese whey yielded 1.7-fold higher BHY with 0.81 mol H_2/mol glucose compared to mesophilic inoculum (Kargi et al., 2012). Thermophilic conditions also amplified the effect of pretreatment of hydrolyzed wheat straw, 1.5-fold higher hydrogen production (333 mL H_2/g Total Solids (TS)), and fermentation of acid-hydrolyzed wheat straw compared to mesophilic conditions (220 mL H_2/g TS), whereas comparable yields in both conditions under no pretreatment (Cakr et al., 2010). Thermophilic and hyperthermophilic xylose-fed fluidized bed reactors with a highly abundant *Thermoanaerobacterales* showed 1.6–2.4-fold higher hydrogen productivity with 1.2 and 0.8 mol H_2/mol xylose at 55°C and 70°C, and 60°C and 65°C, respectively (Dessì et al., 2018). Preincubation and augmentation are crucial for microorganisms to shift from mesophilic to thermophilic conditions, which may require months to years, depending on microbial adaptability.

4.4.3 Additives

Several additives can be added for biohydrogen production and can be classified based on their characteristics into macronutrients, micronutrients, antioxidants, and conductive materials. Dark fermentation microorganisms require sufficient N, P, and S macronutrients to sustain their metabolism, growth, coenzymes, and cellular components. Optimal ranges of macronutrients have been suggested at C/N around 18–50 and C/P around 125–340 (Gbiete et al., 2026; Oztekin et al., 2008; Pérez-Rangel et al., 2020; Sotelo-Navarro & Poggi-Varaldo, 2021).

Micronutrients and trace elements are important for hydrogen production because of their components in coenzymes and hydrogenase. Mg, Fe, Na, and Zn were suggested as crucial metals for hydrogen production. When the medium composition contained 120 mg/L $MgCl_{2h}2H_2O$, 1000 mg/L NaCl, 0.5 mg/L $ZnCl_2$, and 3 mg/L $FeSO_{4h}7H_2O$ resulted in the highest hydrogen production in mixed flora (Qu et al., 2022). Adding trace elements (Fe, Ni, Zn, Co, Cu, Mn, Al, B, Se, Mo, and W) could result in 1.1–2.8-fold increases in biohydrogen production. Fe, Zn, Cu, and Ni were found to highly contribute to hydrogen production from fruit and vegetable waste (Keskin et al., 2018). Fe is important for ferredoxin, which catalyzes the production of hydrogen, a component of hydrogenase, which controls the pathways of hydrogen utilization. In addition, Fe also affects the synthesis of formate lyase (Qu et al., 2022). $FeSO_4$ at Fe^{2+} of 20 mg/L resulted in 2.2-fold higher biohydrogen production (Bao et al., 2013).

In some cases, antioxidants (i.e., L-cysteine and ascorbic acid) could improve hydrogen yield under anaerobic conditions. L-cysteine (0.1–1.0 mM) could scavenge oxygen, provide sulfur, and enhance hydrogen production from 15% to 70%, depending on the operating temperature (Yuan et al., 2008). In dark fermentation of saccharified corn scrap, among 0.6–80 mg/L, 5 mg/L of ascorbic acid resulted in 2.20 mol H_2/mol glucose with a reduced lag phase by 65.6% and increased hydrogen yield by 40.9% compared to the control (Pistekova et al., 2025).

Agricultural residue-based biochar, including pinewood, corn stalk, rice straw, sawdust, coffee ground, and corn barn, can potentially improve 1.2–4.3-fold hydrogen production, which is owing to their physicochemical properties, including surface area and porosity, functional groups, micronutrient content, and electroconductivity (Roychowdhury & Ghosh, 2024). The surface area allows hydrogen-producing biofilm formation, as well as its extracellular (Extracellular Electron Transfer, EET) and interspecies electron transfers (IET). For example, 10 g/L of biochar addition improved 4.2-fold in hydrogen production by sugarcane bagasse owing to synergistic interactions and EET between electroactive bacteria (such as Pseudomonadaceae, Clostridiaceae, and Peptostreptococcaceae) (Bu et al., 2021). Softwood pellet biochar at 10 g/L improved 2.17-fold in hydrogen production, owing to lactate flux reduction and a remarkable improvement in acetate flux (Yang et al., 2024).

4.4.4 Other inhibitors

Several inhibitors also hindered hydrogen production in dark fermentation. Heavy metals play a beneficial role as trace elements but could be inhibited in higher concentrations. Although Cu is very toxic to biohydrogen production, Cd and Zn improved hydrogen production at lower than 20 and 0.5 mg/L, respectively, but inhibited it at higher than 40 and 1 mg/L (Cho & Lee, 2011; Openshaw & Bagnato, 2026). In addition, other than the organic acids aforementioned, organic inhibitors, including phenolic compounds and furan derivatives, should also be considered. Lignocellulosic-derived compounds include phenolic compounds and furan derivatives. At the same concentration (1 g/L), furan derivatives, including furfural and 5-HMF, showed 70%–76% reduction, whereas phenolic compounds (vanillin, syringaldehyde, and phenol) showed 17%–23% reduction (Quéméneur et al., 2012). Phenolic compounds, primarily membrane-active molecules, can interact with and disrupt cell membranes, selectively altering their structure, permeability, and integrity (Basak et al., 2020). Furans can readily cross the cell membrane and accumulate in the cytosol, where they disrupt central metabolic pathways through the inhibition of Nicotinamide Adenine Dinucleotide (NADH)-dependent enzymes. (e.g., glyceraldehyde 3-phosphate dehydrogenase and pyruvate dehydrogenase), their hydrophobicity and low molecular weight (Basak et al., 2020).

4.5 Reactors

Bioreactor systems are central to dark fermentative biohydrogen production, as their design and configuration directly impact microbial activity, substrate conversion efficiency, and process stability. The selection of an appropriate reactor configuration is critical for optimizing hydrogen yield and facilitating scale-up for industrial applications. Dark fermentation reactors are typically anaerobic systems that operate under controlled conditions to promote the growth of HPB while inhibiting hydrogen-consuming microorganisms (Goren et al., 2025; Park et al., 2021). A variety of reactor configurations have been developed and evaluated for biohydrogen production, including continuous stirred tank reactors (CSTRs), anaerobic sequencing batch reactors (ASBRs), upflow anaerobic sludge blanket (UASB) reactors, fixed-bed reactors (FBR), membrane bioreactors (MBRs), and anaerobic baffled reactors (ABRs) (Goren et al., 2025; Park et al., 2021). The choice of reactor depends on factors such as the nature of the substrate, microbial ecology, hydrodynamic behavior, and operational parameters.

Recent research has emphasized the importance of attached-growth systems or biofilm-based reactors for achieving high-rate hydrogen production. These systems prevent washout of slow-growing microorganisms, maintain high cell densities, and enhance resilience to environmental perturbations (Park et al., 2021). FBRs and MBRs, for instance, have demonstrated superior biohydrogen performance compared to suspended-growth systems because of improved biomass retention and higher volumetric loading rates. Similarly the incorporation of baffles in ABRs creates compartmentalized zones that support distinct microbial communities and metabolic phases, thereby improving substrate conversion and hydrogen yield (Grobicki & Stuckey, 1992).

The hydrodynamic characteristics of reactors, such as mixing intensity, flow patterns, and retention times, significantly influence mass transfer, substrate-microbe contact, and hydrogen recovery. Computational fluid dynamics (CFD) simulations have emerged as a valuable tool for visualizing and optimizing these parameters. For example, studies using FLUENT software have revealed how hydraulic retention time (HRT) affects velocity fields and vortex formation in dark fermentation reactors, which, in turn, impact hydrogen production rates.

In addition to configuration the techno-economic feasibility of dark fermentative biohydrogen production is a key consideration. Although reactors like CSTRs are cost-effective and widely used, advanced systems such as MBRs offer higher performance but incur greater energy and capital costs (Goren et al., 2025). Future development efforts should focus on optimizing reactor designs to mitigate these challenges, for instance, by developing low-energy membrane modules or hybrid systems that combine the advantages of multiple configurations.

4.5.1 Reactor structure

The physical structure of a bioreactor determines its flow dynamics, mixing efficiency, and ability to retain biomass, all of which are critical for maintaining an active hydrogen-producing microbial consortium. The structural design must accommodate the specific requirements of dark fermentation, including strict anaerobic conditions, controlled temperature and pH, and efficient gas–liquid separation (Table 4.1).

1. CSTR: The CSTR is one of the most commonly used reactor configurations because of its simplicity and ease of operation. It consists of a cylindrical vessel with a mechanical agitator that provides continuous mixing to ensure uniform distribution of substrate, nutrients, and microorganisms. The reactor is typically equipped with ports for feeding, gas collection, and sampling, and it operates in a continuous mode. The primary advantage of the CSTR is its ability to minimize substrate gradients and enhance mass transfer. However, its suspended-growth nature makes it prone to biomass washout at short HRTs, which can limit the achievable cell density and hydrogen production rate (Goren et al., 2025).
2. UASB reactor: The UASB reactor is characterized by its upflow regime, where wastewater enters from the bottom and passes through a dense blanket of granular sludge. The granular sludge aggregates are self-immobilized microbial consortia that exhibit excellent settling properties, allowing for high biomass retention without the need for supporting materials (Park et al., 2021). The UASB includes a gas–solid–liquid separator (GLS) at the top, which facilitates the collection of biogas and prevents the washout of biomass. This design is highly effective for high-strength wastewaters and has been successfully applied for biohydrogen production from palm oil mill effluent and other agro-industrial wastes (Goren et al., 2025).
3. FBR/packed-bed reactor: FBRs are attached-growth systems packed with inert support materials, such as ceramic beads, activated carbon, plastic rings, or porous materials, which provide a large surface area for microbial adhesion and biofilm formation (Goren et al., 2025). The substrate flows through the packed bed, where it comes into contact with the immobilized biomass. The structure typically includes a feed distribution system at the inlet and a gas–liquid separation system at the outlet. The packed bed can be arranged vertically or horizontally. The biofilm mode of operation in FBRs results in high cell densities, extended microbial residence times, and increased resistance to environmental shocks, leading to enhanced hydrogen productivity. However, challenges, such as channeling, clogging, and pressure drop, need to be managed through appropriate support material selection and flow control (Park et al., 2021).
4. MBR: MBRs integrate a biological reaction unit with a membrane filtration system. The membrane module, which can be based on microfiltration or ultrafiltration, is used to completely retain biomass and suspended solids, resulting in very high mixed liquor suspended solids

TABLE 4.1 Performance comparison of different dark fermentative biohydrogen reactor configurations.

Reactor configuration	Key features	Advantages	Limitations	Typical H_2 yield range
CSTR	Continuous stirring; suspended growth	Simple design; easy operation; effective mixing	Biomass washout; low cell density	Moderate
UASB	Upflow regime; granular sludge	Good biomass retention; high loading rates	Long start-up; granular stability issues	Moderate to high
Fixed-bed reactor	Packed with inert support material	High cell density; prevents washout; good stability	Clogging risk; channeling possible	High
Membrane bioreactor	Integrated filtration membranes	Excellent biomass retention; high effluent quality	High energy consumption; fouling issues	High
ASBR	Fill-and-draw cycles	Flexible operation; good settling	Cyclic operation; more complex control	Moderate

ASBR, *anaerobic sequencing batch reactor;* CSTR, *continuous stirred tank reactor;* UASB, *upflow anaerobic sludge blanket.*

(MLSS) concentrations and superior effluent quality. The MBR structure includes a reaction tank and an internal or external membrane unit. The membranes can be configured as flat-sheet, hollow-fiber, or tubular. Although MBRs achieve some of the highest biohydrogen yields and are excellent for treating low-strength wastewaters, their structure is complex, and membrane fouling remains a significant operational challenge, often requiring frequent chemical cleaning or backwashing (Goren et al., 2025).

5. ABR: The ABR is a compartmentalized reactor with a series of vertical baffles that force the wastewater to flow under and over them as it moves from the inlet to the outlet. This design creates distinct sequential chambers, each of which can develop specialized microbial communities adapted to the changing environmental conditions along the reactor length. The first compartments typically host acidogenic and HPB, whereas subsequent compartments may support other metabolic processes if integrated with posttreatment steps. The baffled structure promotes phase separation, reduces sludge washout, and enhances the reactor's stability against hydraulic and organic shock loads (Grobicki & Stuckey, 1992).
6. Double-chamber plug-flow reactor: Some specialized designs, such as the double-chamber plug-flow reactor, have been developed for specific feedstocks. This reactor features a cylindrical main body with an inverted conical bottom and an internal metal cylinder, creating two isolated chambers. The inner chamber can be used for heating or other processes, whereas the annular space between the inner cylinder and the reactor wall is where the primary digestion occurs. This design is intended for high-solid content feedstocks like livestock manure and crop straw, and it aims to achieve more thorough digestion and higher biogas yield by increasing the substrate's path length and preventing "short-circuiting" of the flow (Changrong, 2013).
7. Materials and construction: Reactor structures are commonly constructed from materials that ensure durability, corrosion resistance, and gas-tight integrity. Stainless steel and carbon steel, often with protective coatings or galvanization, are widely used (Olaniyan et al., 2017). For instance a reported biogas reactor used galvanized steel for the main tank and galvanized aluminum for pipes and valves (Olaniyan et al., 2017). Construction methods can vary, with large-scale fermenters sometimes being built using a cage-and-shell method, where a rotating fermenter cage is lined with gas-tight shell plates (Yousuf et al., 2020).

4.5.2 Reactor operation

The operational strategy of a dark fermentation reactor is pivotal in maximizing hydrogen yield and process efficiency. Key operational parameters must be carefully controlled to maintain an optimal environment for HPB while suppressing methanogens and other competitive microbes.

1. HRT: HRT is a critical parameter that influences microbial population dynamics, substrate contact time, and hydrogen production rate. Shorter HRTs are generally employed to suppress the growth of slow-growing methanogenic archaea and promote acidogenic bacteria. However, an excessively short HRT can lead to washout of active biomass, particularly in suspended-growth systems like CSTRs. Research has shown that an optimal HRT exists for each reactor configuration. For example, in a continuous-flow dark fermentation system, an HRT of 24 hours resulted in the maximum hydrogen production rate of 40.45 mol m^{-3}/d, with longer or shorter HRTs yielding lower rates. Similarly, in a dark–photo circular baffled reactor (DP-CBR), an HRT of 24 hours combined with an initial pH of 6.5 gave the highest hydrogen yield of 0.4 L/g COD removed (Grobicki & Stuckey, 1992). Operation at lower HRTs (e.g., 6–12 hours) can be feasible in attached-growth systems like fixed-bed or MBRs, where biomass retention is independent of HRT (Park et al., 2021).
2. pH and temperature control: Maintaining the reactor pH within the acidic to neutral range (typically 5.0–6.5) is essential for favoring HPB and inhibiting hydrogen-consuming microorganisms (Ghimire et al., 2015). The initial pH of the feed and the inherent buffering capacity of the system must be monitored and adjusted, often using NaOH, KOH, or HCl. For instance a study found that lowering the initial pH to 5.5 deteriorated the performance of the dark fermentation stage in a DP-CBR (Grobicki & Stuckey, 1992). Temperature also profoundly affects microbial metabolism and hydrogenase enzyme activity. Most dark fermentation processes are operated under mesophilic (30°C–40°C) or thermophilic (50°C–60°C) conditions. A lab-scale biogas reactor operated successfully at a mesophilic temperature of 32°C. Temperature control is typically achieved using internal heaters or external heat exchangers (Olaniyan et al., 2017).
3. Substrate type and organic loading rate (OLR): The nature of the organic substrate and its concentration significantly impact hydrogen production. Simple sugars (e.g., glucose and sucrose) are readily converted, whereas complex substrates (e.g., lignocellulosic biomass) require pretreatment (Ghimire et al., 2015). The OLR, which is the amount of organic substrate fed to the reactor per unit volume per day, must be optimized. High OLRs can lead to acid accumulation and pH drop, inhibiting the process, whereas low OLRs may not fully utilize the reactor's capacity. In continuous-flow dark fermentation a substrate concentration of 30 g/L yielded the maximum hydrogen production

rate. Similarly, in photo-fermentation, the OLR was found to significantly influence the hydrogen production rate and yield.

4. Nutrient and trace element supplementation: A balanced nutrient supply is vital for microbial growth and hydrogenase enzyme synthesis. Essential nutrients include nitrogen (e.g., as NH_4Cl), phosphorus, and trace metals like iron, nickel, and magnesium (Ghimire et al., 2015). Iron is particularly important as a key component of [Fe–Fe]-hydrogenases. The C/N ratio should be optimized, as excess nitrogen can divert electron flow toward ammonia production instead of hydrogen.
5. Inoculum and microbial community management: A successful operation begins with a properly developed inoculum, often pretreated to suppress hydrogen consumers. Pretreatment methods include heat shock, acidification, alkalization, or aeration to eliminate methanogens and enrich for spore-forming HPB (Ghimire et al., 2015). During long-term operation the microbial community evolves. Modern high-throughput sequencing techniques (e.g., 16S rRNA gene sequencing) are used to monitor community shifts and identify key players, such as the symbiotic relationship between HPB and biofilm-forming bacteria in attached-growth systems. Maintaining a stable and resilient microbial consortium is key to sustainable operation.
6. Mixing and agitation: Adequate mixing ensures uniform distribution of substrate, temperature, and pH and prevents the formation of dead zones or scum layers. In CSTRs, this is achieved with mechanical impellers (Goren et al., 2025; Olaniyan et al., 2017). In other reactors, mixing can be provided by liquid recirculation, gas sparging, or the upflow velocity in UASB reactors. However, excessive mixing can cause shear stress that disrupts microbial flocs or biofilms.
7. Gas monitoring and process control: Continuous monitoring of the produced gas (volume, composition, and rate) is essential for assessing reactor performance. Hydrogen content in the biogas is typically monitored using gas chromatography. Online sensors for pH, temperature, and pressure provide real-time data for automated control systems. Advanced monitoring may also involve tracking oxidation–reduction potential (ORP), which has been observed to stabilize in specific ranges (e.g., −380 to −455 mV) during efficient hydrogen production.
8. Integration and scale-up considerations: For practical application, operating strategies must address scalability and economic viability. This includes integrating energy-efficient mixing, heat recovery systems, and low-cost nutrient sources. The use of renewable energy, such as solar power for heating, mixing, and lighting (in photo-fermentation), has been explored to improve sustainability and reduce operational costs. Techno-economic assessments are crucial for comparing the operational costs (e.g., energy for mixing, heating, and membrane filtration) against

the value of the hydrogen produced and the environmental benefits achieved (Grobicki & Stuckey, 1992).

4.6 Net-zero effect

Dark fermentation biohydrogen production presents a compelling pathway toward achieving net-zero energy and waste objectives by transforming organic wastes into a carbon-neutral fuel. This process directly contributes to greenhouse gas mitigation by displacing fossil fuel consumption with clean hydrogen and by diverting biodegradable waste from landfills, thereby avoiding methane emissions. Furthermore the process can approach a closed-loop carbon cycle when the residual VFAs from fermentation are valorized, for instance, in subsequent biorefinery stages or for bioenergy production. Thus when integrated within a circular bioeconomy framework, dark fermentation not only generates renewable energy but also promotes sustainable waste management, offering a dual strategy for reducing our carbon footprint and advancing a net-zero future.

4.6.1 Biogenic carbon neutrality

The production of biohydrogen for dark fermentation integrally aligns with net-zero objectives by operating within the biogenic carbon cycle (Nelly López-Hernández et al., 2024). It is widely regarded as a sustainable and low-carbon approach with the potential to contribute to net-zero carbon goals. Furthermore, its carbon neutrality is because the organic waste or biomass utilized in advance already incorporates the atmospheric CO_2 throughout its lifetime, so the emissions that should otherwise originate from conventional hydrogen production are eliminated (Rosa & Mazzotti, 2022). Nonetheless, realizing a truly net-zero effect through dark fermentation is highly dependent on different interconnected factors (Goren et al., 2023). The selection of feedstock plays a significant role. For example, waste-derived materials, such as food waste, agricultural residues, or municipal organic waste, lead to the greatest carbon benefits by diverting organic waste and the utilization of renewable carbon sources (Goren et al., 2023).

This neutrality is influenced by the choice of substrate, pretreatment energy, and fermentation efficiency. Wet wastes, such as food scraps or cheese whey, remove the need for energy-intensive drying and grinding, which marginally lowers Global Warming Potential (GWP) values. When Ganguly et al. (2025) modeled two 50 t/day biorefineries, one of which processed cheese whey and the other solid food waste, they discovered that, when powered by renewable energy sources and outfitted with moderate CO_2 collection, the net life-cycle emissions were −8.6 and −8.0 kg

CO_2-eq/kg H_2 (Ganguly et al., 2025). In comparison, lignocellulosic feeds might need mechanical pretreatment and dilute acid hydrolysis, increasing upstream energy consumption and making GWP more positive (Lui et al., 2022). Table 4.2 shows the emissions reported by several studies on dark fermentation. However, even with less favorable energy balances, the inherent biogenic origin of CO_2 guarantees that dark fermentation cannot surpass the carbon intensity of fossil-derived hydrogen on a cradle-to-gate basis.

4.6.2 Displacement of fossil-derived hydrogen and carbon-negative potential

The most significant climate benefit of biohydrogen arises when it substitutes hydrogen from conventional steam methane reforming (SMR), which emits 9–13 kg CO_2-eq/kg H_2 (Khademi et al., 2021), or hydrogen from other fossil sources. Benchmarks show that, on its own, dark fermentation can emit 6.4–6.6 kg CO_2-eq/kgH_2 when processing wet wastes, all of which is biogenic, and as little as 2.48 kg CO_2-eq/kg H_2 when accounting for efficient downstream anaerobic digestion of effluents (Lee et al., 2024). When renewables, such as solar power auxiliary systems (mixing and gas separation), cradle-to-gate emissions fall below 5 kg CO_2-eq/kg H_2, rivaling the cleanest electrolysis routes (Barghash et al., 2022). Crucially, integrated carbon capture on the CO_2 exhaust stream can push net emissions into negative territory. Ganguly et al. (2025) assumed 50% capture of biogenic CO_2 and obtained −8.6 and −8.0 kg CO_2-eq/kg H_2 for cheese whey and solid food waste, respectively. Boshagh et al. (2025) reported global-warming potentials of only 0.79 kg CO_2-eq/kg H_2 when credits for carbonate buffering agents and process steam were included, over 14-fold lower than SMR.

However, not all biological hydrogen methods displace fossil impacts equally. Goren et al. (2023) found photo-fermentation to have the lowest GWP and acidification potential, whereas dark fermentation and microbial electrolysis cell (DF–MEC) hybrids peaked at 14.6 kg CO_2-eq/kg H_2 because of high electricity demands and water footprints. Such contrasts highlight the importance of process selection, electrification strategies, and water management in realizing a truly carbon-negative potential.

Biohydrogen from dark fermentation (and its integrated variants) was benchmarked against two conventional hydrogen routes (conducted by this study): 1 kg H_2 from a petroleum refinery (1.75 kg CO_2-eq over a 20-year horizon, Ecoinvent 3.6) and 1 kg H_2 via steam-methane reforming (6.2 kg CO_2-eq, openLCA/Ecoinvent 3.6). When benchmarked against hydrogen production from a petroleum refinery operation (1.75 kg CO_2-eq/kg H_2), most dark-fermentation routes emit more GHGs unless they treat the lowest impact substrates (e.g., wastewater) or integrate anaerobic

TABLE 4.2 Summary of key environmental impact assessment studies on the dark fermentation of biohydrogen.

Feedstock	Mode of operation	CO_2 emission	Conditions	Remark	References
Cheese whey	DF + MEC	9 kg CO_2-eq/kg H_2	50 t/day plant; renewable electricity; 50% CO_2 capture	Eligible for \$3/kg H_2 Inflation Reduction Act (IRA) tax credit	Ganguly et al. (2025)
Solid food waste	DF + MEC	11.3 kg CO_2-eq/kg H_2	50 t/day plant; renewable electricity; 50% CO_2 capture		
Mixed biomass resources	DF + MEC	14.6 kg CO_2-eq/kg H_2	Lab-scale DF–MEC; high electricity/heat demand	Highest GWP, water consumption, and scarcity footprint	Goren et al. (2023)
Rice husk	DF only	10.92 kg CO_2-eq/kg H_2	Cogeneration from rice-husk combustion supplying process heat	Lowest environmental impact among gasification scenarios	Reaño (2020)
Wet food waste	DF only	7.25 kg CO_2-eq/kg H_2	Lab-scale DF on wet waste	Wet wastes yield lower GWP than lignocellulosic feeds	Lui et al. (2022)

(*continued on next page*)

TABLE 4.2 Summary of key environmental impact assessment studies on the dark fermentation of biohydrogen—cont'd

Feedstock	Mode of operation	CO_2 emission	Conditions	Remark	References
Corn stover	DF only	9.8 kg CO_2-eq/kg H_2	Lab-scale DF on corn stover	Substrate choice strongly influences GWP	
Food waste	DF + Anaerobic Digestion (AD)	2.48 kg CO_2-eq/t FW treated	Two-stage DF then AD; 1 t FW basis; includes displaced energy	Net avoided GHGs per ton surpass single-stage AD	Lee et al. (2024)
Coffee-manufacturing wastewater	DF + AD	0.79 kg CO_2-eq/kg H_2	Mesophilic/thermophilic DF–AD; credits for buffer chemicals/steam	GWP ~14 × lower than SMR	Boshagh et al. (2025)

DF–MEC, *dark fermentation and microbial electrolysis cell;* SMR, *steam methane reforming.*

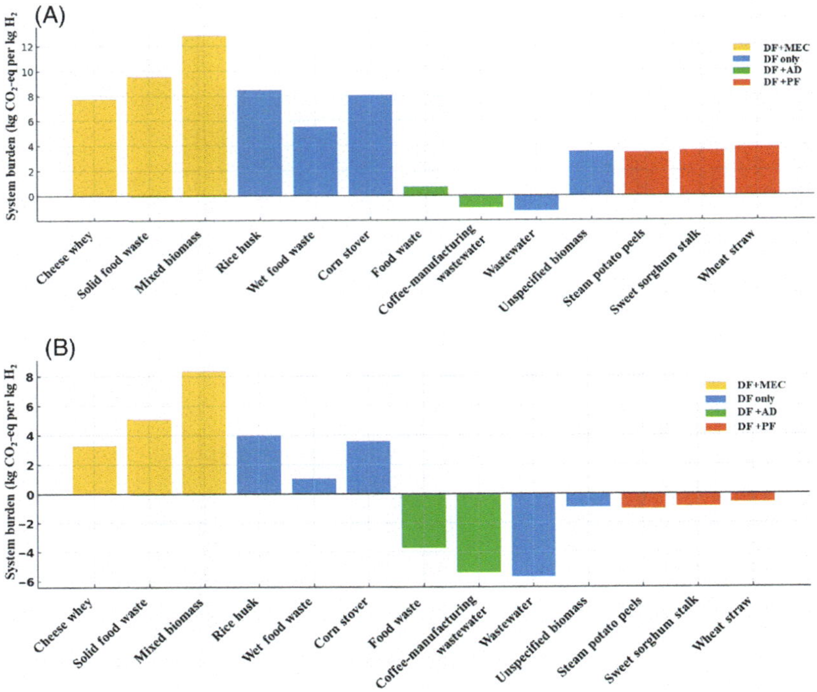

FIGURE 4.1 Net emission from DF biohydrogen production compared to (A) 1 kg H_2 production from a petroleum refinery and (B) via steam methane reforming. *DF*, Dark fermentation.

digestion of residues, as shown in Fig. 4.1A. In contrast, using steam-methane reforming as the reference (6.2 kg CO_2-eq/kg H_2) flips the picture: Every DF–AD configuration and even DF-only on wastewater yields net avoided emissions, with DF + MEC and DF–PF approaches on higher impact feedstocks also delivering carbon savings (Fig. 4.1B). Thus the climate benefits of dark fermentation depend strongly on both the chosen fossil-H_2 benchmark, the choice of feedstock, the scope of the lifecycle assessment (cradle to cradle, cradle to gate, gate to gate), and the degree of process integration.

4.6.3 Integration and covalorization of effluents

When renewable energy sources are combined with effluent valorization, the potential for net-zero emissions is increased. In most cases, combining dark fermentation with other renewable energy systems reduces greenhouse gas emissions when compared to traditional biohydrogen production methods (Boshagh et al., 2025). In other cases, however, the integrated approach may have higher emissions depending on system configuration, energy inputs, or by-product management, demonstrating that system-specific factors are crucial in determining overall environmental performance (Goren et al., 2023). Several studies had demonstrated these findings, with Reaño (2020) showing that coupling dark fermentation of rice husk with cogeneration eliminates nearly all process emissions (10.92 vs. 20.10 kg CO_2-eq/kg H_2 for gasification scenarios), whereas Barghash et al. (2022) found that adding solar electricity to a biohydrogen plant shifts total GHG emissions from 31,300 kg CO_2-eq down to −11,200 kg CO_2-eq per batch. When combining dark fermentation and anaerobic digestion, Lee et al. (2024) measured gross emissions of 2.48 kg CO_2-eq/kg H_2, but the net avoided emissions were −237.7 kg CO_2-eq/t of food waste treated. This result was caused by the fact that 88.5% of GHGs were linked to electricity for purification and mixing, which can be easily reduced by using renewable resources (Lee et al., 2024). According to Ramírez-Díaz et al. (2023), per-kilogram H_2 emissions from sugarcane mill-based dark fermentation can be reduced from over 300 kg CO_2-eq to less than 20 kg CO_2-eq by combining steam surplus recovery and energy integration.

Despite these promising results, not every biorefinery design performs better than fossil fuel approaches. Because of high energy requirements and limited credits for biogenic carbon, Gamero et al. (2024) found a net GWP of 24.4 kg CO_2-eq/kg H_2, which is approximately double that of gray hydrogen from SMR. This was based on a full biorefinery that produced biohydrogen along with several coproducts. These results highlight how crucial it is to do thorough, site-specific techno-economic and Life Cycle Assessment (LCA) studies that take into consideration local energy mixes, feedstock logistics, and process efficiency.

Generally, dark fermentation, whether done alone or in conjunction with renewable systems, can produce net-zero or even net-negative biohydrogen. Substrate conversion and energy efficiency could be maximized by ongoing developments in microbial strain engineering, reactor intensification, and process integration, such as DF–PF, DF–MEC, and hybrid DF–AD designs. To close the gap to commercially viable, climate-neutral hydrogen, regulatory incentives such as hydrogen production tax credits and the strategic use of solar or wind power for auxiliary operations would be essential.

4.7 Projects of dark fermentation

4.7.1 Rotational drum reactor with milli-magnetite for food waste

This research was driven by the dual challenges of managing the ever-increasing volume of global food waste and the urgent need for clean energy sources (Liu et al., 2025). The study focused on dark fermentation,

TABLE 4.3 The performance of the experimental system (RD-M) against the blank control (RD-C) across several key metrics.

Performance metric	RD-C (blank control)	RD-M (with milli-magnetite)
Hydrogen production	Baseline	**+97.04% increase**
VS degradation rate	Baseline	**+23.6% improvement**
Substrate particle size (μm)	1500–4500 (initial)	**100–500 (final)**
Acetate concentration	Not specified	**4.8 g/L**
Butyrate concentration	Not specified	**3.2 g/L**
Clostridium butyricum	Baseline	**41.7% relative abundance (2.3× control)**

VS, *volatile solids.*

a biological process that can convert organic waste into biohydrogen. However, this process is notoriously hampered by significant limitations, primarily slow hydrolysis rates and low overall hydrogen production efficiency. A major issue is the complex microbial community within the reactor, which often shifts toward nonhydrogen-producing metabolic pathways, such as the lactic acid pathway, or even hydrogen-consuming pathways like propionic acid production. Although previous studies have shown that additives like nanosized magnetite can enhance hydrogen yield by facilitating electron transfer, these materials are costly, difficult to recover, and potentially toxic, making them impractical for large-scale application.

To overcome these challenges the research team developed and tested an innovative system: a rotational drum reactor (RDR) integrated with milli-sized magnetite particles (the RD-M system). This design was conceived to synergize mechanical grinding with biochemical stimulation. The substrate used for all experiments was potato peel, a common and representative type of food waste. The experimental design was rigorous, involving three distinct setups for a clear comparison: the main RD-M group (with milli-magnetite), a control group where sand was used instead of magnetite (RD-S), and a blank control group with just the basic RDR (RD-C). The researchers employed advanced analytical techniques, including laser particle size analysis to monitor the physical breakdown of the waste and metagenomic sequencing to delve deep into the microbial community structure and the expression of key functional genes (Table 4.3).

The performance results of the integrated RD-M system were markedly superior. The system achieved a staggering 97.04% increase in hydrogen production compared to the blank control (RD-C). Furthermore the VS degradation rate, a key indicator of waste treatment efficiency, was enhanced by 23.6%. The mechanical action of the rotating drum and the milli-magnetite was highly effective, reducing the substrate particle size from an initial range of 1500–4500 μm down to a final, much finer range of 100–500 μm. This drastic increase in surface area facilitated better microbial access. Metabolically the system favored the high-hydrogen-yield pathways, leading to the accumulation of 4.8 g/L of acetate and 3.2 g/L of butyrate.

The profound improvement was fundamentally linked to a dramatic shift in the microbial community, as revealed by 16S rRNA sequencing. The relative abundance of the key hydrogen-producing bacterium *C. butyricum* soared to 41.7% in the RD-M system, a value that was 2.3 times higher than what was observed in the control group. Metagenomic analysis provided the mechanistic insight: The milli-magnetite acted as an excellent electron conductor, optimizing the entire electron flow within the system. It achieved this by upregulating genes for flagellin proteins to promote direct IET (DIET), activating cytochrome c oxidase for better transmembrane electron transfer, and channeling electrons via pyruvate ferredoxin toward the productive butyrate pathway. This "mechanical-biochemical synergistic enhancement" strategy effectively steered the metabolic processes away from wasteful routes and toward efficient hydrogen generation.

This research successfully demonstrates that the integration of an RDR with milli-magnetite is a highly effective and innovative strategy for significantly boosting biohydrogen production from food waste. By combining physical pretreatment with biochemical enhancement the system overcame key bottlenecks of dark fermentation, resulting in a near-doubling of hydrogen yield, improved waste degradation, and a favorable shift in the microbial consortium. This technology presents a promising and sustainable solution, transforming an energy-consuming waste treatment process into an energy-producing one. It provides a critical theoretical foundation for the development of cost-effective and easily scalable biohydrogen production systems, marking a significant advancement toward a circular bioeconomy.

4.7.2 Integrated bioelectrochemical system (IBES) for wastewater

This research paper presents a comprehensive study on an innovative integrated bioelectrochemical system (IBES) designed to synergistically enhance hydrogen gas production and wastewater treatment (Vinodh Kumar et al., 2025). The system fundamentally addresses two major limitations in the field of bioenergy: the thermodynamic barrier in dark fermentation, known as the "Thauer limit," which caps hydrogen yield at around 4 mol H_2/mol glucose, and the high energy cost associated with conventional MECs that require an external power supply. The proposed solution is an elegant integration of a biological process (dark fermentation) with a spontaneously electrochemical one (a galvanic cell), creating a self-sustaining cycle for energy and resource recovery from organic wastewater.

The core of the methodology involved the design and operation of a multichamber reactor system. The feedstock used was cheese whey, a high-strength organic wastewater with a substantial initial COD of 10,000 mg/L, representing a real-world industrial effluent. The system was structured around an H-shaped reactor divided into three chambers by a Nafion 117 proton exchange membrane. This configuration allowed for the separation of processes: the dark fermentation reactor broke down the complex organics, and the subsequent galvanic cell, equipped with graphite electrodes and a 100 Ω external resistor, utilized the resulting metabolites. A critical aspect of the experimental design was the direct comparison between the performance of a standalone dark fermentation process and the fully integrated IBES, which allowed for a clear quantification of the integration's benefits.

The results demonstrated a remarkable enhancement in hydrogen production when using the integrated system. The standalone dark fermentation process yielded approximately 25 normal liters of hydrogen per kilogram of total organic carbon (NL H_2/kg TOC). In stark contrast the IBES achieved a yield between 75.5 and 78.8 NL H_2/kg TOC. This represents an approximate threefold increase, decisively overcoming the theoretical Thauer limit. This boost is attributed to the galvanic cell's ability to further oxidize the VFAs—such as acetic and butyric acid—produced during fermentation, extracting additional hydrogen that would otherwise remain locked in these by-products. Concurrently the system exhibited exceptional efficacy in wastewater treatment. The integrated process achieved a COD removal efficiency of 78%, significantly reducing the organic pollutant load of the wastewater. Furthermore, it demonstrated a remarkably high removal rate for ammonia (NH_4^+), reaching 94%. This dual action of organic and nutrient removal highlights the system's potential not just for energy production but also for advanced wastewater purification, making it a compelling solution for industries facing strict discharge regulations. From an energy perspective the system proved to be highly efficient. The galvanic cell component generated a peak power density of 62.7 mW/m^3, producing electricity directly from the waste-derived chemicals. More importantly the entire IBES operated with a minimal energy input of only 6.6 kJ/kg TOC. This is a pivotal finding, as it is estimated to be roughly 90% lower than the energy required to operate traditional MECs that rely on an external power source. This energy frugality is the key to the system's economic viability and scalability.

In conclusion, this study successfully demonstrates a proof-of-concept for a synergistic bioelectrochemical platform. The integration of dark fermentation with a galvanic cell creates a virtuous cycle where the waste products of one process become the fuel for the next. This leads to a triple output: a substantial increase in hydrogen yield, effective wastewater treatment with high nutrient removal, and the generation of electrical energy, all achieved with minimal operational energy input. The research positions this IBES as a promising and sustainable technology for the valorization of high-strength organic wastewaters, offering a simultaneous solution for renewable energy generation and environmental remediation.

4.7.3 Carbon felt enhancement for whey wastewater treatment

This research addresses the significant environmental challenge posed by whey wastewater, a by-product of the global dairy industry that generates approximately 180 million t annually, nearly half of which is discharged untreated (). This wastewater is characterized by a high organic load, with COD concentrations reaching up to 20,000 ± 50 mg/L, making it a serious pollutant. The study investigates dark fermentation as a solution, a process that can convert this organic waste into clean biohydrogen. However, the widespread adoption of this technology is limited by inherently low hydrogen yields and incomplete breakdown of the substrate. To overcome these bottlenecks a research team from the Indian Institute of Technology pioneered an innovative strategy of integrating a highly conductive carbonaceous material, specifically carbon felt (CF), directly into the dark fermentation process to enhance the microbial electron transfer mechanisms that are fundamental to hydrogen production.

The experimental methodology was designed to comprehensively evaluate the impact of CF augmentation. The substrate was real whey wastewater collected from a local dairy plant. A critical component of the approach was the use of response surface methodology (RSM) with a central composite design (CCD) to rigorously model and optimize key operational parameters, including temperature, pH, and the initial COD concentration of the wastewater. The researchers conducted a comparative performance analysis of three different sizes of CF—small (CF-S), medium (CF-M), and large (CF-L)—against a control system that contained no additive. To uncover the underlying biochemical and microbiological mechanisms for any observed enhancements the team employed a suite of advanced analytical techniques, including metabolic profiling of VFAs, quantification of electron transfer system activity (ETSA), measurement of extracellular polymeric substances (EPS), and scanning electron microscopy (SEM) to visualize microbial colonization on the CF.

The results demonstrated that the addition of CF, particularly the medium-sized variant (CF-M), led to a dramatic improvement in system performance. The RSM model, which was statistically highly significant

($P < .0001$), identified the optimal operational conditions as a pH of 6.5, a temperature of 35°C, and an initial COD load of 10 g/L. Under these conditions the CF-M-augmented reactor achieved a peak hydrogen production of 1542.8 ± 77 mL/L, which was a remarkable threefold increase compared to the control reactor without CF. Simultaneously the wastewater treatment efficiency was significantly enhanced, with the CF-M system achieving a COD removal rate of 65%, outperforming the control and other CF sizes. Metabolically the process was dominated by the butyrate-type fermentation pathway, which is known for its high hydrogen yield, with butyrate constituting 72.3% of the detected metabolites.

The superior performance was fundamentally linked to the unique properties of the CF. The three-dimensional, porous structure of the material provided an ideal scaffold for microbial colonization, facilitating the formation of a dense and robust biofilm. This was evidenced by the highest recorded concentrations of EPS in the CF-M system, with proteins at 1172.6 ± 58.6 μg/mL and polysaccharides at 198.5 ± 9.9 μg/mL. Furthermore the CF-M reactor exhibited the highest ETSA at 394.08 ± 19.7 μg/mg h, indicating that the CF served as an excellent conductive conduit, drastically enhancing the rate of electron transfer within the microbial community. The study identified a critical "size effect," where the medium-sized CF, with its pore sizes ranging from 0.5 to 1.2 μm, offered the perfect balance between providing ample surface area for microbial attachment and maintaining efficient mass transfer of nutrients and metabolites, thereby overcoming the limitations posed by the smaller and larger variants.

In conclusion, this study successfully established that augmenting a dark fermentation system with medium-sized CF is a highly effective strategy for the synergistic production of biohydrogen and treatment of whey wastewater. The technology achieved a threefold increase in hydrogen yield and a substantial 65% reduction in organic pollutant load. The innovation lies in creating a synergistic environment where the physical structure of the CF supports robust microbial growth, while its superior electrical properties optimize the fundamental biochemical processes. This research, supported by the Indian Department of Science and Technology (DST), provides a viable and sustainable blueprint for the dairy industry, transforming a costly waste disposal problem into a valuable opportunity for clean energy generation and representing a significant advancement toward the practical, industrial-scale application of dark fermentation technology.

4.8 Conclusion

Dark fermentation biohydrogen production constitutes a technologically significant pathway for sustainable energy generation, aligning with circular economy principles through the valorization of organic waste streams. This biological process leverages the metabolic capabilities of anaerobic microorganisms to catalyze the decomposition of complex organic substrates into molecular hydrogen, a carbon-neutral energy vector, under light-independent and anoxic conditions. The transformation from a laboratory-scale phenomenon to a commercially viable technology necessitates a sophisticated integration of microbiological, biochemical, and engineering disciplines, addressing multifaceted challenges in substrate utilization, microbial community management, process parameter optimization, and bioreactor design.

The foundational appeal of this technology resides in its feedstock flexibility, utilizing nonspecialized and often problematic organic residues, including agricultural lignocellulosic waste, food processing effluents, and other high-organic-content streams. This substrate versatility enables a dual environmental benefit: diversion of biodegradable waste from landfill deposition, thereby mitigating fugitive methane emissions, and concomitant production of renewable hydrogen. A critical prerequisite for efficient bioconversion involves the pretreatment of recalcitrant biomasses to hydrolyze complex polymers such as cellulose and lignin into fermentable monosaccharides and oligosaccharides. The selection of pretreatment methodology—whether thermochemical, physicochemical, or enzymatic—profoundly influences substrate bioavailability and may generate inhibitory compounds that impair subsequent microbial metabolic activity. Consequently the initial process stage requires meticulous optimization to maximize saccharification yield while minimizing the formation of microbial growth inhibitors.

The core catalytic machinery of dark fermentation is embodied by specialized microbial consortia, predominantly dominated by obligate anaerobes from the Clostridiaceae family, with supplementary contributions from facultative anaerobes such as *Enterobacter*. These microorganisms facilitate substrate catabolism through glycolytic pathways, culminating in hydrogen evolution mediated by the hydrogenase enzyme complex. The inherent biochemical constraint of this process is the partial oxidation of substrates, resulting in coproduction of molecular hydrogen and a spectrum of soluble metabolic by-products, notably short-chain VFAs like acetate and butyrate. This metabolic partitioning constitutes the fundamental limitation on achievable hydrogen yield, as a substantial fraction of the substrate's chemical energy remains sequestered in organic acid metabolites. Furthermore, VFA accumulation induces systemic acidification, potentially exceeding the pH optimum for hydrogenogenic activity and promoting metabolic shifts toward solventogenesis. Process stability, therefore, mandates precise ecological steering of the microbial community to enrich hydrogenogenic phenotypes while suppressing hydrogenotrophic archaea, including methanogens and homoacetogens, which compete for available reductants.

Process optimization is achieved through strategic manipulation of key operational parameters. Temperature regime selection exerts a deterministic influence on microbial community structure, with thermophilic operation generally enhancing specific hydrogen production rates and providing effective suppression of methanogenic activity. pH maintenance within a moderately acidic range is critical for both inhibiting hydrogen-consuming metabolic pathways and preserving hydrogenase enzyme functionality. HRT requires careful calibration to prevent washout of slow-growing hydrogenogenic taxa while facilitating selective removal of competing microorganisms. Similarly, OLR must be balanced to maximize volumetric production without inducing substrate inhibition or catastrophic acidification from VFA accumulation. The strong interdependence of these parameters necessitates a systems-level optimization approach tailored to specific substrate-inoculum combinations.

Bioreactor configuration serves as the engineering framework determining process intensification potential. Although continuous stirred-tank reactors provide operational simplicity for fundamental investigations, their inability to maintain high biomass density limits volumetric productivity. This has motivated the development of high-rate anaerobic bioreactors, including anaerobic MBRs, expanded granular sludge bed reactors, and fixed-film systems, which achieve biomass retention through cell immobilization or physical separation. Such configurations decouple solid retention time from HRT, enabling higher organic loading capacities and improved process resilience. Contemporary research explores advanced reactor designs incorporating in situ product recovery mechanisms and the integration of conductive materials to potentially facilitate DIET, thereby addressing intrinsic metabolic limitations.

Notwithstanding these advancements the trajectory toward commercial implementation confronts persistent technical barriers, primarily the constrained hydrogen yield resulting from incomplete substrate conversion. However, this apparent limitation reveals a strategic opportunity when reconceptualizing process architecture. The most technologically and economically viable pathway forward involves integrating dark fermentation as the primary unit operation within a cascading biorefinery framework. The metabolite-rich effluent, containing significant concentrations of VFAs, represents a valuable intermediate feedstock for subsequent valorization processes. Potential secondary stages include photofermentative hydrogen production, microbial electrolysis, or conventional anaerobic digestion for biomethane generation, yielding an enriched biohythane mixture. This integrated paradigm dramatically enhances overall energy recovery efficiency from the original biomass, transforms process inhibitors into valuable intermediates, and significantly improves economic viability through polyproduction.

In prospective terms, dark fermentation biohydrogen production represents a technologically robust platform for concurrent waste management and renewable energy generation. Its commercial realization will be predicated on continued multidisciplinary research addressing several critical domains: the development of robust, waste-adapted microbial consortia; advanced dynamic process control strategies for handling feedstock heterogeneity; innovative bioreactor designs for enhanced mass transfer and biomass retention; and, crucially, the demonstration of techno-economic feasibility for integrated biorefinery configurations at pilot and commercial scales. By embracing this integrated systems approach, dark fermentation is positioned to make a substantive contribution to the global sustainable energy portfolio, effectively transforming lignocellulosic and organic waste streams into a clean, renewable hydrogen source.

References

Ali, H. M., Tyagi, V. K., Kazmi, A. A., & Ojha, C. S. P. (2024). Biohythane production from organic fraction of municipal solid waste: Clean and green energy generation under circular economy framework. *International Journal of Hydrogen Energy, 95*, 1080–1094. https://www.sciencedirect.com/science/journal/03603199. doi:10.1016/j.ijhydene.2024.11.251.

Alvarado-Cuevas, Z. D., López-Hidalgo, A. M., Ordoñez, L. G., Oceguera-Contreras, E., Ornelas-Salas, J. T., & De León-Rodríguez, A. (2015). Biohydrogen production using psychrophilic bacteria isolated from Antarctica. *International Journal of Hydrogen Energy, 40*(24), 7586–7592. http://www.journals.elsevier.com/international-journal-of-hydrogen-energy/. doi:10.1016/j.ijhydene.2014.10.063.

Alvarez-Guzmán, C. L., Balderas-Hernández, V. E., González-García, R., Ornelas-Salas, J. T., Vidal-Limón, A. M., Cisneros-de la Cueva, S., & De Leon-Rodriguez, A. (2017). Optimization of hydrogen production by the psychrophilic strain G088. *International Journal of Hydrogen Energy, 42*(6), 3630–3640. http://www.journals.elsevier.com/international-journal-of-hydrogen-energy/. doi:10.1016/j.ijhydene.2016.08.098.

Bao, M. D., Su, H. J., & Tan, T. W. (2013). Dark fermentative bio-hydrogen production: Effects of substrate pre-treatment and addition of metal ions or L-cysteine. *Fuel, 112*, 38–44. doi:10.1016/j.fuel.2013.04.063.

Barghash, H., AlRashdi, Z., Okedu, K. E., & Desmond, P. (2022). Life-cycle assessment study for bio-hydrogen gas production from sewage treatment plants using solar PVs. *Energies, 15*(21). http://www.mdpi.com/journal/energies/. doi:10.3390/en15218056.

Basak, B., Jeon, B. H., Kim, T. H., Lee, J. C., Chatterjee, P. K., & Lim, H. (2020). Dark fermentative hydrogen production from pretreated lignocellulosic biomass: Effects of inhibitory byproducts and recent trends in mitigation strategies. *Renewable and Sustainable Energy Reviews, 133*. https://www.journals.elsevier.com/renewable-and-sustainable-energy-reviews. doi:10.1016/j.rser.2020.110338.

Boshagh, F., van Niel, E., & Lee, C. J. (2025). Techno-economic and life-cycle analyses of dark fermentative hydrogen production integrated with anaerobic digestion from coffee-manufacturing wastewater under thermophilic and mesophilic conditions. *Bioresource Technology, 416*. https://www.sciencedirect.com/science/journal/09608524. doi:10.1016/j.biortech.2024.131737.

Bu, J., Wei, H. L., Wang, Y. T., Cheng, J. R., & Zhu, M. J. (2021). Biochar boosts dark fermentative H_2 production from sugarcane bagasse by selective enrichment/colonization of functional bacteria and enhancing extracellular electron transfer. *Water Research, 202*. www.elsevier.com/locate/watres. doi:10.1016/j.watres.2021.117440.

Cabrol, L., Marone, A., Tapia-Venegas, E., Steyer, J. P., Ruiz-Filippi, G., & Trably, E. (2017). Microbial ecology of fermentative hydrogen producing bioprocesses: Useful insights for driving the ecosystem function. *FEMS Microbiology Reviews, 41*(2), 158–181. http://femsre.oxfordjournals.org/. doi:10.1093/femsre/fuw043.

Cakr, A., Ozmihci, S., & Kargi, F. (2010). Comparison of bio-hydrogen production from hydrolyzed wheat starch by mesophilic and thermophilic dark fermentation. *International Journal of Hydrogen Energy, 35*(24), 13214–13218. doi:10.1016/j.ijhydene.2010.09.029.

Castelló, E., Nunes Ferraz-Junior, A. D., Andreani, C., del Pilar Anzola-Rojas, M., Borzacconi, L., Buitrón, G., Carrillo-Reyes, J., Gomes, S. D., Maintinguer, S. I., Moreno-Andrade, I., Palomo-Briones, R., Razo-Flores, E., Schiappacasse-Dasati, M., Tapia-Venegas, E., Valdez-Vázquez, I., Vesga-Baron, A., Zaiat, M., & Etchebehere, C. (2020). Stability problems in the hydrogen production by dark fermentation: Possible causes and solutions. *Renewable and Sustainable Energy Reviews, 119*. https://www.journals.elsevier.com/renewable-and-sustainable-energy-reviews. doi:10.1016/j.rser.2019.109602.

Cho, Y., & Lee, T. (2011). Variations of hydrogen production and microbial community with heavy metals during fermentative hydrogen production. *Journal of Industrial and Engineering Chemistry, 17*(2), 340–345. doi:10.1016/j.jiec.2011.02.036.

Dessì, P., Porca, E., Waters, N. R., Lakaniemi, A. M., Collins, G., & Lens, P. N. L. (2018). Thermophilic versus mesophilic dark fermentation in xylose-fed fluidised bed reactors: Biohydrogen production and active microbial community. *International Journal of Hydrogen Energy, 43*(11), 5473–5485. http://www.journals.elsevier.com/international-journal-of-hydrogen-energy/. doi:10.1016/j.ijhydene.2018.01.158.

Elbeshbishy, E., Dhar, B. R., Nakhla, G., & Lee, H. S. (2017). A critical review on inhibition of dark biohydrogen fermentation. *Renewable and Sustainable Energy Reviews, 79*, 656–668. https://www.journals.elsevier.com/renewable-and-sustainable-energy-reviews. doi:10.1016/j.rser.2017.05.075.

Gamero, E., Ruppert, S., Miehe, R., & Sauer, A. (2024). Process model and life cycle assessment of biorefinery concept using agricultural and industrial residues for biohydrogen production. *Germany Energies, 17*(17). http://www.mdpi.com/journal/energies/. doi:10.3390/en17174282.

Ganguly, A., Sun, P., Liu, X., Delgado, H. E., Sun, L., & Elgowainy, A. (2025). Techno-economic and life cycle analysis of bio-hydrogen production using bio-based waste streams through the integration of dark fermentation and microbial electrolysis. *Green Chemistry, 27*(21), 6213–6231. http://pubs.rsc.org/en/journals/journal/gc. doi:10.1039/d4gc05020g.

Gbiete, D., Brügging, E., Narra, S., & Kongnine, D. M. (2026). Comprehensive characterization and assessment of biohydrogen production potential of restaurant food waste for scalable dark fermentation systems. *Biomass and Bioenergy, 204*. https://www.sciencedirect.com/science/journal/09619534. doi:10.1016/j.biombioe.2025.108444.

Ghaffar, S. H., Madyan, O. A., Fan, M., & Corker, J. (2018). The influence of additives on the interfacial bonding mechanisms between natural fibre and biopolymer composites. *Macromolecular Research, 26*(10), 851–863. http://www.springerlink.com/content/121720/. doi:10.1007/s13233-018-6119-8.

Ghimire, A., Frunzo, L., Pirozzi, F., Trably, E., Escudie, R., Lens, P. N. L., & Esposito, G. (2015). A review on dark fermentative biohydrogen production from organic biomass: Process parameters and use of by-products. *Applied Energy, 144*, 73–95. http://www.elsevier.com/inca/publications/store/4/0/5/8/9/1/index.htt. doi:10.1016/j.apenergy.2015.01.045.

Glick, B.R., & Pasternak, J.J. (1995). Molecular biotechnology: Principles and applications of recombinant DNA (2nd ed.). American Society for Microbiology Press.

Goren, A. Y., Dincer, I., & Khalvati, A. (2023). Comparative environmental sustainability assessment of biohydrogen production methods. *Science of the Total Environment, 904*. www.elsevier.com/locate/scitotenv. doi:10.1016/j.scitotenv.2023.166613.

Goren, A. Y., Dincer, I., & Khalvati, A. (2025). A comparative evaluation of dark fermentative bioreactor configurations for enhanced hydrogen production. *Environmental Science and Pollution Research, 32*(5), 2182–2209. https://www.springer.com/journal/11356. doi:10.1007/s11356-024-35873-4.

Górska, S., Buda, B., Brzozowska, E., Schwarzer, M., Srutkova, D., Kozakova, H., & Gamian, A. (2016). Identification of *Lactobacillus* proteins with different recognition patterns between immune rabbit sera and nonimmune mice or human sera. *BMC Microbiology, 16*(1). http://www.biomedcentral.com/bmcmicrobiol/. doi:10.1186/s12866-016-0631-9.

Goud, R. K., & Mohan, S. V. (2012). Acidic and alkaline shock pretreatment to enrich acidogenic biohydrogen producing mixed culture: Long term synergetic evaluation of microbial inventory, dehydrogenase activity and bio-electro kinetics. *RSC Advances, 2*(15), 6336–6353. http://pubs.rsc.org/en/journals/journalissues. doi:10.1039/c2ra20526b.

Grobicki, A., & Stuckey, D. C. (1992). Hydrodynamic characteristics of the anaerobic baffled reactor. *Water Research, 26*(3), 371–378. doi:10.1016/0043-1354(92)90034-2.

Gull, U., Lens, P. N. L., & Logan, M. (2025). Review on dark fermentative biohydrogen production: Current status and future perspectives on next-generation strategies. *Energy & Fuels, 39*(45), 21697–21716. doi:10.1021/acs.energyfuels.5c03998.

Guo, S., Lu, C., Wang, K., Wang, J., Zhang, Z., Liu, H., Jing, Y., & Zhang, Q. (2022). Effect of citrate buffer on hydrogen production by photosynthetic bacteria. *Bioresource Technology, 347*, 126636. doi:10.1016/j.biortech.2021.126636.

Jain, R., Panwar, N. L., Jain, S. K., Gupta, T., Agarwal, C., & Meena, S. S. (2024). Biohydrogen production through dark fermentation: An overview. *Biomass Conversion and Biorefinery, 14*(12), 12699–12724. https://www.springer.com/journal/13399. doi:10.1007/s13399-022-03282-7.

Kamalaskar, L. B., Dhakephalkar, P. K., Meher, K. K., & Ranade, D. R. (2010). High biohydrogen yielding *Clostridium* sp. DMHC-10 isolated from sludge of distillery waste treatment plant. *International Journal of Hydrogen Energy, 35*(19), 10639–10644. doi:10.1016/j.ijhydene.2010.05.020.

Kargi, F., Eren, N. S., & Ozmihci, S. (2012). Bio-hydrogen production from cheese whey powder (CWP) solution: Comparison of thermophilic and mesophilic dark fermentations. *International Journal of Hydrogen Energy, 37*(10), 8338–8342. doi:10.1016/j.ijhydene.2012.02.162.

Keskin, T., Arslan, K., Nalakth Abubackar, H., Vural, C., Eroglu, D., Karaalp, D., Yanik, J., Ozdemir, G., & Azbar, N. (2018). Determining the effect of trace elements on biohydrogen production from fruit and vegetable wastes. *International Journal of Hydrogen Energy, 43*(23), 10666–10677. http://www.journals.elsevier.com/international-journal-of-hydrogen-energy/. doi:10.1016/j.ijhydene.2018.01.028.

Khademi, M. H., Lotfi-Varnoosfaderani, M., & Moghadasin, M. H. (2021). Catalytic conversion of bio-renewable glycerol to pure hydrogen and syngas: Energy management and mitigation of environmental pollution. *Energy Conversion and Management, 247*. https://www.journals.elsevier.com/energy-conversion-and-management. doi:10.1016/j.enconman.2021.114719.

Kim, D. H., Shin, H. S., & Kim, S. H. (2012). Enhanced H_2 fermentation of organic waste by CO_2 sparging. *International Journal of Hydrogen Energy, 37*(20), 15563–15568. doi:10.1016/j.ijhydene.2012.03.071.

Kim, G., Yang, H., Lee, J., & Cho, K. S. (2024). Comparative analysis of hydrogen production and bacterial communities in mesophilic and thermophilic consortia using multiple inoculum sources. *Chemosphere, 350*. https://www.sciencedirect.com/science/journal/00456535. doi:10.1016/j.chemosphere.2024.141144.

Kumar, N., Saini, H., Aggarwal, N. K., Jangra, N., Dhiman, K., & Sangwan, I. (2025). Biological pretreatment of lignocellulosic biomass as a sustainable option for biofuel production. *Next Sustainability, 5*, 100133. doi:10.1016/j.nxsust.2025.100133.

Lee, J. Y., Sim, Y. B., Jung, J. H., Pandey, A. K., Kyung, D., & Kim, S. H. (2024). Greenhouse gas emissions and net energy production of dark fermentation from food waste followed by anaerobic digestion. *Energy, 312*. https://www.sciencedirect.com/science/journal/03605442. doi:10.1016/j.energy.2024.133559.

Liu, I. C., Whang, L. M., Ren, W. J., & Lin, P. Y. (2011). The effect of pH on the production of biohydrogen by *Clostridia*: Thermodynamic and metabolic considerations. *International Journal of Hydrogen Energy, 36*(1), 439–449. doi:10.1016/j.ijhydene.2010.10.045.

Liu, Y., Yu, P., Song, X., & Qu, Y. (2008). Hydrogen production from cellulose by co-culture of *Clostridium thermocellum* JN4 and *Thermoanaerobacterium thermosaccharolyticum* GD17. *International Journal of Hydrogen Energy, 33*(12), 2927–2933. doi:10.1016/j.ijhydene.2008.04.004.

Liu, Z., Tang, S., Ren, Y. i, Chen, P., Ma, D., Si, B., Jiang, W., Lu, H., & Zhang, Y. (2025). Biohydrogen production from food waste using a novel rotational drum reactor integrated with milli-magnetite. *Bioresource Technology, 434*, 132822. doi:10.1016/j.biortech.2025.132822.

Lui, J., Sloan, W., Paul, M. C., Flynn, D., & You, S. (2022). Life cycle assessment of waste-to-hydrogen systems for fuel cell electric buses in Glasgow, Scotland. *Bioresource Technology, 359*. www.elsevier.com/locate/biortech. doi:10.1016/j.biortech.2022.127464.

Martínez-Fraile, C., Muñoz, R., Teresa Simorte, M., Sanz, I., & García-Depraect, O. (2024). Biohydrogen production by lactate-driven dark fermentation of real organic wastes derived from solid waste treatment plants. *Bioresource Technology, 403*. https://www.sciencedirect.com/science/journal/09608524. doi:10.1016/j.biortech.2024.130846.

Martínez-Mendoza, L. J., Lebrero, R., Muñoz, R., & García-Depraect, O. (2022). Influence of key operational parameters on biohydrogen production from fruit and vegetable waste via lactate-driven dark fermentation. *Bioresource Technology, 364*. www.elsevier.com/locate/biortech. doi:10.1016/j.biortech.2022.128070.

Meyer, S. T., Ptacnik, R., Hillebrand, H., Bessler, H., Buchmann, N., Ebeling, A., Eisenhauer, N., Engels, C., Fischer, M., Halle, S., Klein, A. M., Oelmann, Y., Roscher, C., Rottstock, T., Scherber, C., Scheu, S., Schmid, B., Schulze, E. D., Temperton, V. M., ... Weisser, W. W. (2018). Biodiversity-multifunctionality relationships depend on identity and number of measured functions. *Nature Ecology and Evolution, 2*(1), 44–49. www.nature.com/natecolevol/. doi:10.1038/s41559-017-0391-4.

Ndayisenga, F., Yu, Z., Zheng, J., Wang, B., Liang, H., Ali Phulpoto, I., Habiyakare, T., & Zhou, D. (2021). Microbial electrohydrogenesis cell and dark fermentation integrated system enhances biohydrogen production from lignocellulosic agricultural wastes: Substrate pretreatment towards optimization. *Renewable and Sustainable Energy Reviews, 145*, 111078. doi:10.1016/j.rser.2021.111078.

Nelly López-Hernández, B., Escamilla-Alvarado, C., Guadalupe Paredes, M., Albalate-Ramírez, A., Rodríguez Valderrama, S., Magnin, J.-P., & Rivas, P. (2024). Environmental impact assessment of biohydrogen production from orange peel waste by lab-scale dark and photofermentation processes. *Revista Internacional de Contaminación Ambiental, 40*, 27–40. doi:10.20937/RICA.55063ï.

Noguer, M. C., Escudié, R., Bernet, N., & Eric, T. (2022). Populational and metabolic shifts induced by acetate, butyrate and lactate in dark fermentation. *International Journal of Hydrogen Energy, 47*(66), 28385–28398. http://www.journals.elsevier.com/international-journal-of-hydrogen-energy/. doi:10.1016/j.ijhydene.2022.06.163.

Olaniyan, A. M., Olawale, T., Alabi, K. P., Adeleke, A. E., & Oyeniyi, S. K. (2017). Design, construction and testing of a biogas reactor for production of biogas using cassava peel and cow dung as biomass. *Arid Zone Journal of Engineering, Technology and Environment, 13*, 478–488. https://api.semanticscholar.org/CorpusID:116380342.

Openshaw, D., & Bagnato, G. (2026). Recent progress in bio-hydrogen production for sustainable energy and chemical production. *Renewable and Sustainable Energy Reviews, 226*, 116307. doi:10.1016/j.rser.2025.116307.

Oztekin, R., Kapdan, I. K., Kargi, F., & Argun, H. (2008). Optimization of media composition for hydrogen gas production from hydrolyzed wheat starch by dark fermentation. *International Journal of Hydrogen Energy, 33*(15), 4083–4090. doi:10.1016/j.ijhydene.2008.05.052.

Palanivel, P. D., Jovita S, A., Babu, K., & Naina Mohamed, S. (2025). Harnessing dark fermentation: Mechanisms, metabolic pathways, and nanoparticle innovations for biohydrogen enhancement. *International Journal of Hydrogen Energy, 109*, 383–396. https://www.sciencedirect.com/science/journal/03603199. doi:10.1016/j.ijhydene.2025.02.033.

Park, J. H., Chandrasekhar, K., Jeon, B. H., Jang, M., Liu, Y., & Kim, S. H. (2021). State-of-the-art technologies for continuous high-rate biohydrogen production. *Bioresource Technology, 320*. www.elsevier.com/locate/biortech. doi:10.1016/j.biortech.2020.124304.

Pérez-Rangel, M., Barboza-Corona, J. E., Buitrón, G., & Valdez-Vazquez, I. (2020). Essential nutrients for improving the direct processing of raw lignocellulosic substrates through the dark fermentation process. *Bioenergy Research, 13*(1), 349–357. http://www.springer.com/life+sci/plant+sciences/journal/12155. doi:10.1007/s12155-019-10083-w.

Pérez-Rangel, M., Villanueva-Galindo, E., Valdez-Vazquez, I., & Moreno-Andrade, I. (2026). Influence of type of organic solid waste and process temperature on lactate-driven hydrogen production and microbial community by self-fermentation. *Process Biochemistry, 160*, 1–10. https://www.sciencedirect.com/science/journal/13595113. doi:10.1016/j.procbio.2025.10.004.

Pistekova, H., Dusankova, M., Sopik, T., Klaban, J., Dostalkova, J., Moucka, R., & Sedlarik, V. (2025). Optimization of the dark fermentation technique for hydrogen production through supplementation with ascorbic acid and/or l-cysteine by *Clostridium butyricum* CCDBC 11. *Journal of Agricultural and Food Chemistry, 73*(22), 13654–13662. http://pubs.acs.org/journal/jafcau. doi:10.1021/acs.jafc.5c03194.

Qu, X., Zeng, H., Gao, Y., Mo, T., & Li, Y. (2022). Bio-hydrogen production by dark anaerobic fermentation of organic wastewater. *Frontiers in Chemistry, 10*. http://journal.frontiersin.org/journal/chemistry. doi:10.3389/fchem.2022.978907.

Quéméneur, M., Hamelin, J., Barakat, A., Steyer, J. P., Carrre, H., & Trably, E. (2012). Inhibition of fermentative hydrogen production by lignocellulose-derived compounds in mixed cultures. *International Journal of Hydrogen Energy, 37*(4), 3150–3159. doi:10.1016/j.ijhydene.2011.11.033.

Ramírez-Díaz, R. C., Prato-Garcia, D., & Vasquez-Medrano, R. (2023). How sustainable is the biohydrogen produced from sugarcane vinasse? An approach based on life cycle assessment. *Biomass Conversion and Biorefinery, 13*(16), 14755–14775. https://www.springer.com/journal/13399. doi:10.1007/s13399-022-03242-1.

Reaño, R. L. (2020). Assessment of environmental impact and energy performance of rice husk utilization in various biohydrogen production pathways. *Bioresource Technology, 299*. www.elsevier.com/locate/biortech. doi:10.1016/j.biortech.2019.122590.

Regueira-Marcos, L., García-Depraect, O., & Muñoz, R. (2023). Elucidating the role of pH and total solids content in the co-production of biohydrogen and carboxylic acids from food waste via lactate-driven dark fermentation. *Fuel, 338*. http://www.journals.elsevier.com/fuel/. doi:10.1016/j.fuel.2022.127238.

Rittmann, S., & Herwig, C. (2012). A comprehensive and quantitative review of dark fermentative biohydrogen production. *Microbial Cell Factories, 11*. doi:10.1186/1475-2859-11-115.

Rosa, L., & Mazzotti, M. (2022). Potential for hydrogen production from sustainable biomass with carbon capture and storage. *Renewable and Sustainable Energy Reviews, 157*, 112123. doi:10.1016/j.rser.2022.112123.

Roychowdhury, S., & Ghosh, S. (2024). State-of-the-art of biochar amended dark fermentative hydrogen production: A sustainable coupling of decarbonization pathways towards low carbon future. *Journal of Cleaner Production, 443*, 141208. doi:10.1016/j.jclepro.2024.141208.

Sharma, A., & Arya, S. K. (2017). Hydrogen from algal biomass: A review of production process. *Biotechnology Reports, 15*, 63–69. http://www.journals.elsevier.com/biotechnology-reports/. doi:10.1016/j.btre.2017.06.001.

Sotelo-Navarro, P. X., & Poggi-Varaldo, H. M. (2021). Hydrogen from dark fermentation of the organic fraction of waste diapers: Optimization based on response surface experiments. *Frontiers in Energy Research, 9*. http://www.frontiersin.org/Energy_Research/. doi:10.3389/fenrg.2021.630212.

Tang, J., Yuan, Y., Guo, W. Q., & Ren, N. Q. (2012). Inhibitory effects of acetate and ethanol on biohydrogen production of *Ethanoligenens harbinese* B49. *International Journal of Hydrogen Energy, 37*(1), 741–747. doi:10.1016/j.ijhydene.2011.04.067.

Tang, T., Chen, Y., Liu, M., Du, Y., & Tan, Y. (2022). Effect of pH on the performance of hydrogen production by dark fermentation coupled denitrification. *Environmental Research, 208*. http://www.elsevier.com/inca/publications/store/6/2/2/8/2/1/index.htt. doi:10.1016/j.envres.2021.112663.

Venkata Mohan, S., Lalit Babu, V., & Sarma, P. N. (2008). Effect of various pretreatment methods on anaerobic mixed microflora to enhance biohydrogen production utilizing dairy wastewater as substrate. *Bioresource Technology, 99*(1), 59–67. doi:10.1016/j.biortech.2006.12.004.

Vinodh Kumar, P., Gobikannan, K., Nandakumar, S., & Manavaalan, G. (2025). Enhancing hydrogen production and wastewater treatment through an integrated bio-electrochemical system combining dark fermentation and galvanic cells. *Biomass and Bioenergy, 201*, 108127. doi:10.1016/j.biombioe.2025.108127.

Wang, B., Wan, W., & Wang, J. (2008). Inhibitory effect of ethanol, acetic acid, propionic acid and butyric acid on fermentative hydrogen production. *International Journal of Hydrogen Energy, 33*(23), 7013–7019. doi:10.1016/j.ijhydene.2008.09.027.

Wang, J., & Wan, W. (2008). Effect of temperature on fermentative hydrogen production by mixed cultures. *International Journal of Hydrogen Energy, 33*(20), 5392–5397. doi:10.1016/j.ijhydene.2008.07.010.

Wu, X., Yao, W., & Zhu, J. (2010). Effect of pH on continuous biohydrogen production from liquid swine manure with glucose supplement using an anaerobic sequencing batch reactor. *International Journal of Hydrogen Energy, 35*(13), 6592–6599. doi:10.1016/j.ijhydene.2010.03.097.

Yang, J., Sim, Y. B., Kim, S. M., Joo, H. H., Mašek, O., & Kim, S. H. (2024). Synergetic effects of conductive materials and bacterial population in inoculum on mixed-culture biohydrogen production. *International Journal of Hydrogen Energy, 53*, 1293–1302. https://www.sciencedirect.com/science/journal/03603199. doi:10.1016/j.ijhydene.2023.12.103.

Yousuf, A., Karim, A., Islam, M. A., Karim, S. U., Khan, M. M. R., & Faizal, C. K. M. (2020). Dry fermenters for biogas production. *Bioreactors: Sustainable design and industrial applications in mitigation of GHG emissions* (pp. 75–87). Elsevier. https://www.elsevier.com/books/bioreactors/singh/978-0-12-821264-6. doi:10.1016/B978-0-12-821264-6.00006-1.

Yuan, Z., Yang, H., Zhi, X., & Shen, J. (2008). Enhancement effect of l-cysteine on dark fermentative hydrogen production. *International Journal of Hydrogen Energy, 33*(22), 6535–6540. doi:10.1016/j.ijhydene.2008.07.065.

Yunusa, U. M., & Karapinar, I. (2025). Comparative optimization of apple pomace pretreatment conditions to maximize sugar generation for dark fermentative biohydrogen production. *Bioresource Technology Reports, 31*. https://www.sciencedirect.com/science/journal/2589014X. doi:10.1016/j.biteb.2025.102191.

Changrong Tang. Semantic Scholar. (2013). *Plug flow type anaerobic biogas reactor*. https://api.semanticscholar.org/CorpusID:135464150.

CHAPTER

5

Photo-fermentation biohydrogen production from organic matters

Yanyan Jing[a], *Wen Cao*[b,*], *Jiyan Lu*[b], *Chaoyang Lu*[a], *Eric Lichtfouse*[b], *and Wenling Yang*[c]

[a]Henan Agricultural University, Zhengzhou, China, [b]Xi'an Jiaotong University, Xi'an, China, [c]Henan Academy of Sciences, Institute of Biology Co. Ltd., Zhengzhou, China

*Corresponding author. E-mail address: caowenxjtu@mail.xjtu.edu.cn

5.1 Introduction

The combustion of fossil fuels releases numerous pollutants, including CO_2, CH_4, and CO, contributing to the greenhouse effect and exerting significant adverse impacts on ecosystems. Furthermore, as fossil fuels constitute a nonrenewable energy source, their excessive utilization inevitably leads to energy scarcity. Against the backdrop of global carbon neutrality, countries worldwide are actively developing renewable energy alternatives, transitioning toward cleaner and more sustainable energy systems.

Among these alternatives, hydrogen energy emerges as a highly promising clean and renewable energy source, offering several distinct advantages over conventional options: (1) Hydrogen possesses an exceptionally high calorific value, releasing three times more energy than gasoline and over five times more than coal per unit mass; (2) hydrogen combustion generates only oxygen and water, producing zero carbon dioxide, nitrogen

Biohydrogen: Advances in Theory and Production
DOI: https://doi.org/10.1016/B978-0-443-27686-6.00002-5

oxides, sulfur compounds, or other harmful emissions; (3) hydrogen exhibits rapid and efficient combustion in the presence of oxygen or air; and (4) hydrogen can be utilized either through direct combustion to generate mechanical energy or as a feedstock for fuel cells, enabling broader and more efficient applications (Van Ginkel & Logan, 2005; Zhao, 2021).

Moreover, hydrogen production can be achieved through various methods, with water electrolysis and petroleum catalytic cracking being the most prevalent. Additional approaches include coal gasification, natural gas reforming, methanol processing, and industrial by-product utilization (Li et al., 2021a). However, conventional methods are energy-intensive and often generate greenhouse gases and environmental pollutants as byproducts. Consequently, researchers have increasingly focused on biohydrogen production technologies, which employ microbial metabolism to convert organic substrates into hydrogen. A notable advantage of this approach is the potential utilization of fermentation residues as fertilizers, adding value to the process. Biohydrogen production leverages widely available organic waste materials, including agricultural residues, forestry byproducts, and organic wastewater, making it both sustainable and economically viable.

Biohydrogen production technologies are broadly categorized into photo-fermentation, dark fermentation, and integrated dark-photo fermentation systems. Photo-fermentation biohydrogen production has attracted significant research interest because of its unique advantages: it utilizes light as an energy source, achieves high theoretical substrate conversion efficiency, minimizes byproduct formation, and offers an environmentally friendly, cost-effective, and efficient hydrogen production pathway (Masilela & Pradhan, 2022; Yue et al., 2021).

The photo-fermentation process is fundamentally a biochemical reaction influenced by multiple factors. Key among these are the photosynthetic activity of bacteria, the synthesis and functionality of hydrogen-producing enzymes, and the availability of light—a critical energy source that drives the conversion of chemical energy (from substrates) into hydrogen. In addition, different photosynthetic bacterial strains exhibit selective substrate utilization patterns, meaning that substrate composition and properties directly impact hydrogen production efficiency. Other operational parameters, such as temperature, pH, the use of additives, and photo-fermentation hydrogen production reactors, also play significant roles in optimizing the hydrogen yield of photosynthetic bacteria.

5.2 Substrate

5.2.1 Substrate type

The feedstocks for photo-fermentation biohydrogen production are diverse, primarily including crop straw, energy grasses, organic wastewater,

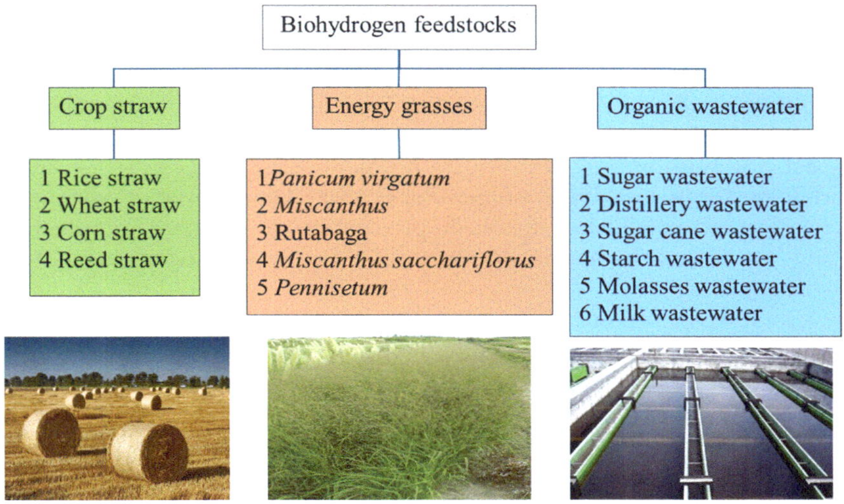

FIGURE 5.1 Sources of feedstock for biological hydrogen production.

and other organic wastes, as illustrated in Fig. 5.1. These substances are rich in essential nutrients that support the metabolic activities of hydrogen-producing microorganisms. During fermentation, these microbes degrade the feedstock and convert it into hydrogen. Consequently the type and properties of the feedstock play a crucial role in determining the efficiency and yield of biohydrogen production via biological fermentation.

5.2.1.1 Crop straw

Crop straw, including rice, wheat, corn, and reed straws, represents a major category of lignocellulosic biomass derived from agricultural by-products. More than half of the photosynthetic output from crops is stored in straw, which is rich in nitrogen, phosphorus, potassium, calcium, magnesium, and organic matter, making it a valuable renewable resource. Because of its high cellulose and hemicellulose content, straw can be enzymatically hydrolyzed into reducing sugars, which are subsequently metabolized by photosynthetic bacteria for hydrogen production. From the economic, energy, environmental, and social perspectives, straw fermentation for biohydrogen production is a highly promising approach (Ren & Ma, 2012). The main straw-based feedstocks used for biohydrogen production are

1. Rice straw: Rice straw, the primary by-product of rice harvesting, is one of the most abundant straw-based raw materials. It consists mainly of cellulose, hemicellulose, and lignin, along with other organic compounds. Because of its relatively low lignin content, rice straw exhibits

high biodegradability, making it an excellent substrate for microbial hydrogen production under optimized conditions (Wang & Yin, 2018).

2. Wheat straw: Wheat straw, a major agricultural residue from wheat cultivation, is a common feedstock for biohydrogen production. It shares similar chemical composition with rice straw. Its abundance, low cost, and high hydrogen production potential make it a suitable candidate for large-scale biohydrogen application. However, its widespread use in feed processing, fertilizer production, and paper manufacturing has limited its application in the field of biohydrogen production (Xiong, 2015).
3. Corn straw: Corn straw, a by-product of maize cultivation, is characterized by its high biomass yield and nutrient value. Like other straws, corn straw is rich in organic substances such as cellulose, hemicellulose, and lignin, which have good potential application value in the process of biological hydrogen production. Owing to its high bioconversion efficiency and hydrogen production potential, corn straw is also one of the most widely utilized feedstocks for biohydrogen production (Sha, 2020).
4. Reed straw: Reed, a perennial aquatic or wet tall grass, thrives in ponds and marshes, riverbanks, and shallow areas along streams worldwide (Cao, 2019). Known for its high adaptability, resilience, and biomass productivity, reed straw contains substantial cellulose content, traditionally used in papermaking and synthetic fiber industries. Given its structural composition, reed straw also holds significant potential as a feedstock for biohydrogen production.

Research has demonstrated that cosubstrate strategies can enhance the efficiency of photo-fermentative hydrogen production. For instance, Zhu (2023) reported that using a mixture of corn straw and soybean residue as a cosubstrate optimized the carbon-to-nitrogen (C/N) ratio, leading to increased hydrogen yield and production rates. Similarly, Xiang et al. investigated the influence of headspace pressure on photo-fermentative hydrogen production using corn straw as the feedstock. Their findings confirmed the excellent suitability of corn straw for hydrogen production, with a maximum yield of 546.57 mL achieved at an optimal headspace pressure of 30 kPa (Xiang et al., 2023).

Beyond corn straw, other agricultural residues such as corncob, soybean stems and leaves, and sorghum stalks have also been explored as potential feedstocks for biohydrogen production. However, different straw-based materials exhibit distinct physicochemical properties, necessitating careful selection and optimization based on their specific characteristics and process conditions. Furthermore, the sustainability and environmental impact of feedstock utilization must be evaluated to ensure the long-term feasibility and scalability of biohydrogen technology. Addressing these factors will be critical for advancing the commercial adoption of straw-based biohydrogen production.

5.2.1.2 Energy grasses

Energy grasses, predominantly tall herbaceous plants, have emerged as preferred energy plants in developed regions such as Europe and the United States. Energy grasses exhibit advantageous biomass composition and remarkable environmental adaptability. Characterized by drought, salinity, and barren soil tolerance, energy grasses demonstrate low requirements for both soil quality and climate conditions. Their robust adaptability enables cultivation in diverse environments, including mountainous regions, semiarid areas, poor-quality soils, and coastal zones. Additional advantages include rapid growth rates, high biomass yields, and extended production cycles, making them cost-effective raw materials for energy production. Commonly utilized energy grasses include mainly *Panicum virgatum*, *Miscanthus* spp., Rutabaga, *Miscanthus sacchariflorus*, and *Pennisetum* spp. (Tan et al., 2014).

These perennial grasses and semishrubs are particularly valuable because of their high cellulose, hemicellulose, and lignin content, coupled with high calorific values and low ash content. Their cultivation and management are relatively straightforward, further enhancing their appeal as energy crops. In the United States the Herbaceous Energy Crops Research Program evaluated 35 herbaceous species, ultimately identifying 18 as most suitable for domestic cultivation. Similarly, European researchers have investigated over 20 energy grass varieties, with *P. virgatum*, *Miscanthus*, and Rutabaga emerging as particularly promising candidates. Zhang et al. (2020a) compared 10 energy grass species for photo-fermentation biohydrogen production, finding that Rutabaga achieved the highest hydrogen production rate of 5.53 mL/(h·g TS (total solids)) after 36 h of enzymatic hydrolysis. Shui et al. (2022) reported that increasing light intensity from 3000 to 6000 lux enhanced both cumulative hydrogen production and maximum production rate by 19.5% and 26.2%, respectively, when using Rutabaga as a substrate. Jiang et al. (2021) investigated Rutabaga-glycerol cosubstrate, showing that the addition 15 g/L glycerol boosted hydrogen production by 294%.

Elephant grass, an energy plant of *Pennisetum purpureum*, discovered by Japanese scientists, stands out as an exceptionally promising energy grass because of its dense foliage, tall culms and stems, vigorous regrowth capacity, high biomass yields, and tolerance to infertile soil. The European Union has allocated approximately 10% of agricultural land to elephant grass cultivation, primarily for power plant feedstock (Song, 2006). Although China's energy grass research initiative began relatively late, significant efforts have been made by institutions, including the Chinese Academy of Sciences, the Chinese Academy of Agricultural Sciences, and the China

Agricultural University. These organizations have focused on energy grass resource collection, germplasm evaluation, and the innovative utilization of genetic resources.

5.2.1.3 Organic wastewater

Recent years have witnessed significant advancements in biological hydrogen production utilizing organic wastewater as feedstock. Various industrial wastewater contains substantial amounts of organic substances, including carbohydrates, proteins, lipids, and lignin, which exist in either suspended or dissolved forms and can be effectively degraded through microbial biochemical processes. The primary industrial sources of organic wastewater suitable for biohydrogen production include sugar manufacturing wastewater, distillery effluent, sugarcane processing wastewater, starch production wastewater, molasses wastewater, and dairy processing wastewater. These waste streams represent valuable resources for sustainable energy production through biological conversion processes.

Biohydrogen production from treated high-concentration organic wastewater represents an innovative approach that simultaneously achieves waste valorization and clean energy generation. This technology enables the conversion of organic pollutants into hydrogen fuel through biological processes, thereby establishing a sustainable pathway for wastewater treatment while producing renewable energy. The resulting biohydrogen can effectively substitute conventional fossil fuels such as coal and petroleum, leading to significant reductions in greenhouse gas emissions and other environmental pollutants. This dual-benefit system not only addresses waste management challenges but also contributes to environmental protection and sustainable energy development (Luo, 2007). Taguchi et al. (1995, 1996) compared the hydrogen production capacity of fibrous disaccharide, glucose, fructose, sucrose, xylose, and other organic wastewater and found that sucrose organic wastewater as a substrate demonstrated the highest hydrogen production efficiency, and starch organic wastewater as a substrate showed the lowest hydrogen-production efficiency. Joshi and Vaidya (2024) simulated the aqueous-phase reforming of low-concentration domestic wastewater by using starch, glucose, carnitine peptone, and sunflower seed oil as representatives of carbohydrates, proteins, and lipids in domestic wastewater, by which 2.13 mmol H_2/g COD (chemical oxygen demand) was obtained and a hydrogen conversion frequency of 3.96 h^{-1} was achieved. Moreover, Wang et al. (2022a) reported the reaction mechanism and workflow for hydrogen production from distillery wastewater and developed an integrated treatment process combining aqueous phase reforming and activated carbon adsorption.

5.2.2 Substrate pretreatment

Lignocellulosic biomass feedstocks, including crop straw and energy grasses, are primarily composed of three polymers, cellulose, hemicellulose, and lignin, which account for 80%–95% of the total biomass mass. Cellulose is synthesized with β-D-glucose as a structural unit linked by 1,4-glycosidic bonds, forming highly ordered crystalline regions through extensive hydrogen bonding networks (Deng, 2011). Hemicellulose is a branched heteropolymer containing both pentose and hexose sugars, characterized by shorter polymer chain lengths and greater structural heterogeneity. Lignin possesses a unique three-dimensional aromatic network structure composed of amorphous phenolic compounds that confer remarkable resistance to biological degradation. In addition to the three polymers, lignocellulosic biomass also contains minor amounts of solvent-extractable organic compounds and inorganic constituents that contribute to its overall structural complexity and recalcitrance (Sathitsuksanoh et al., 2011; Weerachanchai et al., 2014). Hydrogen bonding of lignocellulosic biomass arranges cellulose molecules together in aggregates that form difficult-to-break cellulose crystals, leading to more difficult degradation of cellulose. Whereas hemicellulose consists of sugar chains with more side chains but shorter ones to form compounds with a lower degree of polymerization, which are easier to degrade into monosaccharides. Lignin is a polymer compound formed by lignin molecules connected by C–C bonds, which is more difficult to use for fermentation by conversion (Karunanithy & Muthukumarappan, 2011). And the lignin entangles the cellulose and hemicellulose and hinders the hydrolysis of both cellulose and hemicellulose. The main purpose of pretreatment is to destroy the structure of lignocellulosic biomass, transform or remove the lignin in the biomass, and reduce the crystallization of cellulose so as to improve the contact area with cellulase and convert cellulose and hemicellulose into more reducing sugars (Ji et al., 2014). As shown in Fig. 5.2 the four pretreatment methods—physical, chemical, biological, and combined methods—are currently the most used and most effective pretreatment methods.

5.2.2.1 Physical pretreatment

Physical pretreatment methods involve processes without the addition of any chemical reagents or microorganisms. The more commonly used methods include mechanical pulverization, radiation pretreatment, steam explosion, and microwave treatment.

Mechanical pulverization pretreatment uses mechanical blades to pulverize crop straw biomass by crushing or grinding. Yue (2011) used a ball mill to pulverize straw to the nanometer scale and treated it with cellulase, obtaining a reducing sugar concentration of 13.26 mg/mL. Mechanical pulverization reduces the particle size of the straw raw material, releasing

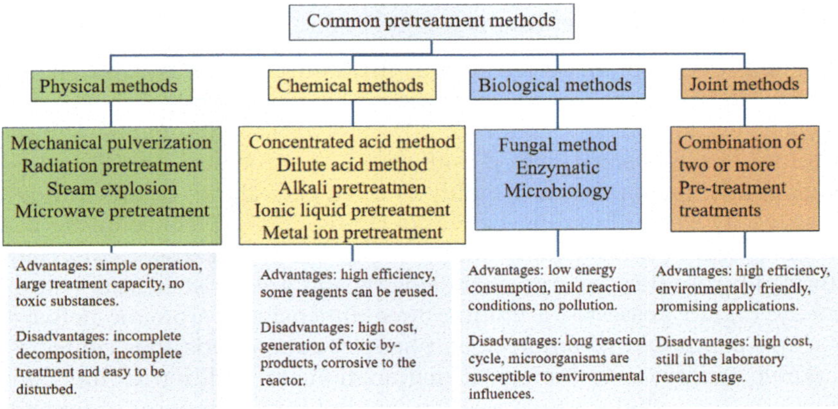

FIGURE 5.2 Common pretreatment methods and characteristics.

more water-soluble components, increasing the cellulose conversion rate. However, mechanical pulverization is energy-intensive and costly, and it cannot remove lignin and hemicellulose from the biomass. Therefore mechanical crushing is usually used before other pretreatment methods to facilitate further processing.

Radiation pretreatment utilizes γ-rays or high-speed electron beams to irradiate crop straw raw materials at specific doses and times, disrupt cellulose molecular chains, improve accessibility, and enhance hydrolysis efficiency. Yang et al. (2008) mechanically crushed wheat straw to 140 mesh and then used γ-rays for radiation treatment at a dose of 500 kGy, obviously damaging the straw's structure. However, radiation pretreatment is not favorable for practical production application because of high costs and energy consumption.

Steam explosion, a widely used method of biomass pretreatment, subjects biomass to saturated water vapor (160°C–260°C, 0.69–4.83 MPa), followed by a rapid drop to atmospheric pressure (Xu, 2011). In the process of steam explosion, high-temperature and high-pressure steam penetrates the raw material's internal structure, fracturing the cellulose framework and disrupting intermolecular hydrogen bonds, releasing more free hydroxyl groups. This method offers significant advantages, including energy efficiency, product purity, and enhanced enzymatic digestibility. However, the process generates fermentation/hydrolysis inhibitors that necessitate posttreatment washing. This washing step not only consumes substantial water resources but also removes soluble sugar from the raw material, consequently reducing total sugar content and affecting the subsequent fermentation. CO_2 and SO_2 steam explosion methods demonstrate similar mechanisms, with CO_2 being more cost-effective and energy-efficient and production of inhibitor-free material that affects enzyme digestion and fermentation.

Microwave pretreatment is an energy-efficient heating technology without thermal gradients. Microwave treatment destroys the hydrogen bonding force between cellulose molecules, thereby enhancing the adaptability of raw materials to enzymes and improving their catalytic efficiency. Hu et al. (2014) treated rice straw with a substrate concentration of 75 g/L for 24 min with 680 W microwave irradiation, resulting in a 30.6% increase in cellulose content, a 30.3% improvement in total sugar conversion rate, and partial lignin removal. However, effective pretreatment requires high temperatures (>160°C), with significantly diminished effects below 100°C. Overall the pretreatment of crop straw biomass using only microwave shows measurable effects, but high-temperature condition may lead to partial lose of feedstock components, and further improvement is needed.

5.2.2.2 Chemical pretreatment

Chemical pretreatment involves adding chemical reagents to the biomass to destroy its structure and make it easier to be degraded. The commonly used chemical pretreatment methods include concentrated acid pretreatment, dilute acid pretreatment, alkali pretreatment, ionic liquid pretreatment, and metal ion pretreatment.

Concentrated acid pretreatment converts biomass feedstock into reducing sugar by treating biomass with concentrated acid (sulfuric acid or hydrochloric acid) with mass fractions exceeding 30% for 2–10 h, generally at temperatures below 100°C under atmospheric pressure. Biomass treated by concentrated acid is easy to be converted into cellulose dextrin, with cellulose and hemicellulose conversion rates exceeding 90% (Sun & Cheng, 2002). However, the hydrolyzed material contains high acid concentrations, which require reasonable processing. And washing after biomass hydrolysis is also cumbersome, and the process is complex, causing environmental impacts and higher recycling costs.

Dilute acid pretreatment is a technique for hydrolyzing biomass to monosaccharides using sulfuric, hydrochloric, or phosphoric acids at concentrations below 10% under temperatures of 100°C–240°C and pressures of 10 atmospheres or more. Song et al. (2006) pretreated corn straw with 0.5% dilute hydrochloric acid at 121°C, achieving an 80.8% conversion rate for both cellulose and hemicellulose. Dilute acid pretreatment has the advantages of fast reaction speed and better effect with less pollution. However, the high temperature and pressure are required, and it produces substances that influence the subsequent fermentation, requiring special equipment.

Alkali pretreatment is a treatment method that utilizes NaOH, $Ca(OH)_2$, and ammonia to treat biomass feedstocks so as to dissolve lignin and obtain reducing sugar. Zheng (2004) investigated the effects of NaOH and $Ca(OH)_2$ pretreatment on straw biohydrogen production, and the results showed that NaOH pretreatment at the same reagent-added concentration was better. During hydrothermal, alkaline, and alkaline-thermal pretreatment on the photosynthetic hydrogen production of corn straw, 2% NaOH-treated corn straw was found to have the greatest hydrogen production (Zhang et al., 2021). Alkali pretreatment damages the force between molecular chains, with mild technological conditions and no acetyl groups generated to affect subsequent fermentation. However, the cost of reagents is relatively high and corrosive, and it dissolves a small amount of cellulose.

Ionic liquid pretreatment uses a type of cellulose solvent called "green solvent," mainly composed of organic cations and inorganic anions, to treat biomass feedstock to achieve the dissolution of high polymer. Ionic liquids are nonvolatile and stable at room temperature to 300°C. By designing and regulating anions the solvent's solubility for polymers can be adjusted (Kilpeläinen et al., 2007). Three ionic liquids, 1-butyl-3-methylimidazole tetrafluoroborate ($[Bmim]BF_4$), 1-butyl-3-methylimidazole hydrogen sulfate ($[Bmim]HSO_4$), and 1-allyl-3-methylimidazole chloride ([Amim]Cl), were used to pretreat Rutabaga (Chen, 2022) to investigate their effects on photo-fermentation hydrogen production. The highest reducing sugar yield of 7.9 g/L and hydrogen yield of 106.1 mL/g TS were obtained from the enzymatic reaction of Rutabaga pretreated with 16 g/L $[Bmim]HSO_4$, which were 68.8% and 35.3% higher than untreated Rutabaga, respectively (Chen, 2022). Li et al. (2019) used synthetic choline acetate ([ChO][OAc]) ionic liquid to dissolve corn straw and remove lignin, indicating the cellulose content of pretreated corn straw recovered up to 98.0%, and the lignin content was reduced from 13.7% to 2.0%. The ionic liquid pretreatment offers significant environmental advantages with minimal pollution, but its widespread application remains constrained by prohibitively high costs that currently limit industrial scalability.

Metal ion pretreatment is a more emerging pretreatment method in recent years, using metal chloride solution at a certain temperature to treat straw to disrupt hydrogen bonding forces between cellulose molecules to reduce crystallinity and polymerization. Zeng (2012) pretreated corn stalk using 0.6 mol/L ferric chloride solution for 15 min at 170°C with the solid–liquid ratio of 10:1 mL/g. After treatment the relative cellulose content and the reducing sugar yield increased by 72.19% and 131.6%, respectively.

1. Biological pretreatment: Biological pretreatment uses fungi, microbial consortia, and enzymes. Compared to physical and chemical methods, biological pretreatment requires lower energy inputs, eliminates the addition of any chemicals, and operates under more environmentally benign conditions, rarely producing harmful inhibitors. However, the main disadvantage of microbial pretreatment is the long pretreatment

cycle and comparatively lower efficiency relative to conventional acid and alkaline pretreatment. The most prevalent biological pretreatment method is fungal pretreatment, such as white rot fungi and brown rot fungi, which have a certain degradation effect on lignocellulose (Zhang & Jun, 2014). Vasco-Correa and Li (2015) investigated the effect of *Ceriporiopsis subvermispora* pretreatment on the solid-state anaerobic fermentation of mangroves, and the results showed that pretreatment with white rot fungi removed 25.7% of lining. *C. subvermispora* pretreated unsterilized yard trimmings, achieving 14.8%–20.2% lignin removal after 30 days of pretreatment with different inoculum/substrate ratios (Zhao et al., 2014).

2. Combined pretreatment method: Because of the low efficiency and high energy consumption of single pretreatment methods the combination of two or more pretreatment methods can effectively improve pretreatment effects. Current research has developed numerous synergistic pretreatment combinations, including a combination of alkali-dilute acid, dilute acid-steam explosion, ultrasonic-alkali, nanobubble-biological joint pretreatment, and other combined systems. Zhang et al. (2023) used ultrasonic-ionic liquid combined pretreatment on Rutabagas, achieving 77.2 mL/g TS of photo-fermentation hydrogen production and improving energy conversion efficiency by 46.4%. The combined pretreatment is the future development trend of biomass degradation and utilization. Low-cost and high-efficiency combined pretreatment will greatly promote biomass fermentation hydrogen production technology toward commercialization.

5.3 Strains

5.3.1 Photosynthetic bacteria

Photosynthetic bacteria are a class of hydrosphere microorganisms widely distributed in lakes, oceans, and soils and are one of the oldest organisms on earth. They are a kind of prokaryotic organism with primitive light energy synthesis system, belonging to bacteria, Eubacteria, and Rhodobacteria, mainly globular, rod-shaped, spiral, or ovoid with the general cell diameter size of 0.5–5 μm. Photosynthetic bacteria contain chlorophyll and carotenoids in their bodies, which can show different colors depending on the species composition and cellular density. Their unique photoheterotrophic metabolism enables simultaneous utilization of light energy and organic carbon sources while performing essential biogeochemical functions including nitrogen fixation, CO_2 assimilation, and hydrogen evolution, making them crucial contributors to global material cycling (Lin et al., 2015).

Gest et al. (1962) first demonstrated that photosynthetic bacteria have the property of converting organic matter to produce hydrogen under light anaerobic conditions and published a related research paper in science, which laid a solid foundation for the subsequent research on hydrogen production by photosynthetic bacteria. Currently, pure strains commonly used in the hydrogen production field include *Rhodospirillum rubrum*, *Rhodobacter sphaeroides*, *Rhodobacter capsulatus*, *Chromatium* spp., *Rhodovulum* spp., and *Rhodopseudomonas palustris*. Feng et al. (2018a) isolated a strain of photosynthetic bacteria, *R. capsulatus* JL1, from the lake water. Zhang et al. (2018a) obtained a highly efficient hydrogen-producing photosynthetic bacterial strain from paper mill activated sludge and identified the strain as belonging to the genus *Rhodobacter capsulatus*, named QW02. Liu (2010) found a photo-fermentation hydrogen production strain RLD-53 (belong to *Rhodopseudomonas faecalis*) from fish pond sediments, which demonstrated a substrate conversion rate of 66%–71% with acetate.

Different bacteria utilize substrates in different ranges and ways, and the various bacteria in a mixed bacterial culture may utilize various components in complex organic waste to produce hydrogen, improving overall substrate utilization efficiency. This consortium has different environmental adaptability. Facing changes in environmental factors such as temperature, pH, and light intensity, various bacteria can collaborate with each other so that the colony can survive and produce hydrogen under a wider range of environmental conditions. For example, some bacteria are more active at lower temperatures, whereas others are more active at higher temperatures, so that the bacterial consortium can maintain relatively stable hydrogen production performance over a range of temperatures. Because of the complementary physiological properties and metabolic pathways of different bacteria, bacterial consortia can establish a relatively stable ecosystem. Compared with a single bacterial strain, a bacterial consortium is less prone to dramatic fluctuations or complete cessation of the hydrogen production process when it is subjected to external disturbances. Henan Agricultural University isolated bacterial consortium for photo-fermentation biohydrogen production from swine fecal wastewater of farms, food factories, and other wastewaters, named HAU-M1, which consisted of *R. rubrum*, *R. capsulate*, *R. palustris*, *R. sphaeroides*, and *R. capsulatus* (Jing et al., 2022).

5.3.2 Mechanism of photo-fermentation hydrogen production

Photo-fermentation hydrogen production is a way for photosynthetic bacteria to regulate residual energy and maintain redox balance in their organism, which is very important for their life activities. There are three types of enzymes involved in hydrogen metabolism in photosynthetic bacteria: nitrogenase, hydrogenase, and reversible hydrogenase. Many

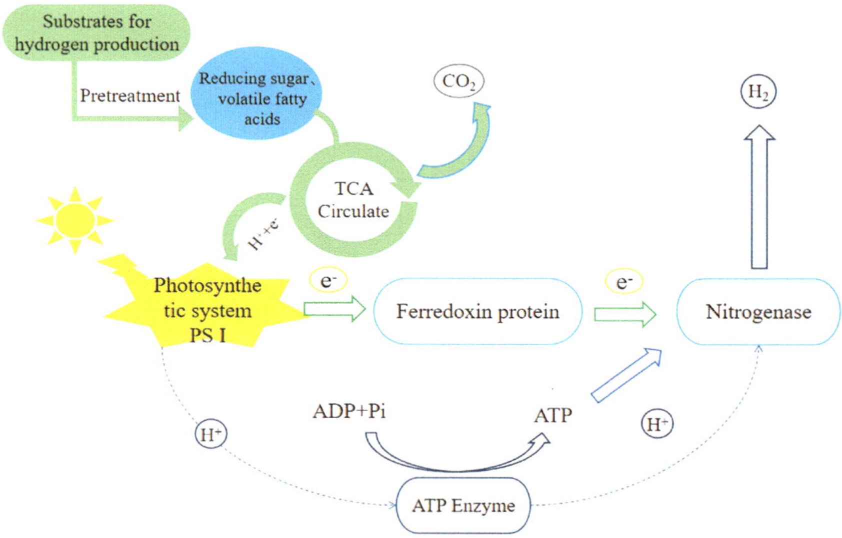

FIGURE 5.3 Mechanism of photo-fermentation hydrogen production. *ATP*, Adenosine triphosphate.

photosynthetic bacteria can also produce hydrogen by anaerobic fermentation under dark conditions, with hydrogenase playing a major role in anaerobic dark fermentation hydrogen production (Basak et al., 2014). Hydrogenases are metal-containing proteins, categorized into hydrogen-absorbing enzymes and reversible hydrogenases. Under light conditions, hydrogenase catalyzes hydrogen oxidation, protons, and electrons and releases adenosine triphosphate (ATP) through reverse electron transfer to compensate for energy demands in nitrogen fixation and cellular redox imbalances (Vignais & Billoud, 2007). Photo-fermentation biological hydrogen production is the process of reducing H^+ to H_2 under anaerobic conditions by utilizing a light source to stimulate the metabolic reaction of photosynthetic bacteria, realizing the decomposition of biomass and energy conversion. The main mechanism is shown in Fig. 5.3. Photosynthetic bacteria contain bacteriochlorophyll and carotenoids, which capture light energy through the photosynthetic system (PSI). The excited high-energy electrons are transferred to the reaction center, where nitrogenase utilizes them along with protons and ATP to catalyze hydrogen generation (Jiang et al., 2015).

5.4 Parameters

Biomass hydrogen production by photo-fermentation is a complex biochemical process, which is affected by various factors, such as pH,

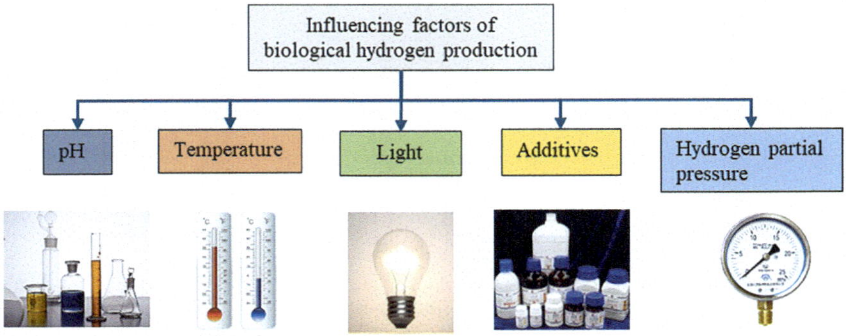

FIGURE 5.4 Parameters of photo-fermentation biological hydrogen production.

temperature, light, additives, and hydrogen partial pressure, as shown in Fig. 5.4. Optimizing these factors to achieve the best hydrogen production is the key to promoting the development of biomass fermentation for energy production.

5.4.1 pH

The pH of the fermentation broth is critical for photosynthetic bacteria to maintain intracellular NADH/ NAD + (nicotinamide adenine dinucleotide) dynamic balance, hydrogenase activity, cellular redox potential, and many other metabolic activities in the process of hydrogen production. The optimal pH for hydrogen production by photosynthetic organisms is usually between 5.0 and 7.0. pH control is essential during operation, as volatile fatty acid (VFA) accumulation can rapidly decrease broth pH to 4.0–5.0 within 24 h. Moreover, pH affects metabolic pathways of hydrogen production bacteria, with butyric acid dominating at pH 4.0, whereas lactic and propionic acids become predominant under neutral conditions (Wang & Wan, 2009). Liu et al. (2015) found that phosphate significantly contributed to hydrogen production by providing nutrient energy to the hydrogen-producing bacteria from leachate fermentation, with the highest hydrogen production at an initial phosphate pH of 5, at 1.58 mol H_2/mol glucose. Khanal et al. (2004) studied the metabolic pathways of hydrogen production in leachate fermentation using sucrose and starch as organic substrates, and it was found that higher initial pH has high productivity but shorter production duration. A rapid pH drop may alter the metabolism of hydrogen-producing bacteria, leading to a shift in the production pathway of intermediates (acetic and butyric acid) and consequently reducing overall hydrogen yield.

In the photo-fermentation hydrogen production process a buffer solution can be added to the medium to reduce the effect of the accumulated acid on photosynthetic bacteria. Zhang and Wang (2013) discovered that white mud produced by adding ammonia-soda to the medium as a buffer in anaerobic fermentation for hydrogen production maintained the medium pH stability. Beyond pH regulation, Salis and Monduzzi (2016) revealed buffers' broader biochemical impacts through modulation of protein and enzyme activities that influence cellular metabolism. Lu et al. (2020a) investigated the effect of the buffer solution Na_2HPO_4/NaH_2PO_4 on the buffering capacity of the fermentation broth and the photosynthetic hydrogen production and showed that the hydrogen production first increased and then decreased with the increase in pH values (5–7) of buffer solution Na_2HPO_4/NaH_2PO_4. Maximum energy conversion efficiency of 9.84%, hydrogen yield of 132.69 mL/g corn stalk, and hydrogen content of 53.88% were achieved at pH value of 6.

5.4.2 Temperature

Temperature is one of the main factors influencing microbial hydrogen production metabolism, and its influence is manifested in the following aspects: (1) temperature change affects the rate of enzymatic reaction, ultimately affecting cell synthesis; (2) high temperature is conducive to the transport of substances in the cell membrane; on the contrary, the fluidity is reduced, which is not conducive to the transport of substances. Therefore temperature affects the absorption of nutrients and the secretion of metabolites; and (3) different photosynthetic bacteria have different optimal hydrogen production temperatures, and changes in temperature will cause changes in substrate utilization, metabolic pathways, and other changes, which in turn cause changes in hydrogen production. Microorganisms are generally very insensitive to low temperatures, and when they reach their minimum growth temperature or below, their metabolic activity gradually decreases, and eventually they are in a state of nonpropagation, but they can still maintain life for a long time, and the rate of death is also significantly reduced. The optimal temperature for the growth and metabolism of most photosynthetic bacteria is 25°C–35°C. When the fermentation temperature is lower than 20°C, photosynthetic bacteria basically do not carry out hydrogen production metabolism. At 20°C–25°C, photosynthetic bacteria need a longer time to adapt to the environment, exhibit slow cellular growth and membrane transport efficiency, and display prolonged hydrogen production lag phases. However, the temperature exceeding 35°C often leads to protein denaturation and inactivation of hydrogen-producing enzymes.

5.4.3 Light

The hydrogen production performance of photosynthetic bacteria depends on the efficiency of carbon source transport and light energy metabolism. Light energy is mainly used in the photosynthetic phosphorylation cycle process of photosynthetic bacteria. The photosynthetic center uses the absorbed and captured light energy for the synthesis of adenosine triphosphate (ATP), the process that facilitates the transmembrane transport of carbon sources, supplying the cell with the required electrons and reducing power. The light conditions in the fermentation system directly influence the efficiency of energy conversion of photosynthetic bacteria, and enhancement light can help to improve the utilization of substrates and hydrogen production by photosynthetic bacteria (Zhang et al., 2020b). Wang et al. (2012) investigated the effects of light wavelength and intensity on the glucose transmembrane transport, cell membrane permeability, and hydrogen production. They observed that both the maximum glucose transmembrane transport rate and permeability coefficient initially increased and then decreased within the wavelength range of 470–630 nm. Syed Jaapa et al. (2009) found that photosynthetic bacteria exhibited better growth and hydrogen production under fluorescent lamps compared to tungsten lamps. When illumination intensity was below 5000 1x, the photo-fermentation hydrogen production increased with illumination intensity but decreased when it exceeded 5000 lx, which identified that photo-fermentation hydrogen production exists a light saturation point similar to that of plant photosynthesis (Yue et al., 2011).

5.4.4 Additives

Adding additives to a fermentation system is a simple and low-cost method to improve the efficiency of biological hydrogen production. The additives used to enhance the effect of fermentation hydrogen production can be mainly classified into nutrients, pH adjusters, photosensitizers, and surfactants. These additives optimize the effect of biomass fermentation hydrogen production in terms of improving the electron transfer, enzyme activity, and microbial growth.

Nutrients such as carbon sources (glucose and sucrose), nitrogen sources (ammonia and nitrates), and trace elements (iron, zinc, and copper) are essential for microbial metabolism and enzyme function. For instance the addition of ascorbic acid (150 mg/L) and zero-valent iron (400 mg/L) significantly increased hydrogen production to 664.0 ± 5.3 mL in soybean residue and corn straw fermentation, improving both yield and production rate by 10.1% and 11.5%, respectively, compared to the control (Zhu, 2023). Vitamins (B_1, B_2, and B_6) also play a critical role in electron transfer during photosynthesis (Fu et al., 2011). Buffers such as phosphate solutions help maintain optimal pH levels, ensuring stable conditions for microbial activity and hydrogen generation (Zhang & Wang, 2013). Titanate photocatalysts (TPCs) like $MgTiO_3$ have been shown to boost hydrogen production by 48.3% in corn straw photo-fermentation, reaching 344 mL (68.8 mL/g TS), while shortening the lag phase by 33.2% (Yang et al., 2023). Surfactants (Tween 20, 60, and 80) enhance cell membrane permeability, facilitating substrate uptake and metabolite release (Fan, 2023). Antifoaming agents can also mitigate foam-related inhibition, increasing hydrogen yield by 23% (Wang et al., 2022b).

5.4.5 Hydrogen partial pressure

Biological hydrogen metabolism is highly sensitive to hydrogen concentration, and the hydrogen produced during fermentation continuously accumulates in the reactor headspace, leading to an increase in hydrogen partial pressure. This, in turn, increases the dissolution of hydrogen and inhibits the precipitation of hydrogen from the fermentation broth. Xiang et al. (2023) investigated the rule of the pressure regulation of the bioreactor headspace on photo-fermentation hydrogen production by maize stover and found that cumulative biohydrogen production increased and then decreased as the reactor headspace pressure decreased (from 100 to 10 kPa), with a maximum hydrogen yield of 546.57 mL achieved at a headspace pressure of 30 kPa. This study suggests that lower hydrogen partial pressures facilitate faster reaction processes and reduce the delay time in hydrogen production. To mitigate the inhibitory effects of hydrogen partial pressure, inert gas purging is usually employed.

5.5 Reactors

In photo-fermentation hydrogen production the reactor's structural design and operational strategies jointly determine the system's overall efficiency and stability. An effective reactor structure should maximize light exposure within a compact space, ensure even light distribution, and enhance gas–liquid mass transfer. However, many current designs face persistent challenges, including light attenuation, heat buildup, and insufficient mixing. These issues significantly limit light utilization and hydrogen yield. Operational conditions further influence reactor performance. Parameters such as operation mode, inoculum volume, hydraulic retention time (HRT), and agitation rate directly affect the metabolic activity of the microorganisms involved. Poor operational conditions can lead to photo-inhibition or hydrogen accumulation-inhibition, both of which reduce productivity and long-term operational stability. To advance the scalability and carbon reduction potential of photo-fermentation hydrogen production, it is essential to understand and optimize the interplay between reactor design and operating strategies. A system-level approach that integrates structural and process engineering holds the key to unlocking the full potential of this sustainable technology. Here, we summarize the advantages and disadvantages of different reactor structures in the process of photo-fermentation hydrogen production, as well as the impacts of operational parameters on system performance. This chapter provides theoretical foundations and practical guidance for the design and operation of photo-fermentation hydrogen production reactors.

5.5.1 Reactor structure

In the process of photo-fermentation hydrogen production the reactor structure plays a crucial role in hydrogen production efficiency. First, the reactor should possess excellent light uniformity and high light utilization efficiency. Light serves as the core energy source, and photosynthetic bacteria typically achieve optimal growth and hydrogen production conditions under specific wavelengths and light intensities. However, in high-density cultures or large-scale reactors, issues such as uneven light distribution and light energy attenuation arise. By shortening the optical path, employing light guiding systems, or using distributed light sources, the penetration depth and uniformity of light within the reactor can be improved. Second, in the gas-liquid two-phase system, the reactor structure should ensure the efficient release and collection of H_2 and CO_2 as well as the uniform distribution of liquid-phase substrates and nutrients, thereby preventing local accumulation, metabolic inhibition, or light shielding phenomena. Therefore an efficient reactor structure should be based on multiparameter coupled regulation and systematic engineering optimization to achieve the integration of light energy input, mass transfer, and metabolic kinetics. This is a key foundation for advancing the photo-fermentation hydrogen production technology toward high efficiency and stability.

Photo-fermentation hydrogen production reactors can be broadly categorized into several structural types, including panel reactors, cylindrical reactors, horizontal tubular reactors, and baffled reactors. Table 5.1 provides an overview of these common reactor configurations, along with hydrogen production rates achieved under optimized conditions.

Among these designs the panel reactor stands out for its uniform light distribution and high surface-area-to-volume ratio (Fig. 5.5). Typically constructed with a thin-layer design the panel reactor minimizes light penetration depth, which enhances both light uniformity and energy utilization efficiency (Singh & Sharma, 2012). By increasing the ratio of the illuminated surface area to the reactor volume, panel reactors can significantly boost photosynthetic activity and hydrogen production. For example,

TABLE 5.1 Reactor structure and operating conditions of photo-fermentation hydrogen production.

Structure	Volume	Materials	Conditions	Illuminance	Performance	Strains	References
Panel	1.4 L	Plexiglas	Substrate: molasses pH value: 7.5 Temperature: 30°C	200 W/m2	14.3 mL $H_2/(L_h h)$	*Rhodobacter capsulatus* YO3	Sagir et al. (2018)
Horizontal tubular	20 L	Glass	Substrate: molasses pH value: 7.5 Temperature: 11°C–40°C	300 W halogen lamps	15.5 mL $H_2/(L_h h)$	*R. capsulatus* YO3	Oflaz and Koku (2021)
Baffled	2 L	/	Substrate: corn stover pH value: 7.0 Temperature: 30°C	3000 lx	191.7 mL $H_2/(L_h h)$	HAU-M1	Li et al. (2021a)
/	1 L	Glass	Substrate: EDF pH value: 7.0 Temperature: 30°C	Halogen lamps 100 W/m_2	8.8–13.2 mL H2/$(L_h h)$	*Rhodopseudomonas palustris* WP3-5	Lo et al. (2008)

(*continued on next page*)

TABLE 5.1 Reactor structure and operating conditions of photo-fermentation hydrogen production—cont'd

Structure	Volume	Materials	Conditions	Illuminance	Performance	Strains	References
Baffled Upflow baffled Upflow tubular	2 L 2 L 0.126 L	/	Substrate: corncob pH value: 6.8 Temperature: 30°C	LED lamps 4000 lx	165.1 mL H2/(L_hh) 151.0 mL H2/(L_hh) 148.5 mL H2/(L_hh)	Mixed bacteria strains	Zhang et al. (2015)
Cylindrical	1.2 L	Glass	Substrate: SMW Temperature: 25.6°C pH value: 7.0	7500 lx	9.1 ml H2/(L_hh)	*Rhodobacter* sp. KKU-PS1	Assawamongkholsiri et al. (2018)
Baffled	8 m^3	/	Substrate: glucose pH value: 5.58 Temperature: 30°C	3000 lx	209.7 mL H_2/(L_hh)	HAU-M1	Lu et al. (2020b)
Tubular	3.34 L	Glass	Substrate: cheese whey pH value: 6.5–7.5 Temperature: 20°C–25°C	4000 lx	3.6 mL H_2/(L_hh)	Mixed bacteria strains	Policastro et al. (2023)

(continued on next page)

Cylindrical	2 L	/	Substrate: malic acid pH value: 6.8–7.0 Temperature: 30°C–37°C	/	/	*Rhodobacter sphaeroides* O·U.001	Sen et al. (2024)[a]
Tubular	80 L	/	Substrate: acetate pH value: 6.4 Temperature:5°C–35 °C	90 W/m^2	6.9 mL $H_2/(L_h h)$	*R. capsulatus* DSM 1710	Boran et al. (2010)
Baffled	4 m^3	/	Substrate: glucose pH value: 7.0 Temperature: 30°C	3000 lx	88.9 mL $H_2/(L_h h)$	HAU-M1	
Horizontal tubular	70 L	Quartz	Substrate: glucose pH value: 7.1 Temperature: 30°C	3000 lx	20.7 mL $H_2/(L_h h)$	*R. sphaeroides* HY01	
Cylindrical	2.7 L	Glass	Substrate: EDF pH value: 7.0 Temperature: 30°C	10,000 lx	1580 mL H2/L	*R. sphaeroides* O·U.001	

EDF, *effluent from dark fermentation;* SMW, *sugar manufacturing plant wastewater.*
[a] *The study involves simulation calculations.*

FIGURE 5.5 The panel reactor is constructed with a PVC frame and Plexiglas sheets, featuring external dimensions of 29 cm × 29 cm × 4 cm and an effective volume of 1.4 L. It has a light exposure area of 0.04 m^2 for immobilized bacteria and is equipped with an internal cooling system. It is suitable for long-term photo-fermentation hydrogen production experiments both indoors and outdoors. *Reproduced with permission from Sagir et al. (2018).*

R. capsulatus YO3 produced hydrogen at a rate of 14.3 mL $H_2/(L_h h)$ in a 1.4 L panel reactor (Sagir et al., 2018). In addition to performance benefits the modular nature of panel reactors allows for scalable assembly and flexible operation in both batch and continuous modes (Yang et al., 2016). However, this design also has limitations. Its relatively low volume efficiency and susceptibility to biofouling can hinder long-term performance (Wang et al., 2014). To overcome these challenges, ongoing research focuses on optimizing parameters such as the reactor's inclination angle and internal fluid dynamics to enhance overall efficiency and reliability.

Cylindrical reactors offer a compact design, making them particularly suitable for settings with limited space and high light intensity requirements (Fig. 5.6A). One of their key operational advantages lies in gas separation: hydrogen produced during photo-fermentation is more readily separated from the culture medium in column-type systems, helping to reduce both gas collection and operational costs (Fig. 5.6B). For example, *Rhodobacter* sp. KKU-PS1 achieved a hydrogen production rate of 9.1 mL $H_2/(L_h h)$ in a 1.2 L cylindrical reactor (Assawamongkholsiri et al., 2018). However, the vertical configuration of cylindrical reactors can lead to uneven light distribution because of light attenuation along the height of the reactor (Singh & Sharma, 2012). To compensate for this, higher light intensities are often applied to maintain microbial hydrogen production efficiency. Notably, hydrogen production by *Rhodobacter* sp. KKU-PS1 and *R. sphaeroides* O.U.001 required illumination levels of 7500 and 10,000 lx, respectively—substantially higher than the typical 3000–4000 lx used in

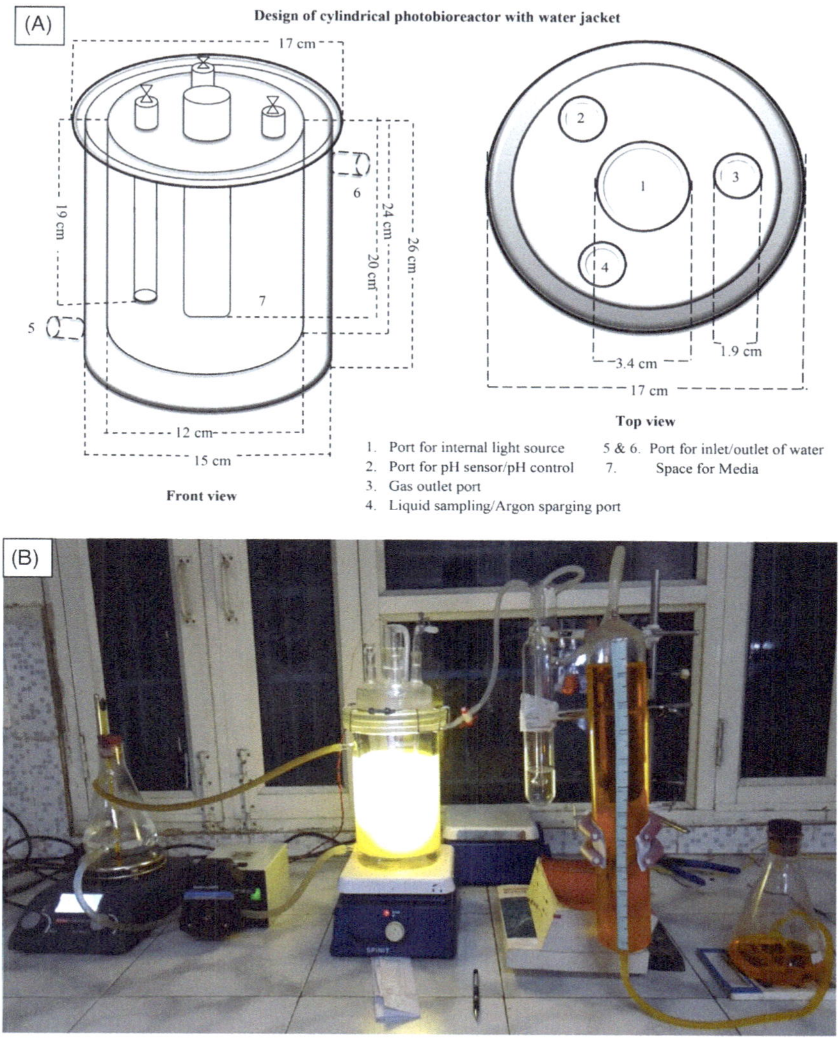

FIGURE 5.6 The cylindrical reactor has a height-to-diameter ratio of 2, with a total volume of 2.7 L, a height of 24 cm, a diameter of 12 cm, and a wall thickness of 0.8 cm. It is equipped with an external water jacket to maintain a constant temperature, and a dual-sided LED light source insertion port is located at the top center to ensure uniform light distribution. The four ports are used for gas outlet, liquid sampling, gas sparging, and pH control, respectively, featuring a compact and versatile design. *Reproduced with permission from Das and Basak (2024).*

other reactor types (Assawamongkholsiri et al., 2018; Das & Basak, 2024). Although effective at small scales, cylindrical reactors face limitations when scaled up. In larger systems the problem of uneven illumination becomes more pronounced, and vertical fluid flow can increase mass transfer

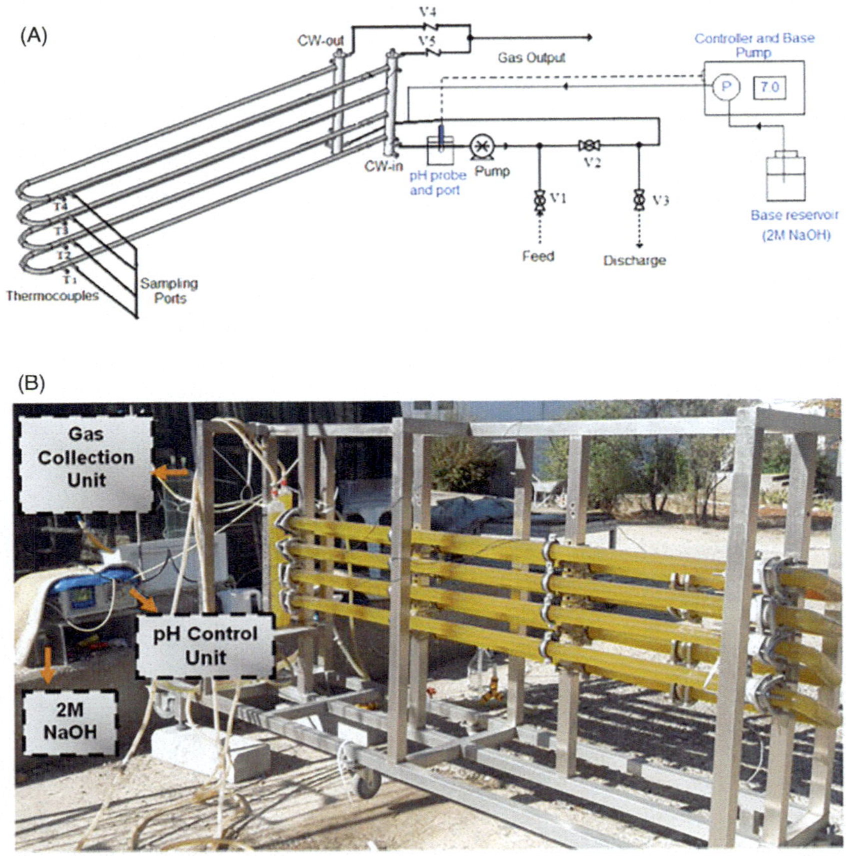

FIGURE 5.7 The horizontal tubular reactor is composed of four parallel glass tubes, each 4 m in length and 4 cm in diameter, arranged vertically with a total volume of 20 L and equipped with sampling ports, pH measurement ports, and temperature control ports to ensure optimal conditions for bacterial growth and hydrogen production. Its design features an inclination to facilitate gas collection from the headspace. *Reproduced with permission from Oflaz and Koku (2021).*

resistance (Katsuda et al., 2000). As summarized in Table 5.1, three cylindrical reactors have been reported with working volumes of 1.2, 2.0, and 2.7 L, further indicating their primary suitability for laboratory- or pilot-scale applications (Assawamongkholsiri et al., 2018; Das & Basak, 2024; Sen et al., 2024).

Horizontal tubular reactors offer significant advantages for photo-fermentation hydrogen production, particularly under natural lighting conditions (Fig. 5.7A). Their design enhances light penetration depth, which improves light energy utilization and ensures a more uniform and sufficient light supply to the photosynthetic microorganisms (Singh & Sharma, 2012). This uniform illumination is critical for maintaining consistent hydrogen production across the reactor volume (Fig. 5.7B). In addition to improved light delivery, horizontal reactors typically support a more stable internal flow environment. This stability promotes efficient interaction between the substrate and microbial cells, enhancing substrate utilization and supporting sustained hydrogen production. These characteristics make horizontal tubular reactors promising candidates for scale-up. Pilot-scale studies have demonstrated the potential of this reactor type. Hydrogen production rates of 15.5 and 20.7 mL $H_2/(L_h h)$ were achieved in 20 and 80 L horizontal tubular reactors, respectively (Oflaz & Koku, 2021; Ren et al., 2024). These results highlight the scalability and performance potential of this configuration. Nevertheless, certain limitations should be addressed before broader application is feasible. Horizontal reactors require a relatively large footprint, which can increase land use and complicate installation planning. Furthermore the internal structure of these reactors makes cleaning and maintenance more difficult, with residual accumulation and access constraints posing practical challenges (Cheng et al., 2019). To support widespread adoption, further pilot-scale investigations are essential to optimize reactor design and operational protocols.

Baffled reactors have demonstrated strong potential for enhancing photo-fermentation hydrogen production by improving internal mixing and mass transfer (Fig. 5.8). One of the key advantages of baffled reactors lies in their ability to improve light utilization. By disrupting laminar flow, baffles help reposition microorganisms into illuminated zones more frequently, reducing the formation of dark regions within the reactor. As a result, light energy is used more effectively, which is crucial for sustaining high levels of hydrogen output in photo-fermentation systems. Moreover, baffled reactors offer structural flexibility and can be adapted to various operational scales and conditions. Their performance has been validated in both laboratory and pilot-scale systems. For example, the HAU-M1 strain achieved hydrogen production rates of 191.7 mL $H_2/(L_h h)$ in 2 L reactors, 88.9 mL $H_2/(L_h h)$ in 4 m^3 systems, and 209.7 mL $H_2/(L_h h)$ in 8 m^3 reactors (Table 5.1), highlighting the scalability and efficiency of this reactor design. Despite these advantages, baffled reactors also present certain challenges. The complexity of the internal structure can increase manufacturing costs and may lead to higher energy consumption because of intensified mixing. In addition, if not properly designed, excessive baffles could create localized light shading, counteracting the benefits of enhanced light distribution. Therefore further optimization is needed to reduce structural complexity, lower energy input, and fine-tune light paths to ensure their practical viability and cost-effectiveness in large-scale applications.

FIGURE 5.8 The baffled reactor presents as an independent chamber structure, whereas inside the chamber, baffle devices are meticulously arranged. These baffles are designed to enhance thorough mixing between the substrate and strains, thereby significantly improving hydrogen production efficiency. *Reproduced with permission from Lu et al. (2020b).*

5.5.2 Reactor operation

Operational parameters play a critical role in determining both the stability and efficiency of photo-fermentation hydrogen production systems. Among them the choice of operating mode forms the foundation of the process, shaping how substrates and metabolic products are distributed and processed within the reactor over time. Following this the inoculum size is essential for ensuring rapid system start-up and effective microbial growth. A suitable inoculum promotes robust establishment of the microbial community, particularly under different operating regimes, which is vital for maintaining consistent hydrogen production. During continuous operation the feeding rate and HRT jointly influence substrate availability and conversion efficiency. These parameters determine how long substrates remain in the reactor and the extent to which they are metabolized, directly impacting hydrogen yield. In addition, stirring plays a key role in maintaining optimal reaction conditions. An appropriate stirring rate improves light penetration and enhances mass transfer, while preventing the formation of biofilms or sedimentation that could disrupt microbial activity and reduce system performance. Together the coordinated optimization of these parameters supports the development of a stable, efficient, and scalable photo-fermentation hydrogen production process.

The choice of operating mode plays a fundamental role in determining the efficiency, stability, and scalability of photo-fermentation hydrogen production systems. Common modes include batch, semicontinuous, and continuous operation, each with distinct advantages and limitations depending on the stage of system development and application scale. Batch operation is straightforward and widely used in laboratory studies and during reactor startup. It allows precise control of experimental variables and facilitates the observation of microbial growth and hydrogen production dynamics. However, because of the need for periodic filling and emptying, its efficiency declines over extended operational periods. As such, batch mode is often employed in early-stage experiments to optimize microbial strains before transitioning to more sustainable operational strategies (Assawamongkholsiri et al., 2018; Li et al., 2021a; Lo et al., 2008; Zhang et al., 2015).

Semicontinuous operation offers a balance between process simplicity and production efficiency. In this mode, part of the fermentation broth is periodically replaced with fresh medium, maintaining microbial activity, and extending production cycles. For example, Policastro et al. (2023) reported a hydrogen yield of 87 mL/(L_hd) using a semicontinuous regime with periodic medium replacement. Oflaz and Koku (2021) further demonstrated that combining semicontinuous feeding with optimized light conditions can significantly enhance both hydrogen output and operational duration.

Continuous operation provides the most stable and scalable solution for industrial applications. With constant feeding and discharge, this mode ensures steady hydrogen production, consistent biomass renewal, and improved substrate utilization (Boran et al., 2010; Zhang et al., 2017). Li et al. (2021b) achieved a hydrogen production rate of 191.7 mL H_2/(L_hh) under continuous conditions. Similarly, Boran et al. (2010) maintained stable hydrogen output over 30 days, with an average production rate of 6.9 mL H_2/(L_hh) during daylight hours. In summary, selecting an appropriate operating mode is crucial for optimizing reactor performance across different phases of development. Although batch and semicontinuous modes are valuable for strain improvement and process refinement, continuous operation is best suited for long-term, large-scale hydrogen production.

The inoculum size is a critical parameter influencing both the efficiency of photo-fermentation hydrogen production and the stability of reactor operation. An appropriate inoculum proportion accelerates system startup, enhances substrate utilization, and improves resilience against environmental fluctuations. Typically, inoculum size is expressed as a percentage of the reactor's total liquid volume, with values commonly ranging from 5% to 25% (Boran et al., 2010; Oflaz & Koku, 2021; Ren et al., 2024; Wang et al., 2019; Zhang et al., 2015, 2018a; Zhang et al., 2018b). Using too little biomass at startup can slow microbial growth, prolong the adaptation period, and increase the system's vulnerability to contamination or environmental shifts. For example, in a study by Policastro et al. (2023), a low inoculum size of 2% resulted in a modest hydrogen production rate of 3.6 mL $H_2/(L_h h)$, illustrating the limitations of underinoculation. However, excessive inoculation may introduce new challenges. A high biomass density can cause light shading, reducing the exposure of photosynthetic bacteria to light, thereby lowering photosynthetic efficiency and increasing operational costs. This can ultimately limit the effective conversion of light energy into hydrogen. Therefore the inoculum size must be carefully optimized based on multiple factors, including bacterial strain characteristics, substrate concentration, illumination conditions, and reactor design. Achieving the right balance is essential for maximizing hydrogen yield while ensuring operational efficiency. For large-scale applications, further research is needed to evaluate different inoculation strategies, such as batch supplementation and continuous inoculation. These approaches may offer pathways to improve long-term performance and system robustness, providing practical insights for the engineering and scale-up of photo-fermentation hydrogen production systems.

The feed rate is a key operational parameter that directly affects microbial activity, substrate utilization, and overall hydrogen production performance in photo-fermentation systems. Maintaining an appropriate feed rate ensures that photosynthetic bacteria remain in a metabolically active state, supporting continuous hydrogen generation. An optimal feed rate helps balance substrate availability with the risk of product inhibition. If the feed rate is too high, unutilized substrates may accumulate, biomass can be washed out, and the system may become overloaded—ultimately reducing hydrogen yield. On the other hand a feed rate that is too low may result in substrate limitation, hinder microbial growth, and lead to the buildup of inhibitory byproducts. To achieve efficient and stable hydrogen production the feed rate must be carefully optimized. This requires consideration of multiple factors, including light intensity, temperature, substrate concentration, reactor configuration, and the growth kinetics of the microbial community. Tailoring the feed strategy to these parameters is essential for maintaining system balance and maximizing hydrogen output.

HRT is a key operational parameter in photo-fermentation hydrogen production, reflecting the average time that substrate fluid remains within the photobioreactor. Defined as the ratio of the reactor's effective volume to the feed flow rate, HRT significantly influences substrate degradation, microbial metabolism, and hydrogen yield. A shorter HRT can increase the system's treatment capacity but may compromise hydrogen production efficiency. Insufficient residence time can lead to incomplete substrate conversion, reduced microbial retention, and biomass washout, all of which diminish metabolic activity and hydrogen output. Conversely, excessively long HRTs may result in overdegradation of substrates and accumulation of inhibitory byproducts—such as organic acids—which suppress the activity of photosynthetic bacteria and reduce system performance. In a study by Zhang et al. (2015), hydrogen production increased with HRT up to 24 h, at which point the system achieved a cumulative yield of 589.21 mmol/L and a substrate conversion efficiency of 40.48%. Beyond this optimal point, performance declined, and biomass loss became significant at HRTs exceeding 48 h (Zhang et al., 2015). The optimal HRT varies depending on the type of substrate. For complex organic matter such as straw hydrolysate, longer HRTs (48–120 h) are generally required to ensure sufficient degradation and hydrogen conversion. When using effluents from dark fermentation or substrates like cheese whey an HRT of around 48 h has shown favorable results (Lo et al., 2008; Policastro et al., 2023; Zhang et al., 2018a). In contrast, simpler carbon sources such as glucose or acetic acid can support high hydrogen production efficiencies at shorter HRTs, typically between 24 and 48 h. Future research should focus on the coordinated regulation of HRT and solids retention time (SRT), particularly under conditions involving variable light regimes and complex feedstocks. This will be essential for enhancing both the operational stability and hydrogen productivity of photo-fermentation systems in real-world applications.

Agitation rate is a critical operational parameter in photo-fermentation hydrogen production, influencing light distribution, mass transfer, and overall reactor performance. Proper agitation ensures even suspension of microbial cells, preventing sedimentation and the formation of anaerobic zones that could hinder hydrogen production. By promoting uniform mixing, agitation improves the distribution of light throughout the reactor, minimizing light-shading effects and enhancing the efficiency of light utilization by photosynthetic bacteria. It also facilitates the transfer of substrates, products, and dissolved gases, such as hydrogen and carbon dioxide, helping to maintain nutrient balance, remove metabolic byproducts, and support continuous microbial activity and hydrogen release. However, the agitation rate should not be too fast. Excessive stirring can impose shear stress on microbial cells, potentially reducing their viability and hydrogen-producing capacity. Moreover, high agitation rates increase energy consumption, negatively impacting the overall efficiency and economic viability of the process. Typical agitation speeds range from 100 to 280 rpm, but the optimal rate depends on reactor design and microbial characteristics. Experimental evaluation is essential to determine the appropriate balance between effective mixing, biological activity, and energy efficiency in each system (Assawamongkholsiri et al., 2018; Lo et al., 2008; Zhang et al., 2015).

5.6 Net-zero effect

Against the backdrop of global energy transition and carbon neutrality goals, photo-fermentation hydrogen production technology has emerged as an advanced research focus in the global energy field, owing to its advantages of utilizing renewable biomass feedstocks, operating under mild reaction conditions, and offering the potential for negative carbon emissions. Photosynthetic bacteria convert organic matter into hydrogen under anaerobic, light-driven conditions via photo-fermentation hydrogen production (Sun et al., 2024). It operates under mild conditions, requires relatively low energy input, and simultaneously enables waste valorization and emission reduction, making it a strong candidate for producing "ultragreen" hydrogen (Jiang et al., 2025). From a carbon reduction perspective, photo-fermentation not only repurposes biomass waste but also contributes to carbon sequestration through biological fixation. Compared to fossil-based methods such as coal gasification (18.6 t CO_2-eq/t of H_2) and microbial electrolysis (18.8 t CO_2-eq/t H_2) (Chen et al., 2019; Goren et al., 2025) the combined process of dark fermentation and photo-fermentation emits significantly less, only 9.4 t CO_2-eq/t H_2 approaching near carbon-neutral performance (Zhang et al., 2024). Its carbon reduction benefits can be further enhanced through advanced strategies such as ethanol coproduction, carbon refixation via the calvin cycle (CBB) cycle, and CO_2 reduction reactions driven by alternative nitrogenase enzymes.

Despite these advantages, realizing true net-zero emissions with photo-fermentation remains challenging. Major bottlenecks include competition between carbon and electron flows within the cell, variability in reactor energy efficiency, and limitations in resource utilization because of the single-function nature of the process (Huang et al., 2025). To overcome these barriers, future efforts should focus on redesigning metabolic pathways through synthetic biology, applying light-responsive nanomaterials to improve light capture, and integrating photo-fermentation with complementary systems such as microalgae cultivation and electrochemical technologies. Together, these innovations will help transition photo-fermentation from lab-scale experimentation to scalable, stable, and low-carbon industrial deployment. Through coordinated advances across these fronts, photo-fermentation hydrogen production has the potential to become a cornerstone of the zero-carbon hydrogen economy. This chapter comprehensively evaluates the contributions of the photo-fermentation hydrogen production process to carbon dioxide reduction from various perspectives. In the outlook section, we emphasize the core scientific challenge of establishing efficient and controllable mechanisms for carbon flow and electron flow distribution and propose system integration pathways that incorporate low-carbon or negative-carbon technologies and enhance energy efficiency. In addition, we discuss the existing issues in the photo-fermentation hydrogen production process and its future development directions. This chapter provides theoretical support and practical pathways for constructing a zero-carbon hydrogen economy based on photo-fermentation biohydrogen production.

TABLE 5.2 Global warming from different hydrogen production technologies.

Hydrogen production method	Global warming t CO_2 eq/t H_2	References
Combined dark- and photo-fermentation	9.4	Zhang et al. (2024)
Coal gasification	18.6	Goren et al. (2025)
Alkaline electrolysis	6.4	Goren et al. (2025)
Microbial electrolysis cell	18.8	Chen et al. (2019)
Two-stage anaerobic fermentation	15.1	Sun et al. (2019)
Water electrolysis	28.6	Siddiqui and Dincer (2019)
Biomass gasification	10.6	Susmozas et al. (2016)
Biogas reforming	5.6	Hajjaji et al. (2016)

5.6.1 Contribution to CO_2 reduction

Photo-fermentation hydrogen production demonstrates significant potential for achieving net-zero effect. It primarily reduces carbon emissions through the following four core pathways:

Photo-fermentation hydrogen production holds strong potential for contributing to net-zero emissions. One of its key advantages lies in its ability to reduce carbon emissions through multiple interconnected pathways. Most notably, photosynthetic bacteria can convert the organic carbon found in agricultural waste into microbial biomass, effectively capturing and storing carbon (Hakobyan et al., 2019). As this organic material originates from atmospheric CO_2 fixed by plants through photosynthesis, the process closes the carbon loop. Although some CO_2 is released during microbial metabolism, the overall balance often results in net carbon uptake. As a result, hydrogen produced via biomass-based photo-fermentation can be classified as carbon-negative and is considered a form of "ultragreen hydrogen" in the hydrogen energy color spectrum (Luo et al., 2025).

Compared to conventional hydrogen production methods, photo-fermentation particularly when integrated with dark fermentation, demonstrates a clear advantage in reducing greenhouse gas emissions. As shown in Table 5.2 the combined dark and photo-fermentation process emits only 9.4 t of CO_2 equivalent per ton of hydrogen produced. This is substantially lower than emissions from fossil-based approaches such as coal gasification (18.6 t CO_2-eq/t H_2) and microbial electrolysis (18.8 t CO_2-eq/t H_2), as well as from two-stage anaerobic fermentation (15.1 t CO_2-eq/t

H_2). Although technologies such as biogas reforming and alkaline electrolysis exhibit relatively low emissions, they often rely on external energy inputs or less sustainable feedstocks. In contrast, photo-fermentation utilizes renewable biomass and solar energy, offering a cleaner and more sustainable pathway for hydrogen production. This positions it as a promising solution for reducing carbon emissions in future energy systems. In addition, measures such as combined heat and power generation and carbon capture, utilization, and storage (CCUS), as well as the resourceful utilization of VFA technologies, have significantly contributed to enhancing the carbon emission reduction benefits of biomass-based manufacturing processes (Luo et al., 2025).

The integration of photo-fermentation hydrogen production with other organic matter coproduction technologies enables the conversion of organic carbon into high-value chemicals. This not only further enhances the conversion rate of organic waste but also reduces carbon emissions and increases hydrogen yield. The coproduction technology of hydrogen and ethanol can increase hydrogen output while generating ethanol. Under the condition of a 55 g/L substrate the one-step coproduction process for hydrogen and ethanol achieved a maximum ethanol concentration of 7.09 g/L (Sun et al., 2024). Meanwhile, the hydrogen yield was 2.76 times higher compared to the efficiency of hydrogen fermentation alone, significantly improving the energy conversion efficiency of corn stover (CS) (Sun et al., 2024). Jin et al. (2024) achieved a mere 0.9 t of CO_2-equivalent emissions per ton of hydrogen produced through the coupling of photo-fermentation, dark fermentation, and methane fermentation.

Photosynthetic bacteria possess CO_2 reduction potential during the process of photo-fermentation. Part of the CO_2 produced during the carbon metabolism of photosynthetic bacteria is fixed into intracellular organic carbon through the carbon fixation and carbon reduction processes of the CBB, respectively (Fig. 5.9). However, the carbon-fixing process of the CBB cycle may compete with the hydrogen-producing process for electrons, thereby leading to a decrease in H_2 yield (Öztürk et al., 2012). The hydrogen production of the strain *R. sphaeroides* HR02, which lacks the cbbR gene, was increased by up to 26.8% compared to the wild-type strain HY01 (Feng et al., 2018b). Therefore balancing the electron allocation between the CBB cycle and nitrogenase is crucial for achieving net-zero emissions and improving hydrogen yield. In nature, apart from the molybdenum (Mo) nitrogenase (encoded by the *nifHDK* genes), there are also two alternative nitrogenases in some photosynthetic bacteria: the vanadium (V) nitrogenase (encoded by the *vnfHDGK* genes) and the iron (Fe) nitrogenase (encoded by the *anfHDGK* genes) (Oehlmann et al., 2024). These alternative nitrogenases only function when the Mo nitrogenase is inactive. Theoretically, alternative nitrogenases have better nitrogen-fixing and hydrogen-producing effects (Oehlmann et al., 2024). However, in

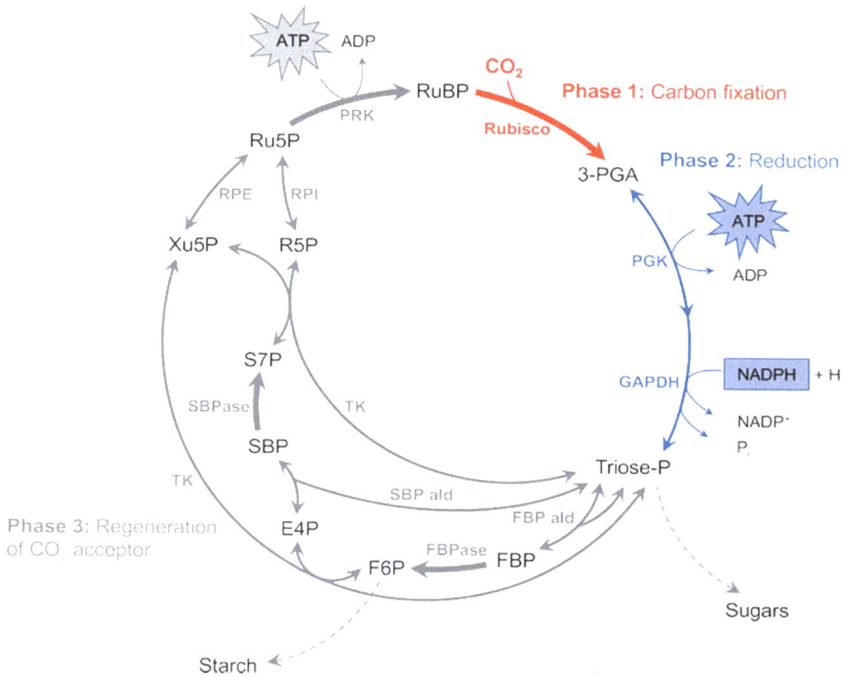

FIGURE 5.9 The Calvin–Benson cycle. *ATP*, adenosine triphosphate. *Reproduced with permission from Schreier and Hibberd (2019)*

experimental processes, the nitrogen-fixing and hydrogen-producing rates of alternative nitrogenases are relatively low (Oehlmann et al., 2024). Nevertheless, the Fe nitrogenase exhibits a stronger CO_2 reduction potential than the Mo nitrogenase, and it is capable of simultaneously reducing both CO_2 and protons to produce high-value chemicals such as formic acid, as well as hydrogen gas (Zheng et al., 2018). Therefore alternative nitrogenases have certain potential in further reducing CO_2 emissions during photo-fermentation hydrogen production. It is noteworthy that the products of CO_2 reduction by alternative nitrogenases include the greenhouse gas CH_4, which may have a certain negative impact on the reduction of CO_2 emissions (Zheng et al., 2018).

5.6.2 Outlook

In the process of photo-fermentation hydrogen production, achieving true zero carbon emissions not only depends on the cleanliness of external energy sources but is more fundamentally constrained by the intracellular carbon and electron flow distribution mechanisms. Carbon flow determines the metabolic fate of carbon elements in the substrate. If a significant amount of carbon is converted into cellular biomass or by-products (such as organic acids and poly-3-hydroxybutyrate), it will markedly reduce the efficiency of carbon flowing toward the hydrogen production pathway, thereby indirectly increasing the system's carbon footprint. Electron flow, on the other hand, directly influences the electron supply capacity for nitrogenase-mediated hydrogen production. If reduced nicotinamide adenine dinucleotide (NADH) is utilized in other competing metabolic pathways, such as the CBB cycle, it will also limit the maximization of hydrogen yield. More importantly, these two types of flows are not only intertwined with each other but are also jointly influenced by light intensity, substrate type, and regulatory mechanisms. Therefore establishing an efficient and controllable synergistic distribution mechanism for carbon and electron flows is one of the core scientific challenges in breaking through the zero-carbon bottleneck of photo-fermentation hydrogen production and achieving carbon neutrality goals.

Although photo-fermentation hydrogen production has demonstrated significant carbon emission reduction potential at the laboratory scale, its energy efficiency fluctuations and resource utilization constraints in a single process may undermine its carbon neutrality advantages in large-scale industrial applications. Therefore constructing a coupled system with other low-carbon or negative-carbon technologies is one of the key pathways to achieving sustained carbon emission reduction benefits. For instance, coupling photo-fermentation with processes such as anaerobic fermentation, electrochemical reduction, or algal photosynthesis can synergistically utilize waste organic matter, light energy, and carbon dioxide resources, achieving synergistic optimization of multiform energy outputs (such as hydrogen, biomethane, and biooil) and carbon sequestration. In addition, by introducing heat integration systems and energy recovery technologies, the overall energy consumption of the system can be further reduced, enhancing the life-cycle carbon efficiency. On a large-scale deployment, cross-technology integration not only helps to address the shortcomings of photo-fermentation in terms of feedstock adaptability, electron supply, and yield fluctuations but also provides a practical pathway for constructing a diversified and highly resilient renewable hydrogen energy system, thereby effectively sustaining its carbon emission reduction advantages.

To further enhance the energy efficiency of photo-fermentation hydrogen production and strengthen its carbon emission reduction capabilities, future research should focus on the synergistic enhancement pathways integrating synthetic biology and nanomaterials. On the one hand, synthetic biology provides powerful tools for reconstructing the metabolic networks of photosynthetic bacteria and optimizing the allocation of carbon and electron flows. Through precise regulation of key hydrogenases, the electron transport chain, and carbon metabolism nodes, efficient guidance of substrates toward the hydrogen production pathway can be achieved, minimizing by-product formation and carbon loss. Meanwhile, by leveraging genetic circuit programming and environmental response regulation, high-yield hydrogen-producing engineered bacterial strains with adaptive characteristics can be constructed. On the other hand the introduction of nanomaterials can further enhance light harvesting, electron transfer, and substrate utilization efficiency at the extracellular or interfacial level. Photosensitive nanomaterials can significantly improve the system's utilization of visible light, whereas functionalized nanocarriers can promote the stability and catalytic activity of hydrogenases. The synergistic optimization of synthetic biological systems and nanomaterial interfaces is expected to break through multiple bottlenecks in current biohydrogen production regarding yield, stability, and carbon emission reduction pathways, propelling it toward a sustainable, efficient, and zero-carbon future.

5.7 Case study

5.7.1 Flow field modeling for photobiological hydrogen generation reactors

When the HRT was set to 24 h, the liquid flow velocity at the inlet of the hydrogen production system was 0.016374 m/s. The flow velocity in the baffle corners ranged approximately between 0.0069 and 0.0104 m/s. At the outlet of each reaction chamber (upper right side) the velocity increased because of the narrowing channel, reaching approximately 0.0242–0.0293 m/s. Meanwhile, at the outlet of the hydrogen production system, the maximum velocity further increased to between 0.0224 and 0.0242 m/s.

In each reaction chamber the average velocity in the downflow zone was significantly higher than that in the upflow zone, and the flow velocity also increased noticeably at the interconnections between chambers (Fig. 5.10). In addition, four large vortices formed in the lower-middle-upper regions of the upflow zones across the four reaction chambers. These vortices were generated as the reaction liquid, after passing beneath the baffles, collided with the bottom plate and the right-side baffle, creating counter-directed flows. Subsequently, under the influence of the left-side baffle, these opposing flows combined to form a vortex. Subsequently the reaction liquid recirculated along the vortex trajectory, reinforcing the previous flow pattern. The formation of three vortices enhanced the mixing between the reaction liquid and bacterial colonies, significantly promoting photo-fermentative hydrogen production.

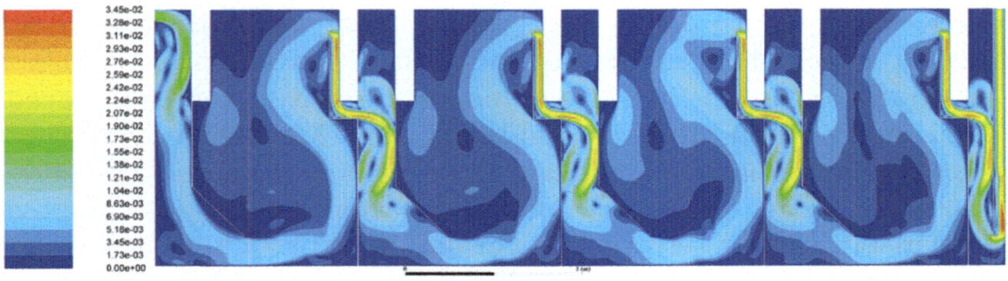

FIGURE 5.10 Simulation of the internal velocity field in a photobiological hydrogen production reactor.

Multiple small vortices were generated in the downflow zones of both Chambers 1 and 2. This phenomenon occurred because shorter HRTs led to increased fluid velocities, which—under the directional guidance of the baffles—more readily induced vortex formation. Notably, under all tested HRT conditions, distinct vortices were consistently observed in the upflow zones of all four reaction chambers. These irregular vortices significantly enhanced liquid-substrate mixing efficiency and bacterial culture homogenization, thereby improving the overall hydrogen production performance of the system.

Computational fluid dynamics simulations of the photobiological hydrogen production reactor revealed that the HRT significantly influences the flow field distribution within the system. The flow field patterns clearly characterize the mixing and distribution of bacterial colonies with the hydrogen-producing culture medium.

5.7.2 Photosynthetic biohydrogen reactor

The photobiological hydrogen production reactor system primarily consists of seven key components: (1) an 8 m^3 photo-fermentation reactor (comprising 8 reaction chambers, each with an effective volume of 1 m^3), (2) a 3 m^3 feed tank, (3) a photosynthetic bacterial culture vessel, (4) a central computer control console, (5) solar heaters with insulation chambers, (6) automated solar concentrators with light-guiding optical fibers, and (7) photovoltaic panels with battery storage systems (Fig. 5.11). The reactor enables individual gas collection and liquid sampling from each reaction chamber, facilitating detailed analysis of fermentation conditions in separate compartments. The baffle design significantly slows horizontal flow velocity, which enhances substrate-microbe contact, improves hydraulic mixing efficiency, optimizes bacterial population distribution, and increases microbial retention capacity. During operation the 135-degrees inclined base of the baffles creates a sudden velocity surge when fluid reaches the sloped interface. This hydrodynamic effect generates impact

FIGURE 5.11 Photosynthetic biohydrogen reactor: (A) gas online analyzer, (B) biological hydrogen production reactor, (C) hot water tank, (D) solar fiber inlet, (E) photovoltaic power generation board, and (F) vacuum collection tube.

forces to resuspend settled substrates and microbes, promotes enhanced mixing at the reactor bottom, and accelerates biochemical reaction rates.

5.7.3 The effect of hydraulic retention time on photosynthetic biohydrogen production reactors

As the HRT decreased from 72 to 24 h: Reactor 1 showed decreased hydrogen production from 181.25 to 70.54 mol/m^3/d; Reactor 2 exhibited rapid increase from 49.55 mol/m^3/d at 72 h to 133.48 mol/m^3/d at 48 h, then declined to 118.30 mol/m^3/d at 24 h; Reactor 3 demonstrated growth from 16.96 to 133.04 mol/m^3/d; and Reactor 4 improved from 8.93 to 96.88 mol/m^3/d. The overall photobiological hydrogen production system increased from 64.29 to 104.91 mol/m^3/d (Fig. 5.12). Under a 72 h HRT in photo-fermentative hydrogen production the feeding rates of both the hydrogen-production medium and photosynthetic bacteria were relatively slow. This extended retention allowed sufficient time for the photosynthetic bacteria in Reactor 1 to effectively degrade substrates, facilitating robust microbial growth and hydrogen generation. Consequently, Reactor 1 demonstrated significantly higher hydrogen production rates compared to other units in the system. As the HRT decreased the feeding rates of both hydrogen-production medium and photosynthetic bacteria were progressively increased. This resulted in rapid delivery of substantial substrates into the photobioreactor system, continuously meeting the nutritional demands for microbial growth and hydrogen production, thereby

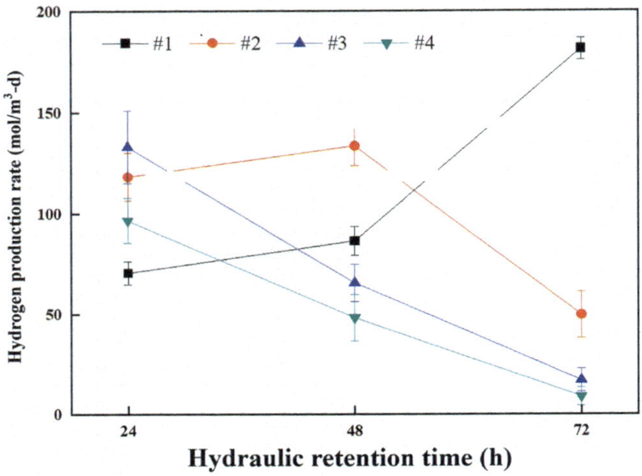

FIGURE 5.12 Effect of hydraulic retention time on hydrogen production rate.

ensuring a significant acceleration in the overall hydrogen generation rate of the photobiological hydrogen production system (Lu et al., 2016, 2019). Concurrently, inadequate retention time for microbial growth in Reactor 1 resulted in the migration of maximum hydrogen production activity from Reactor 1 to subsequent Reactors 2 and 3.

When the HRT was reduced from 72 to 24 h: Reactor 1 showed a pH increase from 4.37 ± 0.01 to 5.52 ± 0.02; Reactor 2 exhibited a pH rise from 4.61 ± 0.09 to 5.29 ± 0.01; and Reactors 3 and 4 demonstrated gradual pH declines (Fig. 5.13). This phenomenon occurs because hydrogen-producing microbes initially degrade glucose into soluble VFAs, whose accumulation progressively decreases the pH of the reaction medium. As the HRT was reduced from 72 to 24 h: The feeding rate of both photosynthetic bacteria and hydrogen-producing culture medium into the bioreactor system increased significantly; photosynthetic hydrogen-producing bacteria transited rapidly through Reactor 1 to subsequent chambers. Consequently, reactors 1 and 2 exhibited pH recovery, and downstream reactors (3 and 4) showed further pH decline. At an HRT of 24 h the pH value in the continuous-flow photo-fermentative biohydrogen production system exhibited a progressive decrease from Reactor 1 to Reactor 3, followed by a recovery trend from Reactor 3 to Reactor 4. This phenomenon may occur because as hydrogen-producing microorganisms completely degrade the available glucose and VFAs accumulate, they begin to utilize these VFAs in the reaction medium for hydrogen production. This metabolic shift consequently leads to a gradual pH recovery in the reaction solution (Lu et al., 2018).

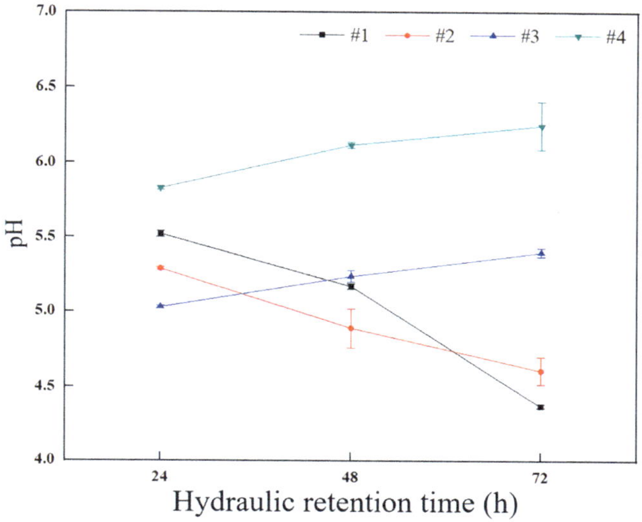

FIGURE 5.13 Effect of hydraulic retention time on pH.

5.8 Conclusion

Photo-fermentation is emerging as a promising and sustainable approach for biohydrogen production from a wide variety of organic feedstocks. The success of this process depends on the careful coordination of several key factors, including substrate type and pretreatment; strains performance; operational conditions, such as pH, temperature, light intensity, and additives; and reactor design and operation. Optimizing these variables is essential to improve hydrogen yield, maintain process stability, and enable system scalability.

Advances in synthetic biology, metabolic engineering, and nanomaterial applications will be crucial for addressing existing limitations, such as imbalances in carbon and electron flow, energy inefficiencies, and limited feedstock flexibility. Integrating photo-fermentation with complementary technologies, including microalgal systems, electrochemical conversion, and carbon capture, presents a promising path toward building resilient, low-emission biohydrogen platforms.

Photo-fermentation offers a scientifically sound and environmentally advantageous route to renewable hydrogen production. With continued innovation and system-level integration, it is well positioned to contribute meaningfully to the global transition toward sustainable energy and net-zero emissions.

References

Assawamongkholsiri, T., Reungsang, A., Plangkang, P., & Sittijunda, S. (2018). Repeated batch fermentation for photo-hydrogen and lipid production from wastewater of a sugar manufacturing plant. *International Journal of Hydrogen Energy, 43*(7), 3605–3617. doi:10.1016/j.ijhydene.2017.12.119.

Basak, N., Jana, A. K., Das, D., & Saikia, D. (2014). Photofermentative molecular biohydrogen production by purple-non-sulfur (PNS) bacteria in various modes: The present progress and future perspective. *International Journal of Hydrogen Energy, 39*(13), 6853–6871. doi:10.1016/j.ijhydene.2014.02.093.

Boran, E., Özgür, E., Van Der Burg, J., Yücel, M., Gündüz, U., & Eroglu, I. (2010). Biological hydrogen production by *Rhodobacter capsulatus* in solar tubular photo bioreactor. *Journal of Cleaner Production, 18*(1), S29. doi:10.1016/j.jclepro.2010.03.018.

Cao, Q. (2019). *Study on wetland protection and landscape construction in Gobi oasis*. University of Architecture and Technology.

Chen, J., Xu, W., Wu, X., Jiaqiang, E., Lu, N., Wang, T., & Zuo, H. (2019). System development and environmental performance analysis of a pilot scale microbial electrolysis cell for hydrogen production using urban wastewater. *Energy Conversion and Management, 193*, 52–63. doi:10.1016/j.enconman.2019.04.060.

Chen, Z. (2022). *Effect of ionic liquid pretreatment process on photosynthetic biohydrogen production from* Arundo donax. Henan Agricultural University.

Cheng, J., Xu, J., Ye, Q., Lai, X., Zhang, X., & Zhou, J. (2019). Strengthening mass transfer of carbon dioxide microbubbles dissolver in a horizontal tubular photo-bioreactor for improving microalgae growth. *Bioresource Technology, 277*, 11–17. doi:10.1016/j.biortech.2019.01.019.

Das, S. R., & Basak, N. (2024). Optimization of process parameters for enhanced biohydrogen production using potato waste as substrate by combined dark and photo fermentation. *Biomass Conversion and Biorefinery, 14*(4), 4791–4811. doi:10.1007/s13399-022-02588-w.

Deng, J. (2011). *Using fruit processing waste as substrate for anaerobic bio-hydrogen production by fermentation*. Northeast Forestry University.

Fan, X. (2023). *Effect of surfactants on photosynthetic biohydrogen production from corncob*. Henan Agricultural University.

Feng, J., Yang, H., & Guo, L. (2018a). The photosynthetic hydrogen production performance of a newly isolated *Rhodobacter capsulatus* JL1 with various carbon sources. *International Journal of Hydrogen Energy, 43*(30), 13860–13868. doi:10.1016/j.ijhydene.2018.03.144.

Feng, J., Yang, H., Wang, X., & Guo, L. (2018b). Enhanced hydrogen production performance of *cbbR* & *pycA* inactived *R. sphaeroides* mutant by improving the ammonium tolerance. *International Journal of Hydrogen Energy, 43*(39), 18142–18150. doi:10.1016/j.ijhydene.2018.07.196.

Fu, C., Zhao, H., & Liu, L. (2011). Effects of B vitamins on hydrogen production and wastewater degradation by *Rhodopseudomonas palustris*. *Food and Fermentation Industries, 37*(06), 25–28.

Gest, H., Ormerod, J. G., & Ormerod, K. S. (1962). Photometabolism of *Rhodospirillum rubrum*: Light-dependent dissimilation of organic compounds to carbon dioxide and molecular hydrogen by an anaerobic citric acid cycle. *Archives of Biochemistry and Biophysics, 97*(1), 21–33. doi:10.1016/0003-9861(62)90039-5.

Goren, A. Y., Dincer, I., & Khalvati, A. (2025). A comprehensive life cycle impact evaluation of hydrogen production processes for cleaner applications. *Energy, 326*. doi:10.1016/j.energy.2025.136182.

Hajjaji, N., Martinez, S., Trably, E., Steyer, J. P., & Helias, A. (2016). Life cycle assessment of hydrogen production from biogas reforming. *International Journal of Hydrogen Energy, 41*(14), 6064–6075. doi:10.1016/j.ijhydene.2016.03.006.

Hakobyan, L., Gabrielyan, L., & Trchounian, A. (2019). Biohydrogen by *Rhodobacter sphaeroides* during photo-fermentation: Mixed vs. sole carbon sources enhance bacterial growth and H_2 production. *International Journal of Hydrogen Energy, 44*(2), 674–679. doi:10.1016/j.ijhydene.2018.11.082.

Hu, H., Li, Y., Wang, X., Zhang, S., & Yang, H. (2014). A review of biomass pretreatment technologies and their effects on pyrolysis products. *Biomass Chemical Engineering, 48*(01), 44–50.

Huang, P., Chen, Y., Yu, S., & Zhou, Y. (2025). Propionic acid enhances H_2 production in purple phototrophic bacteria: Insights into carbon and reducing equivalent allocation. *Water Research, 269*. doi:10.1016/j.watres.2024.122799.

Ji, Z., Ling, Z., Zhang, X., Ma, J., & Xu, F. (2014). Effect of pretreatment on topochemical and ultrastructural changes of lignocellulose plant cell walls: A review. *Shengwu Gongcheng Xuebao/Chinese Journal of Biotechnology, 30*(5), 707–715. doi:10.13345/j.cjb.140078.

Jiang, D., Han, B., & Wang, Y. (2015). Physiological characteristics and hydrogen production properties of HAU-M1 photosynthetic hydrogen-producing bacteria. *Acta Energiae Solaris Sinica, 36*(02), 289–294.

Jiang, D., Zhang, L., Pei, W., Song, Z., Zhang, T., Jing, Y., Chen, Z., Li, W., Zhang, Q., & Lu, C. (2025). Correlation between ethanol addition and photo-fermentative biohydrogen production of *Arundo donax* L. with a consortium. *Renewable Energy, 248*. doi:10.1016/j.renene.2025.123077.

Jiang, D., Zhang, X., Ge, X., Yue, T., Zhang, T., Zhang, Y., Zhang, Z., He, C., Lu, C., & Zhang, Q. (2021). Insights into correlation between hydrogen yield improvement and glycerol addition in photo-fermentation of *Arundo donax* L. *Bioresource Technology, 321*. doi:10.1016/j.biortech.2020.124467.

Jin, M., Wei, X., Mu, X., Ren, W., Zhang, S., Tang, C., & Cao, W. (2024). Life-cycle analysis of biohydrogen production via dark-photo fermentation from wheat straw. *Bioresource Technology, 396*. doi:10.1016/j.biortech.2024.130429.

Jing, Y., Li, F., Li, Y., Jiang, D., Lu, C., Zhang, Z., & Zhang, Q. (2022). Biohydrogen production by deep eutectic solvent delignification-driven enzymatic hydrolysis and photo-fermentation: Effect of liquid–solid ratio. *Bioresource Technology, 349*, 126867. doi:10.1016/j.biortech.2022.126867.

Joshi, A. N., & Vaidya, P. D. (2024). Hydrogen production by aqueous phase reforming of synthetic sewage using Pt/C catalyst: Effect of reaction parameters and pre-treatment strategies. *Waste and Biomass Valorization, 15*(2), 805–819. doi:10.1007/s12649-023-02187-4.

Karunanithy, C., & Muthukumarappan, K. (2011). Influence of extruder and feedstock variables on torque requirement during pretreatment of different types of biomass—A response surface analysis. *Biosystems Engineering, 109*(1), 37–51. doi:10.1016/j.biosystemseng.2011.02.001.

Katsuda, T., Arimoto, T., Igarashi, K., Azuma, M., Kato, J., Takakuwa, S., & Ooshima, H. (2000). Light intensity distribution in the externally illuminated cylindrical photo-bioreactor and its application to hydrogen production by *Rhodobacter capsulatus*. *Biochemical Engineering Journal, 5*(2), 157–164. doi:10.1016/S1369-703X(00)00054-1.

Khanal, S. K., Chen, W. H., Li, L., & Sung, S. (2004). Biological hydrogen production: Effects of pH and intermediate products. *International Journal of Hydrogen Energy, 29*(11), 1123–1131. doi:10.1016/j.ijhydene.2003.11.002.

Kilpeläinen, I., Xie, H., King, A., Granstrom, M., Heikkinen, S., & Argyropoulos, D. S. (2007). Dissolution of wood in ionic liquids. *Journal of Agricultural and Food Chemistry, 55*(22), 9142–9148. doi:10.1021/jf071692e.

Li, J., Li, G., & Ma, S. (2021a). Overview of the progress and development prospects of key technologies for hydrogen production under the goal of carbon neutrality. *Thermal Power Generation, 50*(06), 1–8.

Li, S., Xu, J., & Kang, Y. (2019). Biodegradable ionic liquids pretreatment of corn straw. *Scientia Sinica Chimica, 49*(07), 1001–1006.

Li, Y., Zhang, Z., Jiang, D., Jing, Y., Lu, C., Zhang, H., & Zhang, Q. (2021b). Continuous dark and photo biohydrogen production in a baffled bioreactor and electrons distribution analysis. *Bioresource Technology, 337*. doi:10.1016/j.biortech.2021.125440.

Lin, Y., Liu, J., & Zhou, T. (2015). Iisolation, identification, and culture characteristics of a purple non-sulfur photosynthetic bacterium. *Anhui Agricultural Sciences, 43*(17), 5–8.

Liu, B. (2010). *Screening of photofermentative bacteria and study on hydrogen production by coupling with dark fermentative bacteria*. Harbin Institute of Technology.

Liu, Q., Chen, W., Zhang, X., Yu, L., Zhou, J., Xu, Y., & Qian, G. (2015). Phosphate enhancing fermentative hydrogen production from substrate with municipal solid waste composting leachate as a nutrient. *Bioresource Technology, 190*, 431–437. doi:10.1016/j.biortech.2015.01.139.

Lo, Y. C., Chen, S. D., Chen, C. Y., Huang, T. I., Lin, C. Y., & Chang, J. S. (2008). Combining enzymatic hydrolysis and dark-photo fermentation processes for hydrogen production from starch feedstock: A feasibility study. *International Journal of Hydrogen Energy, 33*(19), 5224–5233. doi:10.1016/j.ijhydene.2008.05.014.

Lu, C., Tahir, N., Li, W., Zhang, Z., Jiang, D., Guo, S., Wang, J., Wang, K., & Zhang, Q. (2020a). Enhanced buffer capacity of fermentation broth and biohydrogen production from corn stalk with Na_2HPO_4/NaH_2PO_4. *Bioresource Technology, 313*, 123783. doi:10.1016/j.biortech.2020.123783.

Lu, C., Wang, Y., Lee, D. J., Zhang, Q., Zhang, H., Tahir, N., Jing, Y., Liu, H., & Zhang, K. (2019). Biohydrogen production in pilot-scale fermenter: Effects of hydraulic retention time and substrate concentration. *Journal of Cleaner Production, 229*, 751–760. doi:10.1016/j.jclepro.2019.04.233.

Lu, C., Zhang, H., Zhang, Q., Chu, C., Tahir, N., Ge, X., Jing, Y., Hu, J., Li, Y., Zhang, Y., & Zhang, T. (2020b). An automated control system for pilot-scale biohydrogen production: Design, operation and validation. *International Journal of Hydrogen Energy, 45*(6), 3795–3806. doi:10.1016/j.ijhydene.2019.04.288.

Lu, C., Zhang, Z., Ge, X., Wang, Y., Zhou, X., You, X., Liu, H., & Zhang, Q. (2016). Bio-hydrogen production from apple waste by photosynthetic bacteria HAU-M1. *International Journal of Hydrogen Energy, 41*(31), 13399–13407. doi:10.1016/j.ijhydene.2016.06.101.

Lu, C., Zhang, Z., Zhou, X., Hu, J., Ge, X., Xia, C., Zhao, J., Wang, Y., Jing, Y., Li, Y., & Zhang, Q. (2018). Effect of substrate concentration on hydrogen production by photo-fermentation in the pilot-scale baffled bioreactor. *Bioresource Technology, 247*, 1173–1176. doi:10.1016/j.biortech.2017.07.122.

Luo, G., Zhao, Z. T., Ding, J., Yang, S. S., Yu, X. L., Bao, M. Y., Sun, H. J., Pang, J. W., Zhang, L. Y., & Ren, N. Q. (2025). Comparative review of carbon emissions from hydrogen production technologies: The hydrogen color characteristic of biomass-based manufacturing. *Renewable and Sustainable Energy Reviews, 217*. doi:10.1016/j.rser.2025.115756.

Luo, H. (2007). *Hydrogen production from high-concentration organic wastewater degradation by immobilized hydrogen-producing sludge*. Kunming University of Science and Technology.

Masilela, P., & Pradhan, A. (2022). Systematic literature review of the sustainability and environmental performance of dark fermentative biohydrogen production. *Journal of Cleaner Production, 372*. doi:10.1016/j.jclepro.2022.133541.

Oehlmann, N. N., Schmidt, F. V., Herzog, M., Goldman, A. L., & Rebelein, J. G. (2024). The iron nitrogenase reduces carbon dioxide to formate and methane under physiological conditions: A route to feedstock chemicals. *Science Advances, 10*(33). doi:10.1126/sciadv.ado7729.

Oflaz, F. B., & Koku, H. (2021). Pilot-scale outdoor photofermentative hydrogen production from molasses using pH control. *International Journal of Hydrogen Energy, 46*(57), 29160–29172. doi:10.1016/j.ijhydene.2020.10.086.

Öztürk, Y., Gökçe, A., Peksel, B., Gürgan, M., Özgür, E., Gündüz, U., Eroğlu, N., & Yücel, M. (2012). Hydrogen production properties of *Rhodobacter capsulatus* with genetically modified redox balancing pathways. *International Journal of Hydrogen Energy, 37*, 2014–2020. doi:10.1016/j.ijhydene.2011.06.078.

Policastro, G., Cesaro, A., & Fabbricino, M. (2023). Photo-fermentative hydrogen production from cheese whey: Engineering of a mixed culture process in a semi-continuous, tubular photo-bioreactor. *International Journal of Hydrogen Energy, 48*(55), 21038–21054. doi:10.1016/j.ijhydene.2022.07.063.

Ren, C., Zhang, S., Li, Q., Jiang, Q., Li, Y., Gao, Z., Cao, W., & Guo, L. (2024). Pilot composite tubular bioreactor for outdoor photo-fermentation hydrogen production: From batch to continuous operation. *Bioresource Technology, 401*. doi:10.1016/j.biortech.2024.130705.

Ren, T., & Ma, X. (2012). Effects of several chemical substances on enzymatic saccharification of rice straw. *Journal of Agro-Environment Science, 31*(01), 206–211.

Sagir, E., Alipour, S., Elkahlout, K., Koku, H., Gunduz, U., Eroglu, I., & Yucel, M. (2018). Biological hydrogen production from sugar beet molasses by agar immobilized *R. capsulatus* in a panel photobioreactor. *International Journal of Hydrogen Energy, 43*(32), 14987–14995. doi:10.1016/j.ijhydene.2018.06.052.

Salis, A., & Monduzzi, M. (2016). Not only pH. Specific buffer effects in biological systems. *Current Opinion in Colloid and Interface Science, 23*, 1–9. doi:10.1016/j.cocis.2016.04.004.

Sathitsuksanoh, N., Zhu, Z., Wi, S., & Percival Zhang, Y. H. (2011). Cellulose solvent-based biomass pretreatment breaks highly ordered hydrogen bonds in cellulose fibers of switchgrass. *Biotechnology and Bioengineering, 108*(3), 521–529. doi:10.1002/bit.22964.

Schreier, T. B., & Hibberd, J. M. (2019). Variations in the Calvin-Benson cycle: selection pressures and optimization? *Journal of Experimental Botany, 70*, 1697–1701. doi:10.1093/jxb/erz078.

Sen, A. S., Rao, R., & Basak, N. (2024). CFD modelling of hydrodynamics and temperature distribution during photo fermentative biohydrogen generation from cheese whey effluent in cylindrical bioreactor. *International Journal of Hydrogen Energy., 60*, 740–755. doi:10.1016/j.ijhydene.2024.02.207.

Sha, H. (2020). *Development and performance optimization of a static magnetic field-enhanced anaerobic fermentation system*. Northeast Electric Power University.

Shui, X., Jiang, D., Zhang, H., Zhang, X., Yang, J., Lei, T., Chen, Z., & Zhang, Q. (2022). Effect of light perturbation on the photo-fermentative hydrogen production of biomass from giant reed. *Journal of Cleaner Production, 351*. doi:10.1016/j.jclepro.2022.131481.

Siddiqui, O., & Dincer, I. (2019). A well to pump life cycle environmental impact assessment of some hydrogen production routes. *International Journal of Hydrogen Energy, 44*(12), 5773–5786. doi:10.1016/j.ijhydene.2019.01.118.

Singh, R. N., & Sharma, S. (2012). Development of suitable photobioreactor for algae production—A review. *Renewable and Sustainable Energy Reviews, 16*(4), 2347–2353. doi:10.1016/j.rser.2012.01.026.

Song, A., Ren, T., Xie, H., Pei, G., & Zhang, B. (2006). *Effect of Chemical Pretreatment on Enzymatic Hydrolysis and Saccharification of Corn Stover. Chemistry and Bioengineering, 8*, 31–33.

Song, Y. (2006). Progress of exploitation and utilization on energy plant. *Biomass Chemical Engineering*, (06), 51–53.

Sun, C., Xia, A., Liao, Q., Fu, Q., Huang, Y., & Zhu, X. (2019). Life-cycle assessment of biohythane production via two-stage anaerobic fermentation from microalgae and food waste. *Renewable and Sustainable Energy Reviews, 112*, 395–410. doi:10.1016/j.rser.2019.05.061.

Sun, P., Lu, C., Liang, X., Wang, G., Song, C., Zhang, Q., Zhang, Z., Wang, H., Alam, M., Liu, H., Wang, K., Xia, C., & Jiang, D. (2024). Enhanced the energy conversion of corn stalk via co-production of photo-fermentation biohydrogen and bioethanol. *Bioresource Technology, 408*, 131196. doi:10.1016/j.biortech.2024.131196.

Sun, Y., & Cheng, J. (2002). Hydrolysis of lignocellulosic materials for ethanol production: A review. *Bioresource Technology, 83*(1), 1–11. doi:10.1016/S0960-8524(01)00212-7.

Susmozas, A., Iribarren, D., Zapp, P., Linβen, J., & Dufour, J. (2016). Life-cycle performance of hydrogen production via indirect biomass gasification with CO_2 capture. *International Journal of Hydrogen Energy, 41*(42), 19484–19491. doi:10.1016/j.ijhydene.2016.02.053.

Syed Jaapa, S. Z., Kalil, M. S., & Anuar, N. (2009). The effect of aeration, agitation and light on biohydrogen production by *Rhodobacter sphaeroides* NCIMB 8253. *Pakistan Journal of Biological Sciences, 12*(18), 1253–1259. doi:10.3923/pjbs.2009.1253.1259.

Taguchi, F., Mizukami, N., Yamada, K., Hasegawa, K., & Saito-Taki, T. (1995). Direct conversion of cellulosic materials to hydrogen by *Clostridium* sp. strain no. 2. *Enzyme and Microbial Technology, 17*(2), 147–150. doi:10.1016/0141-0229(94)00058-Y.

Taguchi, F., Yamada, K., Hasegawa, K., Taki-Saito, T., & Hara, K. (1996). Continuous hydrogen production by *Clostridium* sp. strain no. 2 from cellulose hydrolysate in an aqueous two-phase system. *Journal of Fermentation and Bioengineering, 82*(1), 80–83. doi:10.1016/0922-338X(96)89460-8.

Tan, F., Wu, B., Dai, L., Zhu, Q., Wang, W., Tang, X., Pan, K., Qin, H., Hu, Q., & He, M. (2014). Research status of cellulose-based herbaceous energy plants. *Chinese Journal of Applied and Environmental Biology, 20*(01), 162–168.

Van Ginkel, S. W., & Logan, B. (2005). Increased biological hydrogen production with reduced organic loading. *Water Research, 39*(16), 3819–3826. doi:10.1016/j.watres.2005.07.021.

Vasco-Correa, J., & Li, Y. (2015). Solid-state anaerobic digestion of fungal pretreated *Miscanthus sinensis* harvested in two different seasons. *Bioresource Technology, 185*, 211–217. doi:10.1016/j.biortech.2015.02.099.

Vignais, P. M., & Billoud, B. (2007). Occurrence, classification, and biological function of hydrogenases: An overview. *Chemical Reviews., 107*(10), 4206–4272. doi:10.1021/cr050196r.

Wang, J., Liu, Z., Liang, R., Yan, B., Tao, J., Su, H., & Chen, G. (2022a). Aqueous phase reforming of distiller's grain derived biogas plant wastewater over α-MoO_3 nanosheets. *Chemical Engineering Journal, 430*. doi:10.1016/j.cej.2021.132735.

Wang, J., & Wan, W. (2009). Factors influencing fermentative hydrogen production: A review. *International Journal of Hydrogen Energy, 34*(2), 799–811. doi:10.1016/j.ijhydene.2008.11.015.

Wang, J., & Yin, Y. (2018). Fermentative hydrogen production using various biomass-based materials as feedstock. *Renewable and Sustainable Energy Reviews, 92*, 284–306. doi:10.1016/j.rser.2018.04.033.

Wang, K., Lu, C., Zhang, H., Guo, S., Ru, G., Wang, J., Hu, J., Zhang, N., & Zhang, Q. (2022b). Enhancement effect of defoamer additives on photo-fermentation biohydrogen production process. *Bioresource Technology, 352*, 127070. doi:10.1016/j.biortech.2022.127070.

Wang, L.-L., Tao, Y., & Mao, X.-Z. (2014). A novel flat plate algal bioreactor with horizontal baffles: Structural optimization and cultivation performance. *Bioresource Technology, 164*, 20–27. doi:10.1016/j.biortech.2014.04.100.

Wang, X., Zhu, X., Liao, Q., Wang, Y., Wang, H., & Ding, Y. (2012). Effects of light conditions on transmembrane transport and hydrogen production characteristics of photosynthetic bacteria. *Acta Energiae Solaris Sinica, 33*(10), 1813–1819.

Wang, Y., Joshee, N., Cao, W., Wu, Q., & Tahir, N. (2019). Continuous hydrogen production by dark and photo co-fermentation using a tubular multi-cycle bio-reactor with *Paulownia* biomass. *Cellulose, 26*(15), 8429–8438. doi:10.1007/s10570-019-02468-z.

Weerachanchai, P., Kwak, S. K., & Lee, J. M. (2014). Effects of solubility properties of solvents and biomass on biomass pretreatment. *Bioresource Technology, 170*, 160–166. doi:10.1016/j.biortech.2014.07.057.

Xiang, G., Zhang, H., Li, Y., Liu, H., Zhang, Z., Lu, C., & Zhang, Q. (2023). Enhancing biohydrogen yield from corn stover by photo fermentation via adjusting photobioreactor headspace pressure. *Bioresource Technology, 369*. doi:10.1016/j.biortech.2022.128388.

Xiong, J. (2015). *Properties and composition of different types of wheat straw*. Nanjing Agricultural University.

Xu, B. (2011). *Treatment of corncob hemicellulose by steam explosion*. Jiangnan University.

Yang, C., Shen, Z., Yu, G., & Wang, J. (2008). Effect and aftereffect of γ radiation pretreatment on enzymatic hydrolysis of wheat straw. *Bioresource Technology, 99*(14), 6240–6245. doi:10.1016/j.biortech.2007.12.008.

Yang, X., Li, Y., Zhang, N., Zhang, H., Liu, H., Xiang, G., & Zhang, Q. (2023). Influence of titanate photocatalyst in biohydrogen yield via photo fermentation from corn stover. *Bioresource Technology, 386*, 129544. doi:10.1016/j.biortech.2023.129544.

Yang, Z., Cheng, J., Xu, X., Zhou, J., & Cen, K. (2016). Enhanced solution velocity between dark and light areas with horizontal tubes and triangular prism baffles to improve microalgal growth in a flat-panel photo-bioreactor. *Bioresource Technology, 211*, 519–526. doi:10.1016/j.biortech.2016.03.145.

Yue, G., Lin, H., & Peng, Y. (2021). Future green hydrogen energy from biomass raw materials. *Chemical Industry and Engineering Progress, 40*(08), 4678–4684.

Yue, J. (2011). *Experimental study on physical properties, microstructure, and photosynthetic biohydrogen production of ultrafine straw powder*. Henan Agricultural University.

Yue, J., Li, G., & Jiao, Y. (2011). The effect of illuminance on photosynthetic hydrogen production from enzymatic hydrolysis of straw. *Transactions of the Chinese Society of Agricultural Engineering, 27*(08), 313–317.

Zeng, G. (2012). *Pretreatment of corn stalk with metal ions and kinetic study of fermentation*. Chongqing University.

Zhang, A., & Jun, X. (2014). Research progress on pretreatment methods for biomass ethanol production. *South China Agricultural University, 35*(04), 77–84.

Zhang, A., Wang, Y., & Wang, X. (2018a). Optimization of hydrogen production by photosynthetic bacteria using straw depolymerization solution. *Journal of Shaanxi University of Science & Technology, 36*(01), 17–22.

Zhang, J., & Wang, Q. (2013). Buffering and nutrient effects of white mud from ammonia-soda process on thermophilic hydrogen fermentation from food waste. *International Journal of Hydrogen Energy, 38*(31), 13564–13571. doi:10.1016/j.ijhydene.2013.08.047.

Zhang, Q., Jiao, Y., He, C., Ruan, R., Hu, J., Ren, J., Toniolo, S., Jiang, D., Lu, C., Li, Y., Man, Y., Zhang, H., Zhang, Z., Xia, C., Wang, Y., Jing, Y., Zhang, X., Lin, R., Li, G., … Tahir, N. (2024). Biological fermentation pilot-scale systems and evaluation for commercial viability towards sustainable biohydrogen production. *Nature Communications, 15*(1). doi:10.1038/s41467-024-48790-4.

Zhang, Q., Lu, C., Lee, D. J., Lee, Y. J., Zhang, Z., Zhou, X., Hu, J., Wang, Y., Jiang, D., He, C., & Zhang, T. (2017). Photo-fermentative hydrogen production in a 4 m^3 baffled reactor: Effects of hydraulic retention time. *Bioresource Technology, 239*, 533–537. doi:10.1016/j.biortech.2017.05.067.

Zhang, Q., Yang, J., Zhang, T., Shui, X., Zhang, H., Chen, Z., He, X., Lei, T., Jiang, D., Elgorban, A. M., Syed, A., & Kumar Solanki, M. (2023). Pretreatment of *Arundo donax* L. for photo-fermentative biohydrogen production by ultrasonication and ionic liquid. *Bioresource Technology, 377*, 128904. doi:10.1016/j.biortech.2023.128904.

Zhang, Q., Zhang, Z., Wang, Y., Lee, D. J., Li, G., Zhou, X., Jiang, D., Xu, B., Lu, C., Li, Y., & Ge, X. (2018b). Sequential dark and photo fermentation hydrogen production from hydrolyzed corn stover: A pilot test using 11 m^3 reactor. *Bioresource Technology, 253*, 382–386. doi:10.1016/j.biortech.2018.01.017.

Zhang, X., Jiang, D., Zhang, H., Wang, Y., Zhang, Z., Lu, C., & Zhang, Q. (2021). Enhancement of the biohydrogen production performance from mixed substrate by photo-fermentation: Effects of initial pH and inoculation volume ratio. *Bioresource Technology, 319*, 124153. doi:10.1016/j.biortech.2020.124153.

Zhang, Y., Zhang, H., Lee, D. J., Zhang, T., Jiang, D., Zhang, Z., & Zhang, Q. (2020a). Effect of enzymolysis time on biohydrogen production from photo-fermentation by using various energy grasses as substrates. *Bioresource Technology, 305*. doi:10.1016/j.biortech.2020.123062.

Zhang, Z., Li, L., Zhu, H., & Ai, Z. (2020b). Investigation of the interaction between lighting and mixing applied during the photo-fermentation biohydrogen production process from agricultural waste. *Bioresource Technology, 312*.

Zhang, Z., Wang, Y., Hu, J., Wu, Q., & Zhang, Q. (2015). Influence of mixing method and hydraulic retention time on hydrogen production through photo-fermentation with mixed strains. *International Journal of Hydrogen Energy, 40*(20), 6521–6529. doi:10.1016/j.ijhydene.2015.03.118.

Zhao, J. (2021). *Analysis and optimization of dissipation rate in the process of photosynthetic biological hydrogen production*. Henan Agricultural University.

Zhao, J., Ge, X., Vasco-Correa, J., & Li, Y. (2014). Fungal pretreatment of unsterilized yard trimmings for enhanced methane production by solid-state anaerobic digestion. *Bioresource Technology, 158*, 248–252. doi:10.1016/j.biortech.2014.02.029.

Zheng, W. (2004). *The effect of hot alkali pretreatment on anaerobic fermentation of straw*. China Agricultural University.

Zheng, Y., Harris, D. F., Yu, Z., Fu, Y., Poudel, S., Ledbetter, R. N., Fixen, K. R., Yang, Z. Y., Boyd, E. S., Lidstrom, M. E., Seefeldt, L. C., & Harwood, C. S. (2018). A pathway for biological methane production using bacterial iron-only nitrogenase. *Nature Microbiology, 3*(3), 281–286. doi:10.1038/s41564-017-0091-5.

Zhu, S. (2023). *The mediating effect and enhancement mechanism of zero-valent iron on photo-fermentative hydrogen production from corn stover*. Henan Agricultural University.

CHAPTER

6

Combined dark–photo fermentation biohydrogen production from organic matters

Yameng Li[a], *Zhiping Zhang*[a,*], *Hang Yan*[a], *Lu Feng*[b], *Ikumi Umetani*[b], *Zhihan Zhang*[c], *and Junyu Tao*[d]

[a]Henan Agricultural University, Zhengzhou, China, [b]Norwegian Institute of Bioeconomy Research (NIBIO), Ås, Norway, [c]Zhengzhou University of Aeronautics, Zhengzhou, China, [d]Henan Rural Social Development Service Center, Zhengzhou, China

*Corresponding author. E-mail address: zhangzhiping715@163.com

6.1 Introduction

With the escalating global demand for renewable energy and increasing awareness of the environmental impacts of fossil fuel consumption, biological hydrogen production has attracted growing attention as a sustainable alternative. This approach utilizes renewable biomass under mild reaction conditions, offering advantages such as reduced emissions and lower ecological footprint. Currently, biohydrogen generation through microbial fermentation is mainly categorized into dark fermentation and photo-fermentation. The integration of these two processes—commonly referred to as dark–photo combined fermentation—has emerged as a key area of investigation owing to its complementary characteristics. In this hybrid system, microorganisms first decompose organic substrates

Biohydrogen: Advances in Theory and Production
DOI: https://doi.org/10.1016/B978-0-443-27686-6.00009-8

anaerobically to produce hydrogen and intermediate metabolites, which are subsequently converted into additional hydrogen under light-driven conditions. The synergistic coupling of these stages enables more complete carbon conversion, improved energy yield, and reduced waste accumulation. As a promising biobased route, dark–photo fermentation not only enhances resource efficiency but also represents a potential technological pathway for addressing global challenges related to energy scarcity and environmental degradation.

6.1.1 Substrate type

In biohydrogen production via combined dark–photo fermentation the choice of substrate plays a fundamental role, as its type and composition largely determine system efficiency and energy yield. A wide range of organic materials can serve as feedstocks for this process, offering abundant and renewable sources. Commonly used substrates include carbohydrates, organic acids, agricultural and forestry residues, and various industrial organic wastes, all of which exhibit strong fermentability (Ghosh et al., 2017; Xu et al., 2022).

Carbohydrates are among the most studied substrates because of their simple structure and high biodegradability. Glucose, for instance, is readily assimilated by microorganisms and rapidly converted into hydrogen through enzymatic reactions that efficiently channel its chemical energy. Sucrose, a plant-derived disaccharide, is first hydrolyzed into glucose and fructose before entering the fermentative pathway, providing a cost-effective and accessible carbon source widely used in both laboratory research and pilot-scale studies (Zheng et al., 2014).

Organic acids also play crucial roles in the metabolic pathways of hydrogenogenic microorganisms. Acetate, a common intermediate, is both a by-product and a substrate that can be reused by specific microbial species, promoting cyclic material flow within the fermentation system. Similarly, lactic acid can act as a hydrogen precursor under anaerobic conditions, where its metabolism releases electrons and protons subsequently converted to hydrogen gas. The application of lactic acid in organic waste treatment thus provides simultaneous benefits for pollution mitigation and renewable energy recovery.

Agricultural by-products such as corn and wheat straw are rich in lignocellulosic compounds, including cellulose, hemicellulose, and lignin. Although these polymers are not directly fermentable, pretreatment methods can depolymerize them into fermentable sugars suitable for microbial utilization (Cheng et al., 2011). Such valorization of crop residues not only alleviates environmental issues associated with open-field burning but also transforms agricultural waste into a sustainable energy source. Livestock manure, containing residual carbohydrates and proteins, can similarly undergo anaerobic digestion to produce hydrogen, achieving both waste stabilization and energy recovery (Safonov et al., 2026).

Industrial wastewater represents another promising substrate category. Effluents from food processing industries, rich in sugars, fats, and proteins, can cause severe water pollution if untreated. Incorporating them into biohydrogen production systems allows microorganisms to convert the organic load into hydrogen while simultaneously purifying the wastewater, achieving dual environmental and energetic benefits (Kapdan & Kargi, 2006). Likewise, pulp and paper mill wastewater, which contains lignin-derived compounds, can be utilized after microbial community adaptation to enhance hydrogen yield and promote cleaner production practices.

Furthermore the physicochemical characteristics of the substrate significantly influence hydrogen yield and the distribution of metabolic by-products. Starch-based feedstocks generally achieve higher hydrogen production efficiency than lignocellulosic materials, which are more resistant to microbial degradation. From a biochemical standpoint, readily biodegradable carbohydrates—such as hexoses (e.g., glucose), pentoses (e.g., xylose), and disaccharides (e.g., sucrose, lactose)—are particularly suitable for high-performance dark–photo fermentation systems.

6.1.2 Substrate pretreatment

In integrated dark–photo fermentation systems for biohydrogen production, substrate pretreatment serves as a crucial preliminary step. It functions to unlock the intrinsic energy within raw materials, improve their accessibility to microbial communities, and consequently enhance both process efficiency and operational stability. Pretreatment facilitates the breakdown of complex organic structures, converting them into simpler and more digestible forms suitable for fermentation. Generally, pretreatment strategies are categorized into three main types: physical, chemical (Machineni et al., 2023), and biological methods, as summarized in Table 6.1.

The following sections are the typical methods used for pretreatment.

6.1.2.1 Physical pretreatment methods

Mechanical pretreatment, such as crushing and grinding, is widely employed for agricultural residues, including straw and other bulky biomass. By reducing particle size, these processes increase the specific surface area and enhance microbial accessibility. For instance, converting corn straw from long strips into fine fragments or powder exposes more reactive sites for extracellular enzymes, thereby accelerating the hydrolysis of cellulose and hemicellulose into fermentable sugars and ultimately boosting hydrogen generation (Ji et al., 2016).

TABLE 6.1 Substrate pretreatment method and characteristics.

Preprocessing type	Method	Merit	Shortcoming
Physical method pretreatment	Mechanical crushing, microwave, ultrasound, thermal pretreatment, etc.	Simple operation, large processing capacity, do not produce toxic substances	The decomposition is not complete, the processing is incomplete and easy to interfere with, increasing additional energy consumption
Chemical method pretreatment	Acid/base pretreatment, organic solvents, ionic liquids, etc.	Good treatment effect, high efficiency, and some reagents can be reused	High cost, producing toxic side crops, poor operating environment, and corrosion to the reactor
Biological method pretreatment	Fungi, bacteria, and mixed microflora	Low energy consumption, mild reaction conditions, no pollution	The reaction cycle is long, and the microorganisms are susceptible to environmental influence

Thermal pretreatment methods, such as roasting, have also been demonstrated to improve the enzymatic digestibility of lignocellulosic biomass. Roasting corn straw can enhance saccharification efficiency and increase hydrogen yield during subsequent photo-fermentation processes (Li et al., 2022).

Ultrasonic treatment offers another efficient approach to improving substrate reactivity. High-intensity ultrasound induces cavitation, and the collapse of microbubbles produces localized shock waves and micro-jets that disrupt biomass structure, loosen fiber matrices, and decrease cellulose crystallinity. These structural modifications facilitate chemical reagent diffusion, improve microbial adhesion, and accelerate enzymatic hydrolysis. Similarly, Azuma et al. (1984) reported that microwave-assisted pretreatment at 170°C–230°C partially degraded lignin and hemicellulose, enhancing cellulose accessibility and significantly increasing enzymatic conversion rates in plant biomass.

6.1.2.2 Chemical pretreatment method

Among various chemical pretreatment strategies, acid and alkali treatments are the most extensively explored approaches for enhancing substrate digestibility.

Dilute acid pretreatment, typically using sulfuric or hydrochloric acid, effectively disrupts the compact matrix of lignocellulosic biomass (Brodeur et al., 2011). This process promotes hemicellulose hydrolysis, facilitates sugar release, and causes partial lignin precipitation, although it may also generate inhibitory by-products (Lopez-Hidalgo et al., 2017). After hydrolysis the cellulose and lignin fractions largely remain intact within the residual solid phase (Guerra-Rodriguez et al., 2012). Under optimized conditions of acid concentration, temperature, and residence time, complex polysaccharides are converted into fermentable monosaccharides such as glucose and xylose, which can directly participate in microbial hydrogen metabolism. For instance, applying dilute acid pretreatment to pulp and paper wastewater can effectively break down lignin, enhance cellulose conversion, and increase substrate bioavailability for hydrogen-producing bacteria (Li et al., 2024). However, overly harsh acid treatment may lead to the formation of toxic compounds—such as furfural, 5-hydroxymethylfurfural (5-HMF), and acetic acid—that inhibit microbial growth and reduce hydrogen yields (Azuma et al., 1984; Brodeur et al., 2011).

Alkaline pretreatment, by contrast, employs agents such as sodium hydroxide (NaOH) or calcium hydroxide ($Ca(OH)_2$) to cleave ester bonds between lignin and hemicellulose and to saponify phenolic esters within lignin. This process effectively reduces lignin's shielding effect and increases the enzymatic accessibility of cellulose. Compared with acid hydrolysis, alkaline methods are milder, generate fewer inhibitors, and improve downstream enzymatic digestion. Nevertheless, substrates treated with alkali generally require neutralization prior to fermentation to prevent pH stress on hydrogen-producing microorganisms.

6.1.2.3 Biological pretreatment methods

Biological pretreatment provides an environmentally benign alternative that relies on microorganisms and their enzymatic systems to degrade lignocellulosic structures under mild conditions (Wu et al., 2022). This method minimizes chemical reagent use and energy input while selectively removing lignin and hemicellulose. Cocultivation of multiple microbial species often enhances pretreatment efficiency through synergistic interactions between fungi and bacteria. In such systems, lignin-degrading fungi, such as white rot species, initially decompose lignin to expose polysaccharides, whereas bacteria metabolize the released carbohydrates into fermentable intermediates that further support fungal activity. For example, coculturing white rot fungi with hydrogen-producing bacteria for straw pretreatment allows fungi to preferentially degrade lignin and loosen the fiber structure, after which bacteria convert liberated sugars into hydrogen. This sequential cooperation improves substrate breakdown and promotes efficient energy recovery.

Each pretreatment approach has specific strengths and constraints. The optimal selection depends on the physicochemical nature of the substrate, process parameters, and overall economic considerations. In many cases, combining two or more pretreatment methods yields synergistic effects, leading to improved substrate accessibility and higher hydrogen production efficiency. Effective pretreatment is thus a pivotal step in enhancing the performance, reliability, and scalability of dark–photo-integrated fermentation systems for biohydrogen generation.

6.2 Reactors

6.2.1 Reactor structure

In dark–photo fermentation systems utilizing organic substrates, reactor configuration plays a pivotal role in achieving high efficiency and operational stability. Different reactor designs exhibit unique structural and functional characteristics, and optimizing them for effective mass and heat transfer while maintaining suitable microbial growth conditions is essential for stable and large-scale hydrogen generation. Broadly, bioreactors are divided into two main categories: suspended-growth systems and immobilized-cell systems.

For biohydrogen production, several reactor types have been widely adopted, including the continuous stirred tank reactor (CSTR), packed-bed reactor (PBR), fluidized bed reactor (FBR), anaerobic sequencing batch reactor (ASBR), and anaerobic packed bed biofilm reactor (APBR) (Kumar et al., 2017; Łukajtis et al., 2018). Each configuration possesses distinct operational advantages and limitations, forming the technological basis for optimizing hydrogen yield and process stability. The comparative advantages of these reactor types are summarized in Table 6.2.

The five reactor configurations mentioned above can all be employed for anaerobic biohydrogen production; however, three are generally considered most practical depending on substrate properties, solids concentration, organic loading rate, and process scale. Specifically, CSTRs are preferred for high-solid slurries and systems requiring straightforward operational control; PBRs are suitable when maximizing attached biomass and substrate conversion is desired; and anaerobic FBRs (AFBRs) are advantageous for achieving high mass transfer rates, maintaining continuous stability, and minimizing clogging. The structural and operational characteristics of these three systems are detailed below.

6.2.1.1 Continuous stirred-tank reactor (CSTR)

The CSTR is a continuously operated, mechanically mixed vessel typically equipped with a central shaft and bottom-mounted impellers.

TABLE 6.2 Full name and advantages of the reactor.

Abbreviations	Full name	Advantage
CSTR	Continuous stirred-tank reactor	The substrate and microorganism are mixed evenly to ensure the stability of the fermentation reaction rate and are conducive to the immediate action of enzyme and substrate
PBR	Packed bed reactor	Optimize the light path and light area; improve the utilization efficiency of light energy; accurately regulate the light quality; and stimulate the hydrogen production potential of photosynthetic microorganisms
AFBR	Anaerobic fluidized bed reactor	High volumetric loading rate, high removal efficiency, small footprint, strong resistance to shock loads, biological particle fluidization promotes the renewal of the interface between the biofilm and sewage, which is conducive to mass transfer, high treatment efficiency, less prone to blockage, and easy to control and operate
ASBR	Anaerobic sequencing batch reactor	Intermittent operation adapts to the fluctuation of water quality and water quantity, traps microorganisms in the precipitation stage, maintains high biomass, and effectively treats high concentration of organic wastewater
APBR	Anaerobic packed bed biofilm reactor	Intermittent operation adapts to the fluctuation of water quality and water quantity, traps microorganisms in the precipitation stage, maintains high biomass, and effectively treats high concentration of organic wastewater
CSBR	Conical spouted bed reactor	It is flexible, has a strong ability to treat complex wastewater, and is widely used in the field of anaerobic fermentation

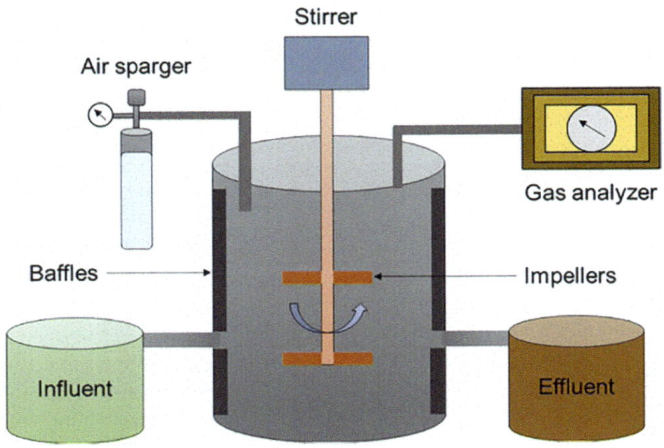

FIGURE 6.1 A continuous stirred tank reactor. *Sahil, S., Singh, R., Masakapalli, S. K., Pareek, N., Kovalev, A. A., Litti, Y. V., Nanda, S., & Vivekanand, V. (2024). Biomass pretreatment, bioprocessing and reactor design for biohydrogen production: A review.* Environmental Chemistry Letters, 22*(4), 1665–1702. https://doi.org/10.1007/s10311-024-01722-6.*

Substrate feed is continuously introduced while maintaining anaerobic conditions. Inert gas (e.g., N_2) may be sparged through the reactor either via a porous distributor or a single-orifice nozzle to facilitate gas stripping and maintain low partial pressures of H_2 and CO_2. Efficient agitation promotes homogeneity within the liquid phase, ensuring uniform distribution of biomass, nutrients, and temperature, while also enhancing gas–liquid mass transfer efficiency.

In biohydrogen production the CSTR design is particularly valuable for its ability to sustain a stable and well-mixed microbial environment. Its operational principle centers on high-speed mechanical mixing driven by a precision motor that rotates internal impellers, ensuring intimate contact between microorganisms and substrates. At the microscopic level, this uniform dispersion allows substrate molecules to continuously reach microbial cell surfaces, where extracellular enzymes rapidly catalyze hydrolysis and conversion reactions leading to hydrogen generation. The reactor structure and mixing dynamics are illustrated in Fig. 6.1 (Sahil et al., 2024).

Taking the dark fermentation of glucose as an illustrative case, continuous agitation in a CSTR maintains a uniform distribution of glucose molecules throughout the reactor, preventing localized substrate concentration gradients. This homogeneity ensures a stable fermentation rate and enhances process reproducibility. Despite these operational benefits the CSTR design also presents certain limitations. The primary concern lies in its relatively high energy consumption—intensive mechanical stirring

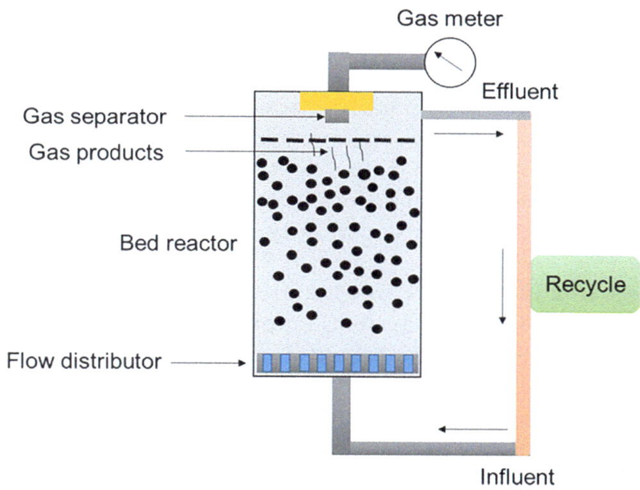

FIGURE 6.2 An anaerobic fluidized bed reactor. *Sahil, S., Singh, R., Masakapalli, S. K., Pareek, N., Kovalev, A. A., Litti, Y. V., Nanda, S., & Vivekanand, V. (2024). Biomass pretreatment, bioprocessing and reactor design for biohydrogen production: A review.* Environmental Chemistry Letters, 22*(4), 1665–1702. https://doi.org/10.1007/s10311-024-01722-6.*

requires substantial power input, which elevates operational costs and diminishes the overall energy conversion efficiency of the system. In large-scale industrial settings the cumulative energy demand associated with continuous mixing becomes a critical barrier, posing challenges to the economic viability and sustainable expansion of CSTR-based biohydrogen production technologies.

6.2.1.2 Anaerobic fluidized bed reactor (AFBR)

The reactor generally features a cylindrical body constructed from durable, corrosion-resistant materials such as stainless steel or reinforced polymers, enabling it to operate stably under diverse and demanding fermentation conditions. Inside the vessel, microparticle carriers—commonly sand, activated carbon, or ceramsite—are loaded to support microbial immobilization. These carriers exhibit fine particle size for fluid mobility, large surface area for microbial colonization, suitable roughness to enhance adhesion, and chemical inertness that prevents interference with microbial activity or substrate conversion (Tseng & Lin, 1994).

This configuration provides excellent hydrodynamic mixing and efficient mass transfer. As illustrated in Fig. 6.2 (Sahil et al., 2024), partially degraded effluents can be recirculated into the reactor, effectively reducing the likelihood of biomass washout (Da Silva et al., 2019) (Table 6.1).

Such systems have been successfully applied to treat diverse wastewaters, including dairy effluents, sugarcane stillage, and synthetic substrates containing glucose or other organics (Barros et al., 2011; Santos et al., 2014). However, under high organic loading rates, archaeal proliferation has been observed, which may consume hydrogen and consequently lower the overall biohydrogen yield (Da Silva et al., 2019). Moreover, periods of elevated hydrogen production are often associated with reduced bacterial diversity, suggesting microbial community shifts that can influence process robustness.

In a representative study, Ramos and Silva (2018) operated an AFBR for a two-stage fermentation system, reporting a 53% increase in energy recovery at an organic loading rate of 21.6 kg chemical oxygen demand (COD)/$m^3{}_h$day (Gois et al., 2021). Similarly, Abud and Amorim evaluated hydrogen generation from sugarcane vinasse using the same design, achieving a maximum yield of 2.4 mol H_2/mol substrate at a hydraulic retention time (HRT) of 4 hours, confirming that vinasse could be utilized without alkalization pretreatment. Ferreira et al. (2019) further demonstrated that shortening the HRT from 8 to 1 hours markedly enhanced hydrogen productivity when sugarcane juice served as the feedstock.

In operation, wastewater rich in organic matter is introduced from the bottom of the reactor and flows upward through the fluidized bed, interacting with microorganisms attached to carrier particles. Under anaerobic conditions, fermentation proceeds, and biogas generated within the reactor is released from the upper outlet.

AFBRs present several distinct operational merits in biohydrogen generation and wastewater treatment:

Superior mass and heat transfer: Upflow contact between liquid and immobilized microbes ensures high transfer efficiency, whereas the fluidized state enhances process stability.

High microbial retention: The carriers' extensive surface area supports dense biofilm formation, maintaining high biomass levels.

Improved substrate utilization: Continuous mixing facilitates efficient microbe–substrate interaction, promoting rapid conversion.

Shock-load resistance: AFBRs tolerate moderate variations in influent composition or organic loading without major performance decline (Dowaidar, 2025; Kaul & Gadakari, 1990).

Low clogging tendency: The fluidized motion of particles prevents blockages common in fixed-bed designs, reducing maintenance.

Efficient three-phase separation: Integrated separators effectively isolate gas, liquid, and solids, improving biogas recovery efficiency.

Compact configuration: High biomass concentration enables substantial throughput in a relatively small volume.

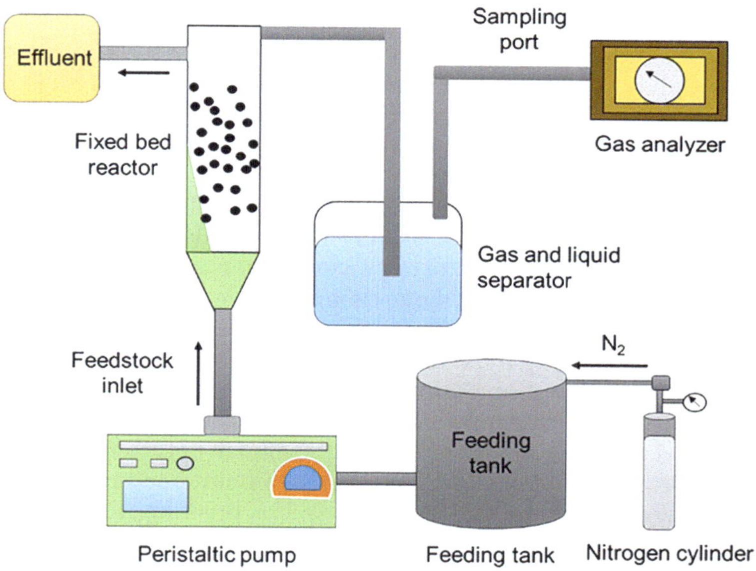

FIGURE 6.3 Packed-bed reactor. *Sahil, S., Singh, R., Masakapalli, S. K., Pareek, N., Kovalev, A. A., Litti, Y. V., Nanda, S., & Vivekanand, V. (2024). Biomass pretreatment, bioprocessing and reactor design for biohydrogen production: A review.* Environmental Chemistry Letters, 22(4), 1665–1702. *https://doi.org/10.1007/s10311-024-01722-6.*

Sustained continuous operation: AFBRs function steadily over extended periods without frequent backwashing, minimizing operational labor and cost.

6.2.1.3 *Packed-bed reactor (PBR)*

PBRs are widely recognized for their operational simplicity and ability to achieve favorable reaction rates. By introducing inert gases such as nitrogen or argon, PBRs can readily maintain anaerobic conditions, as illustrated schematically in Fig. 6.3 (Sahil et al., 2024). These reactors are particularly effective for processing biomass with high organic content.

Reactor body: PBRs generally consist of a columnar, pressure-resistant vessel constructed from materials such as stainless steel, glass, or plastics. Stainless steel is preferred for industrial-scale applications because of its robustness and resistance to high pressures and temperatures. Glass reactors, with high optical transparency, are commonly used in laboratory studies, allowing direct observation of internal processes. Plastic vessels, being lightweight and cost-efficient, are suitable for small-scale setups where mechanical strength is less critical (Andrigo et al., 1999).

Packing carrier: The carrier material inside the reactor plays a central role in microbial immobilization and biofilm formation. Common carriers include ceramics, activated carbon, polyurethane foam, polystyrene, expanded clay, glass beads, perlite, and plastic-based media. These carriers generally exhibit large specific surface area, biocompatibility, chemical stability, and suitable porosity, facilitating microbial colonization, substrate diffusion, and mass transfer, thereby enhancing fermentation performance (Andrigo et al., 1999).

Inlet and outlet configuration: PBRs are equipped with dedicated feed and effluent ports. The feed inlet, usually located at the top or side, ensures even distribution of the substrate to prevent local concentration extremes. The effluent outlet, typically at the bottom, discharges treated liquid while gases generated during fermentation can exit through dedicated ports. Inert gases can also be sparged to maintain anaerobic conditions.

Microbial attachment and growth: During reactor start-up, microorganisms adhere to carrier surfaces through physical adsorption, chemical bonding, and biofilm formation, with extracellular polymeric substances (EPS) supporting immobilization. Progressive biofilm thickening enhances microbial retention, promotes cooperative interactions, improves substrate utilization, and strengthens hydrogen production.

During operation, gaseous products such as hydrogen partially dissolve in the liquid phase and are carried out with the effluent, whereas the remainder is released through the gas outlet. Immobilized microbes are retained in the system, preventing washout. As long as substrate feed and favorable conditions (e.g., pH, temperature, anaerobic environment) are maintained, the microbial community sustains activity, enabling continuous hydrogen production.

PBRs, owing to their inert carriers, high surface area, and porosity, allow extended biomass retention and elevated hydrogen yields, even at relatively short HRTs (Fuess et al., 2016; Namdarimonfared et al., 2023; Ramasamy et al., 2004). By maintaining proper substrate concentrations, high conversion efficiency can be achieved. However, compared with CSTRs, PBRs provide less intensive mixing, potentially limiting mass transfer and hydrogen productivity. To address this, recirculation loops are often implemented to improve substrate–microbe contact. Despite this limitation, PBRs benefit from a simpler design with lower construction and maintenance costs, as no mechanical agitation or external sedimentation units are required (Gomes et al., 2015).

6.2.2 Reactor operation

In combined dark–photo fermentation of organic matter, precise regulation of reactor operation is essential to ensure efficient and stable hydrogen production. Given the diversity of reactor types, each type shows different characteristics in structure and function, and its operation points naturally have different focuses. Operating parameters, therefore, play important role in optimizing the fermentation process and improving the product's yield and quality.

6.2.2.1 Continuous stirred-tank reactor (CSTR) operation

In a CSTR the feed rate is a critical parameter that directly affects microbial metabolism and hydrogen production. Proper control requires consideration of substrate concentration, microbial growth rate, and reactor volume. The optimal feed rate should provide a continuous, stable supply of substrate to meet microbial metabolic demands while avoiding excessive accumulation. For example, when using a high-concentration glucose solution, a feed rate that is too fast can cause substrate concentrations to exceed microbial metabolic capacity, resulting in substrate inhibition and decreased hydrogen yield. Conversely an overly slow feed rate may underutilize reactor volume and limit microbial growth, also reducing hydrogen productivity (Arooj et al., 2008; Maayedukondalu, 2015).

Determining the optimal feeding strategy typically requires experimental data to establish mathematical models relating substrate concentration, feed rate, and hydrogen output. In practice, feed rates should be adjusted according to microbial growth phases. During the logarithmic growth phase the feed rate can be increased to accommodate rapid microbial proliferation and elevated hydrogen production.

Agitation is a central operational factor in CSTRs. The stirring speed influences the homogeneity of substrate distribution, cell–substrate contact, and enzymatic degradation efficiency. Proper agitation ensures uniform dispersion of substrates and promotes effective microbial activity. However, excessively high stirring rates generate strong shear forces that can damage microbial cells, disrupt physiological functions, and ultimately reduce hydrogen production (Aubin et al., 2006).

Optimal stirring conditions are usually determined experimentally. Starting at a low speed, agitation is gradually increased while monitoring hydrogen production rate and microbial activity. The speed at which hydrogen yield peaks without compromising microbial function is considered optimal. In addition, agitator design (e.g., flat-blade vs. inclined-blade impellers) influences mixing efficiency and should be selected based on reactor geometry and specific operational needs (Aubin et al., 2006).

HRT governs the residence of substrate in the reactor and directly affects microbial contact time. Its duration significantly impacts hydrogen production: An excessively long HRT may lead to product accumulation, inhibiting microbial metabolism, increasing endogenous respiration, and wasting energy and substrate. Conversely a short HRT may result in incomplete substrate utilization and suboptimal hydrogen yields.

HRT optimization should consider substrate characteristics, microbial metabolic capacity, and reactor design. In practice the influent flow rate can be adjusted experimentally to identify the HRT that maximizes hydrogen production. Continuous monitoring and timely adjustments are necessary to account for variations in substrate concentration, flow rate, and microbial adaptation, ensuring process stability and efficiency.

Recent studies have focused on HRT-related issues such as biomass washout and optimal hydrolysis time (Moreno-Andrade et al., 2015). Extended HRTs may favor the growth of hydrogen-consuming microorganisms, including nonacidogenic and methanogenic species (Santiago et al., 2019). Consequently, sequencing batch reactors (SBRs) often outperform continuous reactors under certain conditions. For instance, in biohydrogen production from food waste, stable microbial communities with low diversity and evenness were observed at HRTs of 24–72 hours, with *Megasphaera* spp. predominating. Genera, such as *Enterobacter* and *Clostridium*, were also commonly detected. *Enterobacter* strains are particularly industrially relevant because of their high substrate conversion efficiency, rapid growth, tolerance to dissolved oxygen, and adaptability to fluctuating pH (Santiago et al., 2019).

6.2.2.2 Anaerobic fluidized bed reactor (AFBR) operation

6.2.2.2.1 Precise control of substrate flow rate

In AFBRs the flow rate of the substrate solution is a critical operational parameter for achieving efficient and stable hydrogen production. Properly designed flow rates ensure that the substrate maintains adequate contact with microbial biofilms, supplying nutrients continuously to support microbial metabolism. At the same time, excessively high flow can create strong washout forces, causing biofilm detachment and compromising reactor performance (Al-Mamun & Idris, 2021).

Optimal flow rate control depends on several factors, including substrate concentration, biofilm characteristics, and carrier properties. Detailed experimental studies allow observation of how variations in flow affect biofilm stability and hydrogen yield. Furthermore, modern online monitoring technologies, such as real-time measurement of substrate concentration and dynamic biofilm thickness, enable rapid flow adjustments to maintain stable reactor operation (Al-Mamun & Idris, 2021).

6.2.2.2.2 Monitoring and maintenance of microbial biofilms

As the biofilm is central to hydrogen production in AFBRs, comprehensive monitoring and maintenance are essential. Routine observations should include biofilm thickness, color, microbial diversity, and activity levels (Al-Mamun & Idris, 2021). Excessive biofilm accumulation may hinder substrate and product transfer, reducing hydrogen production efficiency. Changes in biofilm color can indicate metabolic imbalance or contamination.

To maintain healthy biofilms, key environmental parameters—such as temperature, pH, and substrate composition—must be strictly controlled. If abnormalities occur, such as localized shedding or reduced microbial activity, corrective actions should be implemented promptly. Measures may include adjusting substrate concentrations, optimizing flow rates, or supplementing nutrients and microbial growth stimulants to promote biofilm recovery and regeneration.

6.2.2.2.3 Implementation of regular backwash operations

During prolonged operation, AFBRs can experience impurity accumulation and biofilm aging, which negatively affect performance. Regular backwashing is an effective strategy to remove aged cells and debris, creating space for new microbial growth, enhancing mass transfer, and improving overall reactor efficiency.

Backwashing must be carefully controlled in terms of intensity and frequency. Excessive force can cause significant biofilm detachment, jeopardizing reactor stability, whereas overly frequent backwashing increases operating costs and may disturb microbial growth (Jankowska et al., 2024). Therefore backwash parameters should be determined based on substrate properties, biofilm development, and reactor operating conditions. Through systematic experimentation and practical experience, backwash strategies can be optimized to maintain long-term stable and efficient operation of AFBR systems.

6.2.2.3 Packed bed reactor (PBR) operation

6.2.2.3.1 Precise control of substrate flow rate

In PBRs the substrate flow rate is a crucial parameter for maintaining efficient and stable hydrogen production. Properly adjusted flow ensures sufficient contact between the substrate and microbial biofilms, supplying nutrients consistently while preventing excessive washout forces that could detach microorganisms and compromise reactor performance (Abu-Reesh, 2025).

Flow rate optimization depends on factors such as substrate concentration, biofilm characteristics, and carrier properties. Systematic experimental studies allow tracking of biofilm stability and hydrogen yield under varying flow conditions. Moreover, integrating real-time monitoring technologies—for instance, continuous observation of substrate concentration and dynamic biofilm thickness—provides a reliable basis for adjusting flow rates promptly to maintain stable operation.

6.2.2.3.2 Monitoring and maintenance of microbial biofilms

As microbial biofilms are central to hydrogen generation in PBRs, continuous monitoring and maintenance are essential. Routine assessments should consider biofilm thickness, color, microbial diversity, and metabolic activity. Overthick biofilms can impede substrate and product transfer, reducing hydrogen output, whereas abnormal coloration may signal metabolic imbalance or contamination.

Maintaining optimal biofilm conditions requires careful control of temperature, pH, and other key environmental parameters. If signs of biofilm deterioration are observed—such as localized shedding or reduced microbial activity—corrective actions should be taken immediately. These may include adjusting substrate concentration, optimizing flow rates, or supplementing nutrients and microbial stimulants to promote biofilm recovery and regeneration.

6.2.2.3.3 Regular backwash operations

During extended operation, PBRs may experience impurity accumulation and biofilm aging, which can reduce reactor efficiency. Scheduled backwashing effectively removes aged microbial cells and debris, facilitating space for new biofilm growth and improving mass transfer between substrates and products.

Backwash intensity and frequency must be carefully managed. Excessive force can cause significant biofilm detachment, threatening operational stability, whereas overly frequent backwashing increases costs and may disrupt microbial growth (Chen et al., 2024; Zhang et al., 2020). Therefore backwash parameters should be determined by considering substrate characteristics, biofilm development, and real-time reactor performance. Through iterative experimentation and operational experience, backwashing strategies can be refined to maintain long-term, stable, and efficient PBR operation.

6.3 Parameters

6.3.1 pH

pH is a key parameter that determines the acidic or alkaline state of the medium and directly influences hydrogen generation. Ma et al. (2015) reported that an optimal pH range of 5.5–6.5 maximizes hydrogen yield while limiting solvent-type by-products. pH strongly affects the activity of (FeFe)-hydrogenase; excessively low pH inhibits enzymatic function, reducing hydrogen output (Liu et al., 2008; Ngo et al., 2012). In addition, protons released during organic acid degradation can penetrate cell membranes, suppress microbial growth, and hinder hydrogen production (Pathak et al., 2017). pH also impacts the activity of by-products, substrate metabolism, protein synthesis, and intracellular material storage, all of which are essential for acidogenic hydrogen-producing bacteria (Venkata Mohan et al., 2008). Deviations from acidic conditions toward neutral can inhibit hydrogen formation (Venkatamohan et al., 2007), whereas highly acidic conditions ($pH < 4.5$) are detrimental. Interestingly, hydrogen production can be enhanced under moderately acidic conditions (pH 5.5) in the presence of elevated sulfate concentrations (3 g SO_4/L) (Lin & Chen, 2006).

6.3.1.1 Mechanisms of pH influence in dark fermentation

During dark fermentation, microorganisms convert organic substrates into hydrogen through intracellular and extracellular enzyme-mediated pathways. The activity of key enzymes—hydrogenase, formate dehydrogenase, and pyruvate decarboxylase—is highly pH-dependent. Hydrogenase, for instance, requires an optimal microenvironment to maintain its active site conformation and enable efficient proton reduction. Deviations from the optimal pH alter proton concentrations, disrupt enzyme-substrate interactions, and decrease reaction rates.

During fermentation, pH often declines because of the accumulation of volatile fatty acids (VFAs) (Argun & Kargi, 2011). At pH < 5, hydrogenase activity is suppressed, halting hydrogen production (Penniston & Gueguim Kana, 2018). Acidic conditions also favor solventogenesis by *Clostridium* spp., producing ethanol, butanol, and acetone instead of hydrogen (Ozyurt et al., 2019). Conversely, alkaline conditions can inhibit acid by-product formation. For example, Ozyurt et al. (2019) demonstrated that maintaining pH 6 increased hydrogen production by *Clostridium butyricum* by 18.4% and extended the production period by 67%. Optimal pH varies by substrate: neutral pH is suitable for animal-processing wastewater, whereas pH 5–6 is preferred for food waste (Argun & Kargi, 2011). Online pH control at 6 in potato juice fermentation increased hydrogen yield 5.4-fold (Ozyurt et al., 2019). pH also regulates cell membrane permeability. Extreme pH values disrupt phospholipid interactions, affecting substrate uptake and metabolite excretion. Highly acidic conditions can enhance cationic substrate uptake but hinder anionic metabolite discharge, impairing energy metabolism and microbial growth, ultimately reducing hydrogen production.

6.3.1.2 pH requirements in photofermentation

In photo-fermentation, pH affects microbial growth, nitrogenase activity, and hydrogenase function, all of which influence substrate utilization and hydrogen generation. Low pH inhibits hydrogen pathways, whereas high pH may shorten the lag phase but reduce maximum hydrogen output (Luo et al., 2010; Phanduang et al., 2017). Photosynthetic pigments, such as chlorophyll and carotenoids, are also pH-sensitive; deviations alter their structure, absorption spectra, and electron transfer efficiency. For example, in purple nonsulfur bacteria, acidic conditions protonate chlorophyll, reducing blue-light absorption and impairing electron transport. This disrupts hydrogenase activity, leading to electron accumulation, oxidative stress, and decreased photo-fermentation efficiency.

6.3.1.3 pH effect in dark–light combined fermentation

During integrated dark–light fermentation, stable pH is crucial for microbial synergy, substrate conversion, and overall hydrogen yield. Fluctuating pH can reduce microbial activity, inhibit enzyme function, or cause cell death. Effective strategies include

Real-time monitoring and feedback control: High-precision pH sensors transmit data to a control system, which adjusts acid/base addition automatically to maintain optimal microbial growth and hydrogen production.

Buffer system construction: Addition of suitable buffers (e.g., phosphate or tris-HCl) stabilizes pH and provides a consistent growth environment.

Phased pH control: Different pH levels are maintained for dark and light fermentation stages—neutral/slightly acidic for dark fermentation and optimal pH for photosynthetic microorganisms during photo-fermentation.

6.3.1.4 Precise pH control strategy

Given the sensitivity of hydrogen-producing systems to pH, continuous real-time monitoring is essential. Modern reactors use high-precision pH sensors (e.g., glass electrode or ion-selective methods) to provide rapid, accurate data for control decisions (Balakrishnan et al., 2023). Buffers help resist pH fluctuations caused by metabolites. When adjustments are required, acid/base addition must be gradual and precisely metered to avoid stressing microbial communities (Penniston & Gueguim Kana, 2018; Pereira et al., 2024).

By combining continuous monitoring, buffer systems, and phased pH control, pH can be precisely regulated, enhancing microbial activity, maximizing hydrogen yield, and facilitating the industrial application of biohydrogen production technologies.

6.3.2 Temperature

Temperature is a key factor affecting microbial hydrogen generation, as it directly regulates metabolic pathways, bacterial growth, and overall hydrogen yield. Numerous studies indicate that moderate increases in temperature can enhance hydrogen production; however, exceeding an optimal threshold leads to decreased output (Arimi et al., 2015; Mohan, 2010; Singh & Rathore, 2017; Wang & Wan, 2008, 2009). Fermentation processes are generally classified by temperature ranges: ambient (15°C–30°C), mesophilic (30°C–39°C), thermophilic (50°C–64°C), and hyperthermophilic (>64°C) (Parihar & Upadhyay, 2016). Fluctuations outside these ranges can disrupt substrate consumption, metabolite formation (e.g., VFAs), and hydrogen yield. Mesophilic conditions are often preferred because of lower operational costs (Vatsala et al., 2008; Venkata Mohan et al., 2007; Yang et al., 2012), although they may also promote the growth of nonhydrogen-producing microorganisms, competing for substrates and reducing efficiency.

Microbial metabolism relies heavily on enzyme-catalyzed reactions, which are highly sensitive to temperature. Within an optimal range, enzymatic reaction rates increase as temperature rises because higher thermal energy enhances molecular motion, increasing the probability of substrate–enzyme interactions. Key enzymes in hydrogen production, such as hydrogenase and amylase, are particularly temperature-sensitive. During dark fermentation, hydrogenases catalyze proton reduction to hydrogen efficiently under optimal conditions (Mukherjee et al., 2025). At low temperatures, enzyme kinetics slow, substrate binding decreases, and reaction rates drop, limiting hydrogen output. Therefore precise temperature control is essential to sustain both biochemical reaction rates and the activity of hydrogen-producing enzymes (Kargi et al., 2012; Łukajtis et al., 2018).

Temperature critically influences microbial proliferation. Each species has an optimal temperature range that promotes rapid cell division and vigorous growth. For mesophilic bacteria, this range typically spans 20°C–45°C. Within this window, cell membranes maintain appropriate fluidity, nutrient uptake and waste excretion are efficient, and DNA replication, transcription, and translation proceed effectively. Temperatures below the minimum growth threshold impede microbial activity, potentially causing intracellular water freezing and structural damage. Conversely, excessive temperatures above the maximum growth limit can denature proteins, compromise membrane integrity, and lead to cell death.

Fermentation reaction rates are closely linked to temperature, as described by chemical kinetics principles. Within an optimal temperature range, increasing temperature accelerates substrate decomposition, intermediate formation, and hydrogen generation. For instance, in glucose-based dark fermentation, moderate temperature increases enhance pyruvate formation and subsequent hydrogen production. However, surpassing a critical thermal limit rapidly decreases reaction rates because of microbial inactivation and enzyme denaturation.

6.3.3 Light exposure

In combined dark–photo fermentation, light is a fundamental factor driving photo-fermentation, profoundly affecting microbial activity and hydrogen yield.

6.3.3.1 Light intensity: Balancing efficiency and saturation

Light intensity directly determines the energy available for photosynthetic microorganisms, which capture photons to drive electron transfer and hydrogen production. Within a suitable range, increasing illumination enhances energy absorption, accelerates photochemical reactions, and improves hydrogen yield. However, beyond a certain threshold, light saturation occurs, limiting additional energy utilization. Effective light intensity for high-efficiency photo-fermentation typically ranges from 6 to 6000 lx (Argun & Kargi, 2010). Excessive light can damage photosynthetic pigments via photooxidation and impair the oxygen-evolving complex, reducing electron transfer efficiency and inhibiting hydrogen generation.

Substrate turbidity caused by pigments and metabolites can further reduce light penetration. Strategies such as photoperturbation and oscillation have proven effective in mitigating these effects (Argun & Kargi, 2010; Cheng et al., 2022). For instance, Shui et al. (2022) demonstrated that introducing light disturbances at 6000 lx increased hydrogen production rate by 19.5% and yield by 26.2%. Similarly, Zhang et al. (2023) found that combining oscillation (150 r/min) with 6000 lx illumination maximized hydrogen production from corn cob hydrolysate, reaching 48 mL H_2/g substrate and a light energy conversion efficiency of 11.04%, representing over 600% improvement compared to static conditions. These findings highlight the synergistic effect of optimal light intensity and mechanical oscillation in enhancing photo-fermentation efficiency.

6.3.3.2 Light quality: Wavelength-specific effects

Light quality, defined by wavelength, selectively interacts with photosynthetic pigments. Chlorophyll *a* absorbs primarily red and blue–violet light, whereas carotenoids absorb mainly in the blue–violet range (Glibert, 2024; Milrad et al., 2024). Different wavelengths not only drive energy capture but also influence metabolic regulation. Red light facilitates electron transfer within the photosystem, improving photosynthetic efficiency, whereas blue–violet light regulates growth, pigment synthesis, and cell differentiation.

Light quality can also steer microbial metabolism, directing more energy toward hydrogen production. For example, blue-light exposure has been reported to enhance hydrogenase activity in certain photosynthetic bacteria, stimulating hydrogen output (Wang et al., 2024). In addition, light wavelength affects carbon metabolism, influencing CO_2 fixation and biosynthesis of organic intermediates, which in turn impacts overall hydrogen production efficiency.

6.3.3.3 Light cycle: Mimicking natural rhythms

Alternating light and dark periods, or light cycles, significantly affect photosynthetic microbial growth and hydrogen production. Mimicking natural day–night cycles allows cells to balance energy capture with repair and metabolic maintenance. During illumination, cells perform photosynthesis and hydrogen generation; during dark periods, they repair damaged pigments, restore energy reserves, and maintain metabolic homeostasis.

Appropriate light–dark regimes enhance both stability and long-term efficiency. For instance, a 12-hour light/12-hour dark cycle has been shown to optimize hydrogen output, whereas continuous or excessively long illumination can cause photoinhibition and cellular fatigue. Moreover, light cycles interact with other environmental factors, such as temperature and pH, influencing microbial growth and hydrogen productivity. Optimizing photo-fermentation thus requires coordinated management of light intensity, wavelength, and temporal patterns.

6.3.4 Additives

Additives play crucial roles in hydrogen generation from organic substrates through combined dark–photo fermentation. They can modulate microbial physiology, improve fermentation performance, and enhance both the efficiency and stability of hydrogen production.

6.3.4.1 *Trace metal ions*

Trace metal ions serve as essential cofactors for key enzymes involved in hydrogen metabolism, maintaining their structural integrity and catalytic activity. Enzymes, such as hydrogenases and oxidoreductases, require specific metal ions at their active sites to function properly. For instance, nickel (Ni) forms part of certain hydrogenases' catalytic centers, stabilizing the enzyme's three-dimensional structure and facilitating electron–proton coupling during proton reduction to hydrogen. Similarly, iron (Fe) ions act as electron carriers in oxidoreductases, participating in cellular redox reactions and supporting metabolic continuity.

Metal ions also influence electron transport and energy metabolism. In dark–photo fermentation, efficient electron transfer links substrate oxidation, intermediate conversion, and hydrogen generation. By mediating redox reactions within the electron transport chain, metal ions accelerate electron flow and improve energy conversion. For example, Fe–S protein complexes in the photosynthetic electron transport chain rely on iron supplementation to enhance light energy capture, conversion to chemical energy, and hydrogen production (Zhou et al., 2024). In addition, metal ions contribute to adenosine triphosphate (ATP) synthesis and proton motive force regulation, optimizing energy utilization and ensuring stable hydrogen-producing metabolism.

6.3.4.2 *Chemical promoters*

Chemical promoters modulate cell membrane permeability, supporting efficient substrate uptake and product release. Certain surfactants can insert into the phospholipid bilayer, increasing membrane fluidity and allowing faster substrate entry (Atukunda & Nasr, 2024). This improves assimilation of macromolecular or poorly soluble compounds while facilitating hydrogen excretion, reducing product inhibition, and enhancing overall hydrogen yield. However, excessive use may disrupt intracellular integrity and impair microbial function.

Chemical promoters also increase substrate solubility. Insoluble organic materials, such as polysaccharides or lignin derivatives, can be made more accessible through solubilizing reagents that form interactions with substrate molecules, improving contact with microbial cells (Mumtha & Mahalingam, 2024). Furthermore, some promoters mitigate product inhibition by binding or chemically modifying accumulated hydrogen or organic acids, maintaining microbial activity and sustaining hydrogen production over time.

6.3.4.3 *Biological additives*

Biological additives, including vitamins, growth factors, and beneficial microbial consortia, support microbial growth and hydrogen generation. Growth factors, such as amino acids and nucleotides, are precursors for protein and nucleic acid synthesis, enhancing cell proliferation and metabolism. Vitamins often act as coenzymes; for example, thiamine (vitamin B1) is required for pyruvate dehydrogenase activity, facilitating pyruvate oxidation and energy generation during dark fermentation. Supplementation of vitamin B1 can therefore improve microbial energy metabolism and hydrogen yield.

The addition of selected microbial consortia can stabilize the fermentation ecology. Symbiotic interactions among strains allow complex substrates to be degraded into smaller molecules, providing readily available carbon sources for hydrogen-producing photosynthetic microorganisms (Abdulsalam et al., 2024). This mutualistic cooperation enhances community stability, improves substrate utilization, and increases overall hydrogen production efficiency.

6.3.5 Other factors

In biohydrogen production via combined dark–photo fermentation, system pressure influences microbial physiology in multiple ways. Moderate pressure can act as a subtle regulator, affecting cell membrane fluidity and permeability and thereby modulating the transport of substrates and metabolites across the membrane. Within an optimal range, increasing pressure compacts the phospholipid bilayer, reducing pore size and altering the efficiency of molecular exchange (Sun et al., 2025).

Pressure also affects enzyme function and protein structure within microbial cells. Certain enzymes may experience slight conformational changes under pressure, which can modify catalytic efficiency and alter metabolic pathways. In particular, piezotolerant microorganisms adapted to high-pressure environments can use pressure as a cue to regulate gene expression, adjusting metabolic strategies to maintain survival and function under stress.

In addition, system pressure directly impacts hydrogen solubility in the medium. According to Henry's law, gas solubility rises with increasing pressure, which can facilitate hydrogen accumulation and improve subsequent separation and recovery. However, this relationship is complex: excessive pressure may inhibit microbial growth and metabolism, whereas higher hydrogen solubility can intensify product inhibition, as elevated hydrogen concentrations near cells hinder its release and disrupt the production process.

Therefore optimizing pressure in biohydrogen systems requires a balance between enhancing hydrogen solubility and maintaining microbial activity. Establishing an appropriate pressure range—high enough to promote dissolution but not so high as to impair microorganisms—is critical for maximizing hydrogen production efficiency.

6.4 Net zero effect

In the global effort to achieve net-zero emissions (NZE) by 2050, hydrogen (H_2) has emerged as a promising energy carrier. The adoption of hydrogen provides a practical route to decarbonize sectors that are difficult to abate, including industry, heating, and transportation (Yan et al., 2020). Meeting the NZE 2050 target requires a comprehensive transformation of energy systems, integrating a range of technological solutions. Core elements of decarbonization include renewable electricity generation, widespread deployment of low-carbon energy technologies, hydrogen production with minimal CO_2 emissions, improvements in energy efficiency, and the implementation of carbon capture, utilization, and storage (CCUS) (King, 2021).

At present, global hydrogen production reaches around 75 million metric tons (Mt) of pure H_2 and an additional 45 Mt in mixed gas forms annually. Under NZE scenarios, approximately 70% of the projected 200 Mt H_2 demand by 2030 will need to come from low-carbon processes, rising to nearly 500 Mt by 2050 through advanced low-emission technologies. Employing low-carbon methods for hydrogen and its derivatives could potentially reduce cumulative CO_2 emissions by 60 gigatons (Gt) by midcentury (Yan et al., 2020).

Many nations are actively incorporating hydrogen into their climate strategies, developing national roadmaps that define market potential, regulatory frameworks, and transition pathways (Gerres et al., 2022). Japan led these efforts by introducing its hydrogen policy in 2017, whereas Germany launched the first EU national hydrogen strategy in 2020, soon followed by France. Rapid expansion of low-carbon hydrogen production is essential to build a sustainable energy system by 2050. Countries with limited domestic resources may rely on imports, whereas those with abundant renewable energy or CO_2 storage potential could become key exporters. Despite the urgency, many critical hydrogen technologies remain in the early stages of development, emphasizing the need for continued innovation (Dagdougui et al., 2018; Nnabuife et al., 2022).

Hydrogen is widely regarded as a sustainable solution to future energy demands and can function as an energy carrier in both electricity generation and transportation. In power systems, hydrogen and fuel cells offer a viable alternative to batteries, with current advancements focusing on direct methanol-fueled cells. Future decentralized electricity systems, whether for large-scale or distributed applications, are expected to rely on improved hydrogen fuel cell efficiency. In transportation, hydrogen's reaction with atmospheric oxygen yields energy with water as the only byproduct, providing a cleaner alternative to conventional fuels (Abdin et al., 2020). Nevertheless, not all hydrogen production methods are environmentally neutral, as associated carbon emissions vary with the production technology and feedstock employed (Arcos & Santos, 2023).

6.4.1 Path to carbon neutrality by 2050

Carbon neutrality seeks to achieve NZE by 2050, with developed countries generally aiming for earlier targets than developing ones. The NZE objective aligns with the Sustainable Development Goals (SDGs), emphasizing energy transition and environmental sustainability by 2030. This target is consistent with the Intergovernmental Panel on Climate Change (IPCC) Sixth Assessment Report, which recommends a 45% reduction in CO_2 emissions by mid-century and attainment of NZE by 2050 to limit global warming to 1.5°C (Intergovernmental Panel on Climate Change, 2022).

The main components of the NZE pathway include

Transition to renewable energy: Expanding the use of renewable technologies to cut CO_2 emissions by 2050 without depending on land-use offsets. The speed of deployment is influenced by market dynamics, policy frameworks, technological maturity, costs, and existing infrastructure.

Resilient energy systems: Ensuring reliable, affordable, and continuous energy supply through coordinated policies, incentives, and regulatory mechanisms that foster sustainable energy practices and support the adoption of clean technologies.

International cooperation: Achieving NZE requires global collaboration, with both developed and developing nations contributing. Developing countries are encouraged to accelerate their transition efforts.

SDG alignment: Promoting widespread access to renewable electricity by 2030 in accordance with SDG objectives.

Methane mitigation: Reducing methane emissions from fossil fuel operations to provide a transitional window for developing nations to balance economic development with greenhouse gas (GHG) reductions.

Technology sharing: Global collaboration facilitates the dissemination of low-carbon technologies, reduces costs, and offers financial support to developing economies in their NZE initiatives.

According to the United Nations Climate Action program, over 1000 cities, 9000 companies, more than 600 organizations, and around 1000 financial and academic institutions are participating in global CO_2 reduction efforts aimed at 2030. Currently, 140 countries targeting NZE by 2050 account for roughly 88% of global emissions. Key emitters, including the EU, the United States, China, India, and Russia, represented about 60% of global GHGs in 2021; including additional nations such as Canada, Japan, and the United Kingdom raises the coverage to 76%. Developing nations and small island states, in comparison, contribute only 3.8% and 1% of emissions, respectively (United Nation, 2023). The 2023 United Nations Environment Program (UNEP) report indicates that although GHG emissions were initially projected to increase by 16% since the 2015 Paris Agreement, recent estimates suggest a smaller rise of approximately 3%. Nevertheless, a 42% reduction in emissions remains necessary to cap global warming at 1.5°C (United Nation, 2023).

6.4.2 Role of hydrogen toward net-zero emission target

Hydrogen (H_2) is increasingly recognized worldwide as a key sustainable energy carrier critical for achieving NZE by 2050. It offers a promising pathway for decarbonizing sectors that are otherwise difficult to abate, including heavy industry, transport, and heating. In recent years the adoption of H_2 as an energy carrier has accelerated, driven by steadily declining costs that are enhancing its competitiveness, despite ongoing technical and economic challenges (Schmidt et al., 2017). H_2 is one of the few energy vectors capable of supplying energy to hard-to-decarbonize sectors with minimal GHG emissions (Greene et al., 2020). Although H_2 can serve as an alternative to fossil fuels, the sustainability of its production depends on the method and feedstock employed. In particular, green H_2 produced via renewable electricity represents a carbon-free and environment-friendly solution that is central to building a sustainable energy system (Hassan et al., 2025). Its versatility and ecoefficiency have positioned green H_2 as a crucial component of the global H_2 supply chain, which was valued at $155 billion in 2022, with rapid industry growth expected in the coming years. Across various sectors, green H_2 can function both as an energy carrier and as an industrial feedstock, enhancing energy use efficiency (Meng et al., 2022).

To achieve NZE by 2050 the global energy system must prioritize energy efficiency, replace fossil fuels with renewable electricity, integrate H_2 as a direct fuel and fuel precursor, and deploy CCUS (Greene et al., 2020). Global hydrogen demand is projected to surpass 200 Mt by 2030, up from roughly 90 Mt in 2020. Meeting NZE targets requires that this expansion rely on low- or zero-emission production pathways (Nnabuife et al., 2023). The market for low-carbon H_2 is expected to grow by 70% by 2030. Incorporating a 15% H_2 blend into the natural gas grid could reduce global CO_2 emissions from gas use by about 6%. Under NZE scenarios, by 2030 approximately 70% of the projected 200 Mt H_2 will be produced via low-emission technologies such as electrolysis, rising to roughly 500 Mt by 2050 using advanced low-carbon methods. Achieving these production levels will require scaling electrolysis capacity from the current 0.3 GW to around 850 GW by 2030 and 3600 GW by 2050. Between 2022 and 2030, CO_2 emissions per kilogram of H_2 are expected to fall from 12–13.5 to 6–7.5 kg. H_2 is particularly important for decarbonizing sectors such as shipping, heavy-duty road transport, aviation, heavy industry, and select power generation processes, while also providing seasonal energy storage and flexibility to balance variable renewable electricity (Schulthoff et al., 2021).

Technological and infrastructural development for H_2 is still progressing (Bataille et al., 2021). Sweden pioneered carbon-free steel production in 2019, followed by a green ammonia plant in Spain in 2021. Several large-scale projects, each with capacities of tens of kilotons, are expected to come online within the next few years, supported by global investments totaling approximately $37 billion in clean H_2 initiatives (Slorach & Stamford, 2021). Industrial deployment in sectors such as glass, cement, and ceramics manufacturing is also underway. Countries, including Korea, France, the Netherlands, and Japan, are promoting fuel cell vehicles powered by clean H_2 to advance sustainable transportation (Capurso et al., 2022). Bataille et al. (2018) highlighted that industrial decarbonization requires alternatives such as electrification and H_2-based fuels, alongside enhanced integration of industry in energy system models and consideration of regional impacts from industrial transformation. Accelerating the adoption of low-carbon H_2 is therefore essential to achieve a decarbonized energy system by 2050.

Global policymakers are supporting this transition through carbon pricing, taxation, trade frameworks for low-carbon H_2, and regulatory measures governing its use. Rapid implementation of these policies, reinforced by comprehensive legislation, is expected to unlock the potential of renewable H_2 demand and supply, necessitating coordinated national strategies for a zero-emission future (Kshirsagar et al., 2023). Currently,

41 governments have enacted H_2 policies, whereas others are formulating national roadmaps. Nevertheless, only 1% of global H_2 production originates from low-carbon sources, meaning production must increase roughly 100-fold by 2030 to align with NZE objectives.

China has established a detailed roadmap for its H_2 sector for 2021–35, emphasizing green H_2. Key initiatives include (1) developing centralized platforms to advance R&D and innovation in green H_2 technologies; (2) reducing electricity and electrolyzer costs; (3) expanding H_2 transportation, storage, and manufacturing infrastructure; (4) promoting H_2 use in heavy industry, energy storage, and transport; and (5) implementing quality, safety, and pricing standards while establishing an integrated H_2 energy system with national and global coordination. The 14th Five-Year Plan for Renewable Energy Development (June 2022) set targets for pioneering an H_2 supply system using industrial by-products and renewable energy. Subsequently, the 2023 Hydrogen Energy Industry Standard System Construction Guide laid the foundation for global collaboration and industry standardization. Targeted outcomes include deploying 50,000 H_2 vehicles by 2025, reducing annual CO_2 emissions by 1–2 million t, and cutting approximately 84 million tons of CO_2 from cement, steel, ammonia, and oil refining industries between 2024 and 2025.

6.5 Case study

6.5.1 Combined photo and dark fermentation mode 11 m^3 baffle photobioreactor

Pilot-scale experiments for sequential dark and photo-fermentation hydrogen (H_2) production were carried out for the first time in an 11 m^3 integrated reactor, consisting of a 3 m^3 dark fermentation compartment and an 8 m^3 photo-fermentation compartment (Fig. 6.4) (Zhang et al., 2018). The system was equipped with a combined solar and light-emitting diode (LED) lighting arrangement, as well as a thermal regulation system, which was tested for operational feasibility. Hydrolyzed corn stover was employed as the substrate for hydrogen generation.

Activated sludge served as the inoculum for the dark fermentation stage, whereas the mixed photosynthetic bacterial consortium HAU-M1 was applied in the photofermentation stage. The experimental procedure included a 60-day adaptation phase followed by a 30-day production phase.

During the initial adaptation phase, 300 kg of activated sludge was introduced into the dark fermentation compartment and mixed with 2 m^3 of cultivation medium for 48 hours. Subsequently, hydrolyzed corn stover with a chemical oxygen demand (COD) of 10 g/L was continuously fed

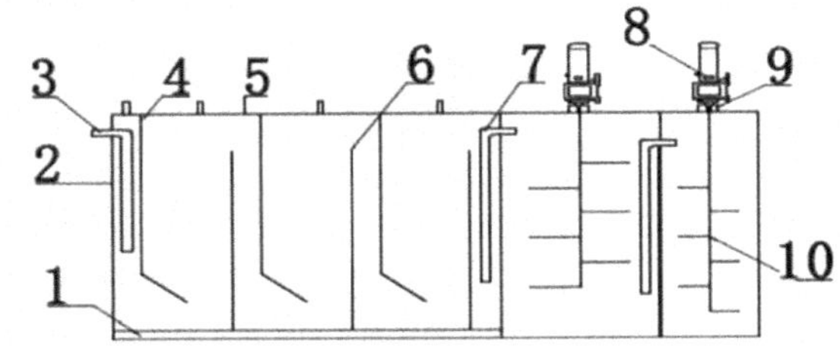

1. Heating layer 2. support plate 3. feed pipe 4. traverse baffle 5. top plate 6. flow baffle 7. discharge tube 8. blender motor 9. support frame 10. Blender

(A)

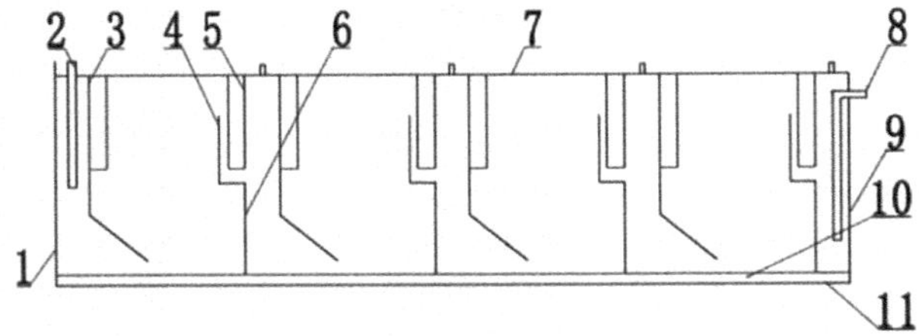

1. Support plate 2. feed pipe 3. traverse baffle 4. flow baffle 5. water seal tank 6. flow baffle 7. top plate 8. discharge tube 9. support plate 10. heating layer 11. baseplate.

(B)

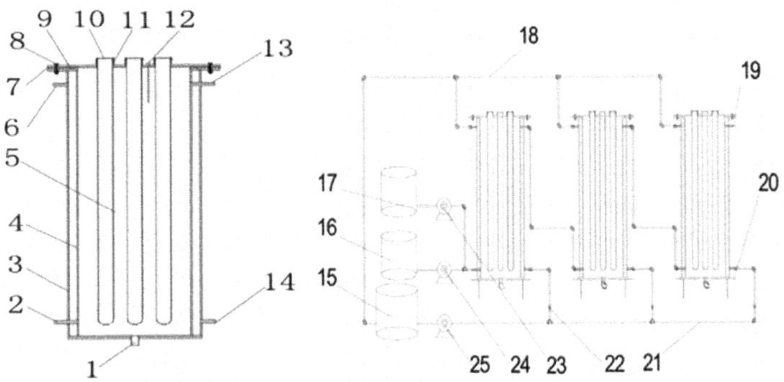

1. discharge pipe 2. feed pipe 3. tank 4. heat pressing cavity 5. lighting pipes 6. outlet pipe 7. seal cover 8. fastening bolts 9. o-rings 10. glass tube head 11. glass pipe joint 12. temperature probe 13. discharge tube 14. water inlet 15. heat storage tank 16. storage tank of the growth medium 17. HAU-M1 18 return line 19. discharge line 20. pedestal 21. hot water circulating pipe line 22. magnet valve 23. feed pump of HAU-M1 24. feed pump of the growth medium 25. heat water circulating pump.

(C)

FIGURE 6.4 Schematic of the experimental setup: (A) dark-fermentation unit, (B) photo-fermentation unit, and (C) inoculum cultivation reactor. *Zhang, Q., Zhang, Z., Wang, Y., Lee, D. J., Li, G., Zhou, X., Jiang, D., Xu, B., Lu, C., Li, Y., & Ge, X. (2018). Sequential dark and photo fermentation hydrogen production from hydrolyzed corn stover: A pilot test using 11 m^3 reactor.* Bioresource Technology, 253, *382–386. https://doi.org/10.1016/j.biortech.2018.01.017.*

into the dark fermenter at an HRT of 16 hours. The dark fermentation conditions were maintained at pH 4.5°C and 35°C.

Effluent from the dark fermenter was then adjusted to a pH of 7.0 and a C/N ratio of 10:1 before entering the photofermentation compartment. The photofermenter was operated at 30°C under an internal illumination of 3000 lx, with an HRT of 48 hours. The inoculum, obtained from the cultivation reactor at a cell concentration of 2.1 g/L, was supplied at 25% (v/v) of the feed volume.

After the 60-day adaptation period the reactor achieved stable operation, and its performance was subsequently monitored and recorded during the 30-day production phase.

During the 30-day production phase, hydrogen (H_2) generation rates remained stable, with values of 6.4, 12.5, and 3.3 $m^3/m^3{}_h$day for compartments 1, 2, and 3, respectively.

In row 1, the production rates were 1.1, 2.9, 5.9, and 7.7 $m^3/m^3{}_h$day for compartments 1–1 to 1–4, respectively. Row 2 exhibited rates of 2.1, 3.9, 5.3, and 9.8 $m^3/m^3{}_h$day for compartments 2–1 to 2–4. The higher H_2 output in row 2 compared to row 1 is likely because of uneven feed and inoculum distribution between the two rows.

The reactor's total H_2 production reached 87.8 ± 3.8 m^3/day, with an H_2 content of 68.0% ± 0.8%. On average, H_2 generation was 59.7 m^3/day, corresponding to 2440 mol-H_2/day. Of this total, 22.4 m^3/day originated from the dark fermentation compartment, whereas 37.3 m^3/day was contributed by the photofermentation compartment, resulting in volumetric production rates of 7.5 and 4.7 $m^3/m^3{}_h$day, respectively.

Notably, H_2 production increased progressively from compartments 1 to 4, indicating potential for further performance enhancement. The observed differences in production rates among compartments suggest that reactor suspension mixing could be optimized. However, any increase in mixing intensity must be balanced against the associated operational costs.

6.5.2 Engineering design of a two-stage dark–photo fermentation process

Chen et al. (2008) reported a two-stage dark–photo fermentation biohydrogen production system (Fig. 6.5). In the first stage, *Clostridium pasteurianum* CH4 was used for dark fermentation, where hydrogen was produced from sucrose as the carbon source, achieving a maximum hydrogen yield of 3.80 mol H_2/mol sucrose. The primary VFAs (such as butyric acid and acetic acid) produced in the dark fermentation phase were subsequently used in the second stage for photo-fermentation with *Rhodopseudomonas palustris* WP3–5. The photo-fermentation was conducted under conditions of approximately 35°C, pH ≈ 7.0, continuous illumination (about

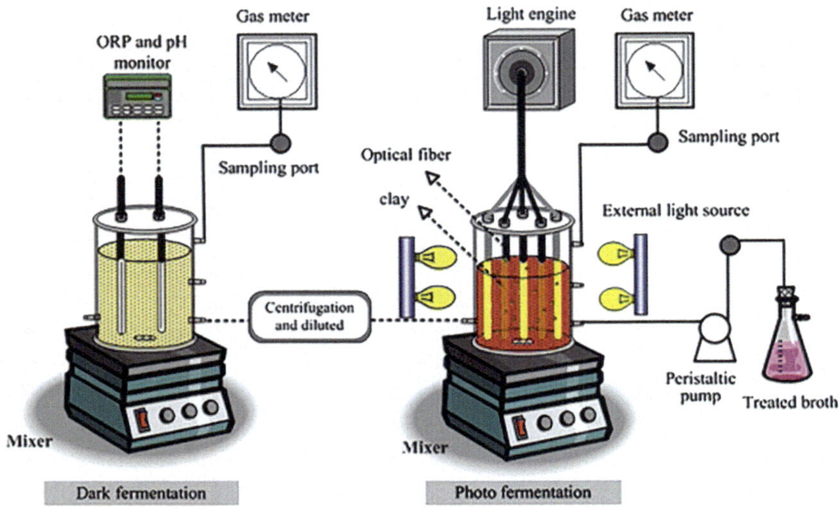

FIGURE 6.5 Schematic description of the two-stage process combining dark and photo fermentation. *Chen, C. Y., Yang, M. H., Yeh, K. L., Liu, C. H., & Chang, J. S. (2008). Biohydrogen production using sequential two-stage dark and photo fermentation processes.* International Journal of Hydrogen Energy, *33(18), 4755–4762. https://doi.org/10.1016/j.ijhydene.2008.06.055.*

10,000 lx using side-light optical fiber), and a 48-hour retention time. Through this combined process, the overall hydrogen yield significantly increased to 10.02 mol H_2/mol sucrose, whereas the COD removal reached approximately 72.0%.

Furthermore, when side-light optical fiber illumination was employed and 2.0% (w/v) clay carriers were added to the photobioreactor, the efficiency of photo-fermentation and microbial attachment improved, further enhancing the hydrogen yield to 14.2 mol H_2/mol sucrose and achieving nearly 90% COD removal. This study demonstrates that the two-stage dark–photo fermentation system not only significantly improves biohydrogen production but also contributes to organic waste removal from wastewater, realizing the dual benefits of energy production and environmental remediation.

6.6 Conclusion

Dark–photo fermentation has emerged as a versatile and promising approach in clean energy production. Its adaptability is evident from the wide range of substrates it can utilize, including simple sugars, organic acids, agricultural and forestry residues, and industrial wastes, with pretreatment methods enhancing substrate accessibility and process efficiency.

Reactor design and operational control are pivotal: CSTRs provide uniform mixing, FBRs improve microbial density and stability, and PBRs optimize light management for photosynthetic microorganisms. Key parameters such as pH, temperature, and illumination critically influence enzyme activity, pigment stability, and electron transfer, whereas trace metal ions, chemical and biological additives, and system pressure further modulate microbial physiology. By integrating substrate management, reactor optimization, and precise parameter control, dark–photo fermentation offers a scalable pathway to efficient and sustainable hydrogen production.

References

Abdin, Z., Zafaranloo, A., Rafiee, A., Mérida, W., Lipiński, W., & Khalilpour, K. R. (2020). Hydrogen as an energy vector. *Renewable and Sustainable Energy Reviews, 120*, 109620. doi:10.1016/j.rser.2019.109620.

Abdulsalam, M., Salihu, A. T., Usman, H. Y., & Yahaya Usman, M. (2024). Protein biosynthesis in microorganisms: Mechanisms, regulation, and biotechnological applications. *World Journal of Advanced Research and Reviews, 21*(1), 869–881. doi:10.30574/wjarr.2024.21.1.2469.

Abu-Reesh, I. M. (2025). Mathematical model-based optimization of continuous flow photobioreactor operating at steady state using MATLAB optimization function. *Arabian Journal for Science and Engineering, 50*(6), 3779–3792. https://www.springer.com/journal/13369. doi:10.1007/s13369-024-09155-7.

Al-Mamun, A., & Idris, A. B. (2021). Anaerobic fluidized bed reactor (AFBR) for the treatment of high solid content agro-industrial wastewater. *Asian Journal of Microbiology, Biotechnology and Environmental Science, 23*(1), 102–105.

Andrigo, P., Bagatin, R., & Pagani, G. (1999). Fixed bed reactors. *Catalysis Today, 52*(2–3), 197–221. http://www.sciencedirect.com/science/journal/09205861. doi:10.1016/S0920-5861(99)00076-0.

Arcos, J. M. M., & Santos, D. M. F. (2023). The hydrogen color spectrum: Techno-economic analysis of the available technologies for hydrogen production. *Gases, 3*(1), 25–46. https://www.mdpi.com/journal/gases. doi:10.3390/gases3010002.

Argun, H., & Kargi, F. (2010). Photo-fermentative hydrogen gas production from dark fermentation effluent of ground wheat solution: Effects of light source and light intensity. *International Journal of Hydrogen Energy, 35*(4), 1595–1603. doi:10.1016/j.ijhydene.2009.12.040.

Argun, H., & Kargi, F. (2011). Bio-hydrogen production by different operational modes of dark and photo-fermentation: An overview. *International Journal of Hydrogen Energy, 36*(13), 7443–7459. doi:10.1016/j.ijhydene.2011.03.116.

Arimi, M. M., Knodel, J., Kiprop, A., Namango, S. S., Zhang, Y., & Geißen, S. U. (2015). Strategies for improvement of biohydrogen production from organic-rich wastewater: A review. *Biomass and Bioenergy, 75*, 101–118. http://www.journals.elsevier.com/biomass-and-bioenergy/. doi:10.1016/j.biombioe.2015.02.011.

Arooj, M. F., Han, S. K., Kim, S. H., Kim, D. H., & Shin, H. S. (2008). Continuous biohydrogen production in a CSTR using starch as a substrate. *International Journal of Hydrogen Energy, 33*(13), 3289–3294. doi:10.1016/j.ijhydene.2008.04.022.

Atukunda, A., & Nasr, M. (2024). Enhancement of biohydrogen production using chemical additives and nanoparticles. *Science and Business Media LLC*, 255–285. doi:10.1007/978-3-031-49818-3_10.

Aubin, J., Kresta, S. M., Bertrand, J., Xuereb, C., & Fletcher, D. F. (2006). Alternate operating methods for improving the performance of continuous stirred tank reactors. *Chemical Engineering Research and Design, 84*(7), 569–582. doi:10.1205/cherd.05216.

Azuma, J.-I., Tanaka, F., & Koshijima, T. (1984). Enhancement of enzymatic susceptibility of lignocellulosic wastes by microwave irradiation. *Journal of Fermentation Technology*.

Balakrishnan, D., Manmai, N., Ponnambalam, S., Unpaprom, Y., Chaichompoo, C., & Ramaraj, R. (2023). Optimized model of fermentable sugar production from Napier grass for biohydrogen generation via dark fermentation. *International Journal of Hydrogen Energy, 48*(55), 21152–21160. http://www.journals.elsevier.com/international-journal-of-hydrogen-energy/. doi:10.1016/j.ijhydene.2022.12.011.

Barros, A. R., Adorno, M. A. T., Sakamoto, I. K., Maintinguer, S. I., Varesche, M. B. A., & Silva, E. L. (2011). Performance evaluation and phylogenetic characterization of anaerobic fluidized bed reactors using ground tire and pet as support materials for biohydrogen production. *Bioresource Technology, 102*(4), 3840–3847. doi:10.1016/j.biortech.2010.12.014.

Bataille, C., Åhman, M., Neuhoff, K., Nilsson, L. J., Fischedick, M., Lechtenböhmer, S., Solano-Rodriquez, B., Denis-Ryan, A., Stiebert, S., Waisman, H., Sartor, O., & Rahbar, S. (2018). A review of technology and policy deep decarbonization pathway options for making energy-intensive industry production consistent with the Paris Agreement. *Journal of Cleaner Production, 187*, 960–973. https://www.journals.elsevier.com/journal-of-cleaner-production. doi:10.1016/j.jclepro.2018.03.107.

Bataille, C., Nilsson, L. J., & Jotzo, F. (2021). Industry in a net-zero emissions world: New mitigation pathways, new supply chains, modelling needs and policy implications. *Energy and Climate Change, 2*, 100059. doi:10.1016/j.egycc.2021.100059.

Brodeur, G., Yau, E., Badal, K., Collier, J., Ramachandran, K. B., & Ramakrishnan, S. (2011). Chemical and physicochemical pretreatment of lignocellulosic biomass: A review. *Enzyme Research, 2011*(1), 1–17. doi:10.4061/2011/787532.

Capurso, T., Stefanizzi, M., Torresi, M., & Camporeale, S. M. (2022). Perspective of the role of hydrogen in the 21st century energy transition. *Energy Conversion and Management, 251*, 114898. doi:10.1016/j.enconman.2021.114898.

Chen, C. Y., Yang, M. H., Yeh, K. L., Liu, C. H., & Chang, J. S. (2008). Biohydrogen production using sequential two-stage dark and photo fermentation processes. *International Journal of Hydrogen Energy, 33*(18), 4755–4762. doi:10.1016/j.ijhydene.2008.06.055.

Chen, L., DeGroot, C. T., & Bassi, A. (2024). Biofilm growth enhancement in microalgae biofilm reactors: Parameters, configurations, and modeling. *Journal of Water Process Engineering, 65*, 105780. doi:10.1016/j.jwpe.2024.105780.

Cheng, C. L., Lo, Y. C., Lee, K. S., Lee, D. J., Lin, C. Y., & Chang, J. S. (2011). Biohydrogen production from lignocellulosic feedstock. *Bioresource Technology, 102*(18), 8514–8523. doi:10.1016/j.biortech.2011.04.059.

Cheng, D., Ngo, H. H., Guo, W., Chang, S. W., Nguyen, D. D., Bui, X. T., Wei, W., Ni, B., Varjani, S., & Hoang, N. B. (2022). Enhanced photo-fermentative biohydrogen production from biowastes: An overview. *Bioresource Technology, 357*. Www.elsevier.com/locate/biortech. doi:10.1016/j.biortech.2022.127341.

Dagdougui, H., Sacile, R., Bersani, C., & Ouammi, A. (2018). *Hydrogen infrastructure for energy applications: Production, storage, distribution and safety*. Academic Press.

Da Silva, A. N., Macedo, W. V., Sakamoto, I. K., de Los Angeles Dalmao Pereyra, D., Mendes, C. O., Maintinguer, S. I., Caffaro Filho, R. A., Zamariolli Damianovic, M. H., Amancio Varesche, M. B., & Lucena Cavalcante de Amorim, E. (2019). Biohydrogen production from dairy industry wastewater in an anaerobic fluidized-bed reactor. *Biomass & bioenergy, 120*, 257–264.

Dowaidar, M. (2025). Microbial pathways for sustainable hydrogen production. *International Journal of Hydrogen Energy, 142*, 825–841. https://www.sciencedirect.com/science/journal/03603199. doi:10.1016/j.ijhydene.2024.12.446.

Ferreira, T. B., Rego, G. C., Ramos, L. R., de Menezes, C. A., Soares, L. A., Sakamoto, I. K., Varesche, M. B. A., & Silva, E. L. (2019). HRT control as a strategy to enhance continuous hydrogen production from sugarcane juice under mesophilic and thermophilic conditions in AFBRs. *International Journal of Hydrogen Energy, 44*(36), 19719–19729. http://www.journals.elsevier.com/international-journal-of-hydrogen-energy/. doi:10.1016/j.ijhydene.2019.06.050.

Fuess, L. T., Mazine Kiyuna, L. S., Garcia, M. L., & Zaiat, M. (2016). Operational strategies for long-term biohydrogen production from sugarcane stillage in a continuous acidogenic packed-bed reactor. *International Journal of Hydrogen Energy, 41*(19), 8132–8145. http://www.journals.elsevier.com/international-journal-of-hydrogen-energy/. doi:10.1016/j.ijhydene.2015.10.143.

Gerres, T., Serna Zuluaga, S., & Cossent, R. (2022). National hydrogen strategies in a global context: Common design elements across country-specific visions. *SSRN Electronic Journal*. doi:10.2139/ssrn.4203346.

Ghosh, S., Dairkee, U. K., Chowdhury, R., & Bhattacharya, P. (2017). Hydrogen from food processing wastes via photofermentation using Purple Non-sulfur Bacteria (PNSB) – A review. *Energy Conversion and Management, 141*, 299–314. doi:10.1016/j.enconman.2016.09.001.

Glibert, P.M. (2024). Capturing light and the diversity of pigments. Phytoplankton whispering: An introduction to the physiology and ecology of microalgae (pp. 85–103). Springer, Cham, Switzerland, 2024, https://doi.org/10.1007/978-3-031-53897-1_5.

Gois, G. N. S. B., Macêdo, W. V., Peiter, F. S., Chaves, T. C., Sales, V. C. R., Barbosap, A. M. A., Abuda, K. S., & Amorim, E. L. C. (2021). Evaluation of biohydrogen production from sugarcane vinasse in an anaerobic fluidized bed reactor without ph control. *Latin American Applied Research, 51*(1), 63–69. https://laar.plapiqui.edu.ar.

Gomes, S. D., Fuess, L. T., Penteado, E. D., Lucas, S. D. M., Gotardo, J. T., & Zaiat, M. (2015). The application of an innovative continuous multiple tube reactor as a strategy to control the specific organic loading rate for biohydrogen production by dark fermentation. *Bioresource Technology, 197*, 201–207. Www.elsevier.com/locate/biortech. doi:10.1016/j.biortech.2015.08.077.

Greene, D. L., Ogden, J. M., & Lin, Z. (2020). Challenges in the designing, planning and deployment of hydrogen refueling infrastructure for fuel cell electric vehicles. *eTransportation, 6*. Www.journals.elsevier.com/etransportation. doi:10.1016/j.etran.2020.100086.

Guerra-Rodriguez, E., Portilla-Rivera, O. M., Jarquin-Enriquez, L., Ramirez, J. A., & Vazquez, M. (2012). Acid hydrolysis of wheat straw: A kinetic study. *Biomass & Bioenergy, 36*, 346–355.

Hassan, Q., Algburi, S., Sameen, A. Z., Salman, H. M., & Jaszczur, M. (2025). RETRACTED: Green hydrogen: A pathway to a sustainable energy future. *International Journal of Hydrogen Energy, 137*, 651. doi:10.1016/j.ijhydene.2025.05.078.

Intergovernmental Panel on Climate Change. (2022). *Mitigating climate change.Contribution of working group III to the sixth assessment report of the Intergovernmental Panel on Climate Change,Cambridge University Press, Cambridge, UK and New York, NY, USA, 2022,* ⟨17:italic ⟩ https://doi.org/10.1017/9781009157926⟨/17:italic⟩.

Jankowska, K., Domingo-Félez, C., Prado-Rubio, O. A., Skiadas, I. V., Woodley, J. M., & Pinelo, M. (2024). Effect of backwash as a strategy for biofouling control in the submerged ceramic membrane bioreactor for high-density cultivations: Process optimization and fouling mechanism at pilot scale. *Separation and Purification Technology, 338*. https://www.sciencedirect.com/science/journal/13835866. doi:10.1016/j.seppur.2024.126428.

Ji, G., Gao, C., Xiao, W., & Han, L. (2016). Mechanical fragmentation of corncob at different plant scales: Impact and mechanism on microstructure features and enzymatic hydrolysis. *Bioresource Technology, 205*, 159–165. www.elsevier.com/locate/biortech. doi:10.1016/j.biortech.2016.01.029.

Kapdan, I. K., & Kargi, F. (2006). Bio-hydrogen production from waste materials. *Enzyme and Microbial Technology, 38*(5), 569–582. doi:10.1016/j.enzmictec.2005.09.015.

Kargi, F., Eren, N. S., & Ozmihci, S. (2012). Bio-hydrogen production from cheese whey powder (CWP) solution: Comparison of thermophilic and mesophilic dark fermentations. *International Journal of Hydrogen Energy, 37*(10), 8338–8342. doi:10.1016/j.ijhydene.2012.02.162.

Kaul, S. N., & Gadakari, S. K. (1990). Fluidized bed reactor for wastewater treatment. *Chemical Engineering World, 25*(2), 25–35.

King, J. (2021). Hydrogen and Net Zero-A crucial role in decarbonisation strategies. *FST Journal, 22*(10), 17–18. doi:10.53289/CNLO5061.

Kshirsagar, S., Mishra, S., Nallapaneni, A., Bhattacharjee, A., & Mulukutla, P. (2023). *Understanding global hydrogen strategies: Strengthening clean hydrogen opportunities.*

Kumar, G., Sivagurunathan, P., Pugazhendhi, A., Thi, N. B. D., Zhen, G., Chandrasekhar, K., & Kadier, A. (2017). A comprehensive overview on light independent fermentative hydrogen production from wastewater feedstock and possible integrative options. *Energy Conversion and Management, 141*, 390–402. doi:10.1016/j.enconman.2016.09.087.

Li, C., Lu, Y., Liu, Y., Yue, J., Zhou, X., & Jing, Y. (2024). Comparative analysis of photo-fermentation hydrogen production from pretreated corn stover by deep eutectic solvents. *BioResources, 19*(1), 1433–1445. https://bioresources.cnr.ncsu.edu/wp-content/uploads/2024/01/BioRes_19_1_1433_Li_LLYZJ_Compar_Photo_Ferment_H2_Pretreat_Cornstover_DES_23123.pdf. doi:10.15376/biores.19.1.1433-1445.

Li, Y., Fan, X., Zhang, H., Ai, F., Jiao, Y., Zhang, Q., & Zhang, Z. (2022). Pretreatment of corn stover by torrefaction for improving reducing sugar and biohydrogen production. *Bioresource Technology, 351*, 126905.

Lin, C., & Chen, H. (2006). Sulfate effect on fermentative hydrogen production using anaerobic mixed microflora. *International Journal of Hydrogen Energy, 31*(7), 953–960. doi:10.1016/j.ijhydene.2005.07.009.

Liu, Y., Yu, P., Song, X., & Qu, Y. (2008). Hydrogen production from cellulose by co-culture of *Clostridium thermocellum* JN4 and *Thermoanaerobacterium thermosaccharolyticum* GD17. *International Journal of Hydrogen Energy, 33*(12), 2927–2933. doi:10.1016/j.ijhydene.2008.04.004.

Lopez-Hidalgo, A. M., Sánchez, A., & De León-Rodríguez, A. (2017). Simultaneous production of bioethanol and biohydrogen by *Escherichia coli* WDHL using wheat straw hydrolysate as substrate. *Fuel, 188*, 19–27. http://www.journals.elsevier.com/fuel/. doi:10.1016/j.fuel.2016.10.022.

Łukajtis, R., Hołowacz, I., Kucharska, K., Glinka, M., Rybarczyk, P., Przyjazny, A., & Kamiński, M. (2018). Hydrogen production from biomass using dark fermentation. *Renewable and Sustainable Energy Reviews, 91*, 665–694. https://www.journals.elsevier.com/renewable-and-sustainable-energy-reviews. doi:10.1016/j.rser.2018.04.043.

Luo, G., Xie, L., Zou, Z., Zhou, Q., & Wang, J. Y. (2010). Fermentative hydrogen production from cassava stillage by mixed anaerobic microflora: Effects of temperature and pH. *Applied Energy, 87*(12), 3710–3717. http://www.elsevier.com/inca/publications/store/4/0/5/8/9/1/index.htt. doi:10.1016/j.apenergy.2010.07.004.

Ma, J., Zhao, Q. B., Laurens, L. L. M., Jarvis, E. E., Nagle, N. J., Chen, S., & Frear, C. S. (2015). Mechanism, kinetics and microbiology of inhibition caused by long-chain fatty acids in anaerobic digestion of algal biomass. *Biotechnology for Biofuels, 8*(1). http://www.biotechnologyforbiofuels.com/. doi:10.1186/s13068-015-0322-z.

Maayedukondalu, T. S. (2015). *Steady state simulation of continuous stirred tank reactor (CSTR) system using aspen plus.*

Machineni, L., Deepanraj, B., Chew, K. W., & Rao, A. G. (2023). Biohydrogen production from lignocellulosic feedstock: Abiotic and biotic methods. *Renewable and Sustainable Energy Reviews, 182*. https://www.journals.elsevier.com/renewable-and-sustainable-energy-reviews. doi:10.1016/j.rser.2023.113344.

Meng, X., Chen, M., Gu, A., Wu, X., Liu, B., Zhou, J., & Mao, Z. (2022). China's hydrogen development strategy in the context of double carbon targets. *Natural Gas Industry B, 9*(6), 521–547. https://www.sciencedirect.com/science/journal/23528540. doi:10.1016/j.ngib.2022.11.004.

Milrad, Y., Mosebach, L., & Buchert, F. (2024). Regulation of microalgal photosynthetic electron transfer. *Plants, 13*(15), 2103. doi:10.3390/plants13152103.

Mohan, S. V. (2010). Waste to renewable energy: A sustainable and green approach towards production of biohydrogen by acidogenic fermentation. *Sustainable biotechnology: Sources of renewable energy* (pp. 129–164). Springer Netherlands. http://www.springerlink.com/openurl.asp?genre=book&isbn=978-90-481-3308-6. doi:10.1007/978-90-481-3295-9_7.

Moreno-Andrade, I., Carrillo-Reyes, J., Santiago, S. G., & Bujanos-Adame, M. C. (2015). Biohydrogen from food waste in a discontinuous process: Effect of HRT and microbial community analysis. *International Journal of Hydrogen Energy*. http://www.journals.elsevier.com/international-journal-of-hydrogen-energy/40. doi:10.1016/j.ijhydene.2015.04.084.

Mukherjee, T., Senevirathne, N., & Kaparaju, P. (2025). Dark fermentative hydrogen production from sugar syrup at different temperatures and inoculum-to-substrate ratios. *International Journal of Hydrogen Energy, 102*, 37–49. https://www.sciencedirect.com/science/journal/03603199. doi:10.1016/j.ijhydene.2025.01.011.

Mumtha, C., & Mahalingam, P. U. (2024). Biohydrogen production from co-substrates through dark fermentation by bacterial consortium. *3 Biotech, 14*(11). https://www.springer.com/journal/13205. doi:10.1007/s13205-024-04106-3.

Namdarimonfared, M., Zilouei, H., & Tondro, H. (2023). Biological hydrogen production from paper mill effluent via dark fermentation in a packed bed biofilm reactor. *Fuel, 338*, 127231. doi:10.1016/j.fuel.2022.127231.

Ngo, T. A., Nguyen, T. H., & Bui, H. T. V. (2012). Thermophilic fermentative hydrogen production from xylose by *Thermotoga neapolitana* DSM 4359. *Renewable Energy, 37*(1), 174–179. doi:10.1016/j.renene.2011.06.015.

Nnabuife, S. G., Oko, E., Kuang, B., Bello, A., Onwualu, A. P., Oyagha, S., & Whidborne, J. (2023). The prospects of hydrogen in achieving net zero emissions by 2050: A critical review. *Sustainable Chemistry for Climate Action, 2*. Www.sciencedirect.com/journal/sustainable-chemistry-for-climate-action. doi:10.1016/j.scca.2023.100024.

Nnabuife, S. G., Ugbeh-Johnson, J., Okeke, N. E., & Ogbonnaya, C. (2022). Present and projected developments in hydrogen production: A technological review. *Carbon Capture Science and Technology, 3*. https://www.journals.elsevier.com/carbon-capture-science-and-technology. doi:10.1016/j.ccst.2022.100042.

Ozyurt, B., Soysal, F., Hitit, Z. Y., Camcioglu, S., Akay, B., & Ertunc, S. (2019). An efficient dark fermentative hydrogen production by GMV control of pH. *International Journal of Hydrogen Energy, 44*(36), 19709–19718. http://www.journals.elsevier.com/international-journal-of-hydrogen-energy/. doi:10.1016/j.ijhydene.2019.06.048.

Parihar, R. K., & Upadhyay, K. (2016). Production of biohydrogen gas from dairy industry wastewater by anaerobic fermentation process. *International Journal of Applied Research, 2*, 512–515.

Pathak, V. V., Tyagi, V. V., Kothari, R., & Ahmad, S. (2017). A critical review on factors influencing fermentative hydrogen production. *Frontiers in Bioscience A: Journal & Virtual Library, 22*(8), 1195–1220.

Penniston, J., & Gueguim Kana, E. B. (2018). Impact of medium pH regulation on biohydrogen production in dark fermentation process using suspended and immobilized microbial cells. *Biotechnology and Biotechnological Equipment, 32*(1), 204–212. http://www.tandfonline.com/loi/tbeq20#.Vx8TFE1f3cs. doi:10.1080/13102818.2017.1408430.

Pereira, J. C. V., Saini, R., & Brar, S. K. (2024). *Biohydrogen production by photosynthetic microorganisms, biohydrogen-advances and processes* (pp. 125–147). Springer.

Phanduang, O., Lunprom, S., Salakkam, A., & Reungsang, A. (2017). Anaerobic solid-state fermentation of bio-hydrogen from microalgal *Chlorella* sp. biomass. *International Journal of Hydrogen Energy, 42*(15), 9650–9659. http://www.journals.elsevier.com/international-journal-of-hydrogen-energy/. doi:10.1016/j.ijhydene.2017.01.084.

Ramasamy, E. V., Gajalakshmi, S., Sanjeevi, R., Jithesh, M. N., & Abbasi, S. A. (2004). Feasibility studies on the treatment of dairy wastewaters with upflow anaerobic sludge blanket reactors. *Bioresource Technology, 93*(2), 209–212. doi:10.1016/j.biortech.2003.11.001.

Ramos, L. R., & Silva, E. L. (2018). Thermophilic hydrogen and methane production from sugarcane stillage in two-stage anaerobic fluidized bed reactors. *World hydrogen energy conference*.

Safonov, A. V., Kovalev, D. A., Dorokhov, A. S., Pavkin, D. Y., Karelina, M. Y., Filatov, V. V., & Kovalev, A. A. (2026). Integration of technologies for utilization of organic agrowaste: Production of feed additives, reduction of carbon dioxide emissions and production of hydrogen for sustainable agriculture. *International Journal of Hydrogen Energy, 197*, 151518. doi:10.1016/j.ijhydene.2025.151518.

Sahil, S., Singh, R., Masakapalli, S. K., Pareek, N., Kovalev, A. A., Litti, Y. V., Nanda, S., & Vivekanand, V. (2024). Biomass pretreatment, bioprocessing and reactor design for biohydrogen production: A review. *Environmental Chemistry Letters, 22*(4), 1665–1702. https://www.springer.com/journal/10311. doi:10.1007/s10311-024-01722-6.

Santiago, S. G., Trably, E., Latrille, E., Buitrón, G., & Moreno-Andrade, I. (2019). The hydraulic retention time influences the abundance of *Enterobacter, Clostridium* and *Lactobacillus* during the hydrogen production from food waste. *Letters in Applied Microbiology, 69*(3), 138–147. doi:10.1111/lam.13191.

Santos, S. C., Rosa, P. R. F., Sakamoto, I. K., Varesche, M. B. A., & Silva, E. L. (2014). Organic loading rate impact on biohydrogen production and microbial communities at anaerobic fluidized thermophilic bed reactors treating sugarcane stillage. *Bioresource Technology, 159*, 55–63. Www.elsevier.com/locate/biortech. doi:10.1016/j.biortech.2014.02.051.

Schmidt, O., Gambhir, A., Staffell, I., Hawkes, A., Nelson, J., & Few, S. (2017). Future cost and performance of water electrolysis: An expert elicitation study. *International Journal of Hydrogen Energy, 42*(52), 30470–30492. doi:10.1016/j.ijhydene.2017.10.045.

Schulthoff, M., Rudnick, I., Bose, A., & Gençer, E. (2021). Role of hydrogen in a low-carbon electric power system: A case study. *Frontiers in Energy Research, 8*. http://www.frontiersin.org/Energy_Research/. doi:10.3389/fenrg.2020.585461.

Shui, X., Jiang, D., Zhang, H., Zhang, X., Yang, J., Lei, T., Chen, Z., & Zhang, Q. (2022). Effect of light perturbation on the photo-fermentative hydrogen production of biomass from giant reed. *Journal of Cleaner Production, 351*, 131481. doi:10.1016/j.jclepro.2022.131481.

Singh, A., & Rathore, D. (2017). *Biohydrogen production: Sustainability of current technology and future perspective*. Springer.

Slorach, P. C., & Stamford, L. (2021). Net zero in the heating sector: Technological options and environmental sustainability from now to 2050. *Energy Conversion and Management, 230*, 113838. doi:10.1016/j.enconman.2021.113838.

Sun, B., Zhang, M., Sun, Q., Zhong, J., & Shao, G. (2025). Review on natural hydrogen wells safety. *Nature Communications, 16*(1). https://www.nature.com/ncomms/. doi:10.1038/s41467-024-55773-y.

Tseng, S. K., & Lin, M. R. (1994). Treatment of organic wastewater by anaerobic biological fluidized bed reactor. *Water Science and Technology, 29*. https://iwaponline.com/wst. doi:10.2166/wst.1994.0602.

United Nation. (2023). *Emissions gap report 2023, United Nation Environment Programme*.

Vatsala, T., Raj, S., & Manimaran, A. (2008). A pilot-scale study of biohydrogen production from distillery effluent using defined bacterial co-culture. *International Journal of Hydrogen Energy, 33*(20), 5404–5415. doi:10.1016/j.ijhydene.2008.07.015.

Venkata Mohan, S., Lalit Babu, V., & Sarma, P. N. (2007). Anaerobic biohydrogen production from dairy wastewater treatment in sequencing batch reactor (AnSBR): Effect of organic loading rate. *Enzyme and Microbial Technology, 41*(4), 506–515. doi:10.1016/j.enzmictec. 2007.04.007.

Venkata Mohan, S., Lalit Babu, V., Srikanth, S., & Sarma, P. N. (2008). Bio-electrochemical evaluation of fermentative hydrogen production process with the function of feeding pH. *International Journal of Hydrogen Energy, 33*(17), 4533–4546. doi:10.1016/j.ijhydene.2008.05. 073.

Venkatamohan, S., Vijayabhaskar, Y., Muralikrishna, P., Chandrasekhararao, N., Lalitbabu, V., & Sarma, P. (2007). Biohydrogen production from chemical wastewater as substrate by selectively enriched anaerobic mixed consortia: Influence of fermentation pH and substrate composition. *International Journal of Hydrogen Energy, 32*(13), 2286–2295. doi:10.1016/ j.ijhydene.2007.03.015.

Wang, J., & Wan, W. (2008). Effect of temperature on fermentative hydrogen production by mixed cultures. *International Journal of Hydrogen Energy, 33*(20), 5392–5397. doi:10.1016/j. ijhydene.2008.07.010.

Wang, J., & Wan, W. (2009). Factors influencing fermentative hydrogen production: A review. *International Journal of Hydrogen Energy, 34*(2), 799–811. doi:10.1016/j.ijhydene.2008. 11.015.

Wang, X., Li, S. Y., chen, Y., Chen, Y. P., Guo, J. S., Liu, S. Y., & Yan, P. (2024). New insights into microalgal photobiological hydrogen production: Potential role of aggregation forms in modulating the hydrogenase activity and metabolic properties of microalgae during hydrogen production. *Chemical Engineering Journal, 500*. https://www.sciencedirect.com/ science/journal/13858947. doi:10.1016/j.cej.2024.156802.

Wu, Z., Peng, K., Zhang, Y., Wang, M., Yong, C., Chen, L., Qu, P., Huang, H., Sun, E., & Pan, M. (2022). Lignocellulose dissociation with biological pretreatment towards the biochemical platform: A review. *Materials Today Bio, 16*. Www.journals.elsevier.com/ materials-today-bio. doi:10.1016/j.mtbio.2022.100445.

Xu, X., Zhou, Q., & Yu, D. (2022). The future of hydrogen energy: Bio-hydrogen production technology. *International Journal of Hydrogen Energy, 47*(79), 33677–33698. http: //www.journals.elsevier.com/international-journal-of-hydrogen-energy/. doi:10.1016/ j.ijhydene.2022.07.261.

Yan, Y., Manovic, V., Anthony, E. J., & Clough, P. T. (2020). Techno-economic analysis of low-carbon hydrogen production by sorption enhanced steam methane reforming (SE-SMR) processes. *Energy Conversion and Management, 226*. https://www.journals.elsevier.com/ energy-conversion-and-management. doi:10.1016/j.enconman.2020.113530.

Yang, F., Hanna, M. A., & Sun, R. (2012). Value-added uses for crude glycerol—A byproduct of biodiesel production. *Biotechnology for Biofuels, 5*. doi:10.1186/1754-6834-5-1322413907.

Zhang, B., Kotsalis, G., Khan, J., Xiong, Z., Igou, T., Lan, G., & Chen, Y. (2020). Backwash sequence optimization of a pilot-scale ultrafiltration membrane system using data-driven modeling for parameter forecasting. *Journal of Membrane Science, 612*, 118464.

Zhang, Q., Zhang, Z., Wang, Y., Lee, D. J., Li, G., Zhou, X., Jiang, D., Xu, B., Lu, C., Li, Y., & Ge, X. (2018). Sequential dark and photo fermentation hydrogen production from hydrolyzed corn stover: A pilot test using 11 m^3 reactor. *Bioresource Technology, 253*, 382–386. Www.elsevier.com/locate/biortech. doi:10.1016/j.biortech.2018.01.017.

Zhang, Z., Ai, F., Zhang, H., Zhang, H., Zhu, S., Zhang, Q., & Li, Y. (2023). Synergetic effect evaluation of light and mass transfer enhancement strategies on photo fermentative biohydrogen production process: Illumination, shake, and high solid level. *Energy, 269*, 126841. doi:10.1016/j.energy.2023.126841.

Zheng, Y., Zhao, J., Xu, F., & Li, Y. (2014). Pretreatment of lignocellulosic biomass for enhanced biogas production. *Progress in Energy and Combustion Science, 42*(1), 35–53. doi:10.1016/j.pecs.2014.01.001.

Zhou, S., Liu, D., Fan, K., Liu, H., & Zhang, X. D. (2024). Atomic-level design of biomimetic iron-sulfur clusters for biocatalysis. *Nanoscale, 16*(40), 18644–18665. http://pubs.rsc.org/en/journals/journal/nr. doi:10.1039/d4nr02883j.

CHAPTER

7

Economic analysis of production and sustainability of biohydrogen

Yangyang Li[a,*], *Peiyu Tian*[a], *Beier Shang*[a], *Nan An*[a], *Bingyan Wu*[a], *Yuao Su*[b], *Hui Sun*[c], *Tian Yuan*[d], *and Tingting Hou*[c]

[a]China Agricultural University, Beijing, China, [b]Seoul National University of Science & Technology, Seoul, South Korea, [c]Henan Agricultural University, Zhengzhou, China, [d]University of Tsukuba, Tsukuba, Japan

*Corresponding author. E-mail address: liyy@cau.edu.cn

7.1 Introduction

Global energy demand remains predominantly dependent on fossil fuels, coal, oil, and natural gas, which account for over 80% of global primary energy consumption. The remaining share is derived from alternative sources: biomass (e.g., agricultural residues such as wheat straw, forestry biomass, and organic waste-derived biogas), nuclear energy (powered by nuclear fission for electricity), solar energy (photovoltaic panels and concentrated solar power), hydropower (run-of-river and reservoir plants), and geothermal energy (Yaashikaa et al., 2022). This fossil-dominated structure has fueled industrial growth since the 18th century but triggers severe environmental crises: greenhouse gas (GHG) emissions (mainly CO_2) drive global warming, sulfur oxides (SO_x) and nitrogen oxides (NO_x) cause smog and acid rain, and overexploitation degrades ecosystems (e.g., oil spills harm marine life).

DOI: https://doi.org/10.1016/B978-0-443-27686-6.00003-7

As a renewable biobased fuel, hydrogen (H_2) emerges as a fossil fuel alternative, with key advantages: high productivity (theoretical yield of 12 mol H_2/mol glucose in biohydrogen processes), zero-pollution combustion (only H_2O as a byproduct), recyclability (storable in closed-loop systems), nontoxicity (safe for humans/ecosystems), high conversion efficiency (>60% in fuel cells), and low heating rate (reducing equipment thermal stress) (Ebrahimian et al., 2022; Yusuf et al., 2021). These align with global carbon neutrality goals (e.g., the Paris Agreement), making H_2 a core of future clean energy. However, commercialization faces dual challenges: ecologically, biomass cultivation may deplete soil nutrients or compete with food crops; economically, high costs persist in production, storage (high-pressure tanks or cryogenic vessels), transportation (pipelines/tankers), and infrastructure (hydrogen refueling stations) (Hassan et al., 2019). Traditional H_2 production is fossil dependent: thermal reforming [e.g., steam methane reforming (SMR)], coal gasification, and heavy oil partial oxidation. SMR, 75% of global H_2 output, uses methane + high-temperature steam (700°C–1000°C) over nickel catalysts to produce H_2 and CO_2. It consumes 28–35 MJ/kg H_2 and emits 8–10 kg CO_2/kg H_2. Coal gasification converts coal to syngas (H_2 + CO) at 1200°C–1500°C and then purifies H_2 via water–gas shift reactions. It uses 40–45 MJ/kg H_2 and releases SO_x, NO_x, and slag, worsening pollution (Jamali et al., 2019). These high energy, high emission methods conflict with clean transition, spurring biohydrogen development.

To solve these issues, biohydrogen production arises as a sustainable option. It uses biomass (sugarcane bagasse, food waste, and microalgae) as feedstock, leveraging microbial/enzymatic catalysis via three processes: photolysis, fermentation, and electrochemical conversion (Chezeau & Vial, 2019). Compared to traditional methods, it has lower energy use [e.g., microbial electrolysis cells (MECs) consume 1/3–1/2 the energy of SMR] and operates at ambient conditions (25°C–35°C, 1atm), avoiding high-temperature/pressure equipment and cutting capital costs (Li et al., 2022). Yet each pathway has pros and cons. Photolysis: Uses photosynthetic microbes (cyanobacteria, green algae) to split H_2O with sunlight. Direct photolysis enables continuous H_2 production but suffers from O_2-induced hydrogenase inhibition, low efficiency (5%–8%), and reliance on high light intensity. Indirect photolysis separates O_2 release and H_2 production (solving O_2 accumulation) but needs higher capital/operational costs. Fermentation: Anaerobic microbes (*Clostridium, Rhodobacter*) degrade carbohydrates. Photo-fermentation (PF) uses sunlight for high yield (10–12 mol H_2/mol glucose); dark fermentation (DF) operates without light, treats organic waste, and has faster rates but lower yield (2–4 mol H_2/mol glucose) and byproducts (organic acids like acetic acid). Electrochemical conversion (MECs): Microbes oxidize organics at the anode, generating electrons/protons that form H_2 at the cathode. MECs have 75% efficiency and treat wastewater but are costly because of platinum catalysts and proton exchange membranes (Safari et al., 2022).

To achieve a sustainable hydrogen economy (integrating production-storage-transportation-utilization), efficient, economical, and eco-friendly H_2 systems are essential (Jin et al., 2024). Life cycle assessment (LCA), a "cradle to grave" tool, evaluates resource use (water and energy), emissions (CO_2 and SO_x), and ecological impacts (land use and biodiversity loss) across biohydrogen's life cycle: feedstock collection, pretreatment (biomass hydrolysis), production, purification (pressure swing adsorption), and waste disposal. LCA identifies hotspots (e.g., fertilizer use in biomass cultivation causes eutrophication) and validates GHG reductions. Li et al. (2020a) used LCA to show biohydrogen cuts GHG emissions by 40%–60% vs. coal-based H_2 (Li et al., 2020b; Sun et al., 2021). Techno-economic analysis (TEA) assesses biohydrogen's economic viability. It models capital expenditure (CAPEX: equipment and plant construction) and operational expenditure (feedstock, labor, and maintenance) to calculate metrics: levelized cost of H_2 (LCOH), return on investment (ROI), payback period (PBP), and net present value (NPV). TEA finds bottlenecks (e.g., MEC catalysts make up 40% of CAPEX) and analyzes sensitivities (feedstock price and subsidies). Han et al. (2016b) studied DF with sugarcane molasses: plants ≥ 40 m^3 have positive ROI (0.63% for 40 m^3 and 9.25% for 50 m^3) and PBPs ≤ 10 years; smaller plants are unviable. Biohydrogen's feasibility depends on tech maturity (PF is lab scale), feedstock cost (<$50/t agricultural waste lowers LCOH), and scale (large production cuts LCOH by 15%–30%) (Ahmad et al., 2016; Ganeshan et al., 2023; Shahbaz et al., 2017). TEA and LCA are key for biohydrogen evaluation. TEA pinpoints cost bottlenecks (e.g., low-cost catalysts) to guide optimization; LCA ensures eco-sustainability via GHG quantification. Together, they support policy-making (subsidies and carbon pricing) and tech breakthroughs (higher fermentation yields), making them vital for assessing biohydrogen's environmental and economic benefits. This integration drives biohydrogen from lab to industry, accelerating the global low carbon energy transition.

Although current research on biohydrogen production has focused on its technical potential and the evaluation value of TEA and LCA, there is a lack of systematic and comparative comprehensive analysis on the environmental and economic performance of different biohydrogen production technical pathways (such as photolysis, fermentation, and electrochemistry), making it difficult to clarify the applicable scenarios and optimization directions of each pathway in practical applications. This chapter summarizes the results of TEA and LCA reported in the literature. It can not only fill the gap in the "technology-environment-economy" linkage analysis in existing studies and provide quantitative basis for technology selection under different scenarios but also identify key constraints of each pathway (such as the efficiency bottleneck of photolysis and the cost structure of fermentation) to offer targeted guidance for technological optimization and industrial implementation. This chapter can not only consolidate the theoretical foundation for the commercialization of biohydrogen production but also provide practical support for promoting its transition from laboratory-scale to large-scale application and facilitating the development of a sustainable hydrogen economy.

7.2 Material flow analysis of biohydrogen technologies

7.2.1 Dark fermentation biohydrogen production technology

DF biohydrogen production technology allows for hydrogen production as microorganisms decompose organic substrates under anaerobic conditions (Rani et al., 2024). It has a core process comprising four stages—pretreatment, strain inoculation, fermentation reaction, and gas separation and purification (Bhatia et al., 2021)—which form an integrated chain via substance and energy flow (Pachapur et al., 2019). The technology can run continuously without light, and it boasts advantages such as simple equipment, low-cost, a high reaction rate, and wide substrate adaptability. Yet, it has limitations such as low hydrogen purity, restricted substrate conversion efficiency, and microorganisms' high sensitivity to environmental factors such as temperature and pH (Mohanakrishna et al., 2023). In pretreatment, organic substrates (e.g., agricultural waste), pretreatment reagents (such as dilute acids), and water are introduced, and physical, chemical, or biological methods are used to break down the substrate structure for easier subsequent microbial metabolism (Rocha-Meneses et al., 2022). For strain inoculation, pretreated products, hydrogen-producing strains, nutritional additives, and water are mixed into an inoculated mixture that reserves functional microorganisms. In the fermentation reaction the inoculated mixture is the key input; electrical energy is used to maintain an anaerobic environment, where substrates are metabolized into crude hydrogen, fermentation broth, and residual substrates (the latter are filtered out). During gas separation and purification, crude hydrogen, fermentation broth, absorbents, and electrical energy are input, and high-purity hydrogen is obtained through adsorption-based purification. Byproducts (e.g., concentrated CO_2 and organic acids) can be recycled, whereas waste liquids need treatment. This technology has great application potential in agricultural waste resource utilization, being especially fit for treating readily available organic substrates such as crop straw and livestock manure. Moreover, with low requirements for operational

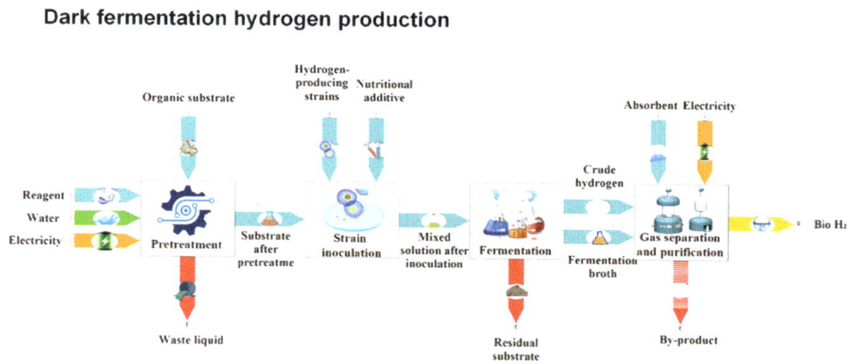

FIGURE 7.1 Material and energy flow diagram of dark fermentation biohydrogen production technology.

space, it enables distributed hydrogen production in small- and medium-sized farms or rural areas, helping upgrade rural energy structures (Fig. 7.1).

7.2.2 Photo fermentation biohydrogen production technology

Photosynthetic bacteria produce hydrogen by decomposing organic substrates under light (Zuo et al., 2025), with the process involving four stages: pretreatment, strain inoculation, fermentation reaction, and gas separation and purification. This technology uses light energy, enables secondary substrate conversion via DF byproducts, and offers high hydrogen purity (low purification cost). However, it has limitations: strict light requirements (increasing equipment costs), slow reaction rate, narrow substrate adaptability (mostly simple organics), and inability to operate around the clock because of light restrictions. Its mechanism relies on photosynthetic bacteria conducting photosynthesis—like reactions under light to drive organic substrate decomposition, releasing electrons and protons to form hydrogen via specific pathways. Current research progresses in strain improvement (enhancing light harvesting efficiency and key enzyme activity), light utilization enhancement (optimizing light sources and photobioreactors), and substrate expansion (using complex organic wastes as carbon sources) (Song et al., 2022). Technological stages: Pretreatment inputs organic substrates, reagents, water, and electricity to generate small molecule organic acids, waste liquids, and residues; strain inoculation mixes pretreated products, hydrogen-producing strains, nutritional additives, and water into an inoculated mixture; fermentation sees the mixture metabolize under light to generate crude hydrogen, fermentation broth, and residual substrates (filtered out); gas separation and purification are

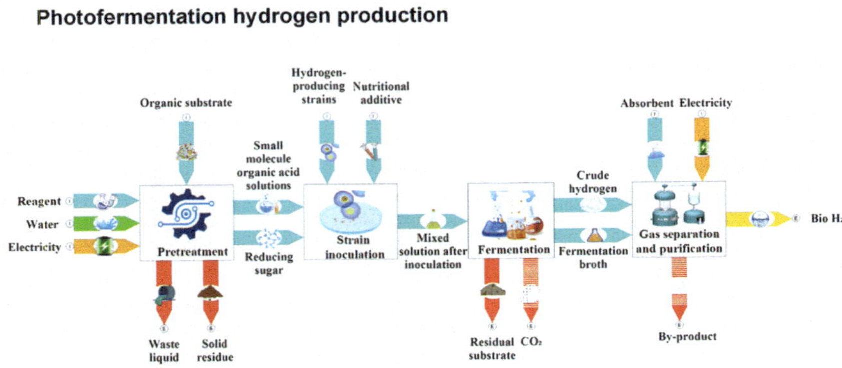

FIGURE 7.2 Material and energy flow diagram of photo fermentation biohydrogen production technology.

same as DF, ultimately yielding high-purity hydrogen, with byproducts and waste liquids requiring treatment (Fig. 7.2).

7.2.3 Dark-photo-coupled fermentation biohydrogen production technology

The dark-photo-coupled fermentation biohydrogen production technology, a pivotal biohydrogen advancement, achieves hydrogen generation via the synergy of dark and photo fermentation (Crespi et al., 2024). Its process consists of four stages (Kucharska et al., 2018): pretreatment, DF, PF, gas separation, and purification. It combines both merits López-Hernández et al. (2024), efficiently using complex substrates to boost total hydrogen production—DF provides PF with essential substrates and eliminates inhibitors, enhancing system stability vs. single PF. Yet it has flaws: Complex system (needing precise condition coordination), high equipment costs, PF limited by light, and strict microbial compatibility demands (Zhang et al., 2024). Pretreatment inputs organic substrates, reagents, water, and electricity, yielding utilizable DF materials (plus waste liquids/residues). DF adds pretreated substrates and dark fermentative bacteria, producing organic acids, alcohols, H_2/CO_2 (PF feedstock) (Hossain et al., 2023). PF uses these products and photosynthetic bacteria (under proper light) to make high-purity H_2 mixed gas. Gas separation and purification give high-purity hydrogen (CO_2 is recyclable, solid residues need harmless disposal) (Muntasir Shovon et al., 2024). For strain inoculation: in DF, inoculating high-efficiency dark fermentative bacteria raises substrate conversion and optimizes organic acid/hydrogen output, benefiting subsequent PF. As for microbial mutual influence, DF's organic acids are ideal photosynthetic bacteria carbon sources, but if it produces

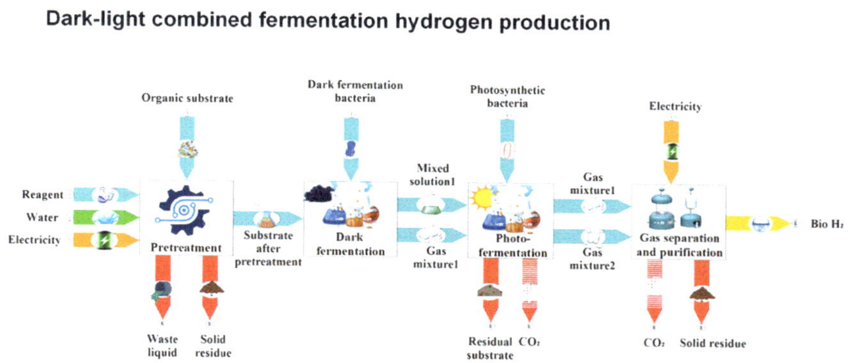

FIGURE 7.3 Material and energy flow diagram of dark-photo-coupled fermentation biohydrogen production technology.

toxins/inhibitors, photosynthetic bacteria growth and hydrogen production will be suppressed. Thus selecting compatible strains and optimizing conditions is key to smooth coupled fermentation (Fig. 7.3).

7.2.4 Pyrolysis biohydrogen production technology

Pyrolysis biohydrogen production technology converts organic substrates into hydrogen via pyrolysis reactions, with a four-stage process: pretreatment, pyrolysis reaction, gas purification, and hydrogen purification. It boasts a fast reaction rate, exceptionally broad substrate adaptability (unconstrained by microbial metabolism and applicable to lignocellulosic biomass such as wood and bark and even plastic wastes), and easy regulation of process parameters (e.g., temperature and pressure) to optimize hydrogen yield. However, it demands high temperatures (500°C–1000°C), relying on electric or fossil energy input that leads to high energy consumption (energy ROI often <1.5 for low calorific substrates); pyrolysis gas contains impurities such as CH_4, CO, and H_2S (increasing purification costs, requiring multistep scrubbing or membrane separation); tar (polycyclic aromatic hydrocarbons) is prone to formation (clogging pipelines and deactivating catalysts); and it cannot utilize low concentration organic wastewater (because of high moisture diluting heat input). The technological stages are as follows (Arfan et al., 2023): Pretreatment inputs organic substrates and electricity, using physical methods (crushing, drying to <15% moisture) to obtain pretreated substrates, with inert solid residues (e.g., ash) produced. Pyrolysis reaction inputs pretreated substrates, electrical energy, and inert gas (N_2/Ar to prevent oxidation), where high-temperature cracking breaks chemical bonds of organic matter into pyrolysis gas (H_2, CO, and CH_4) and liquid tar (with concurrent

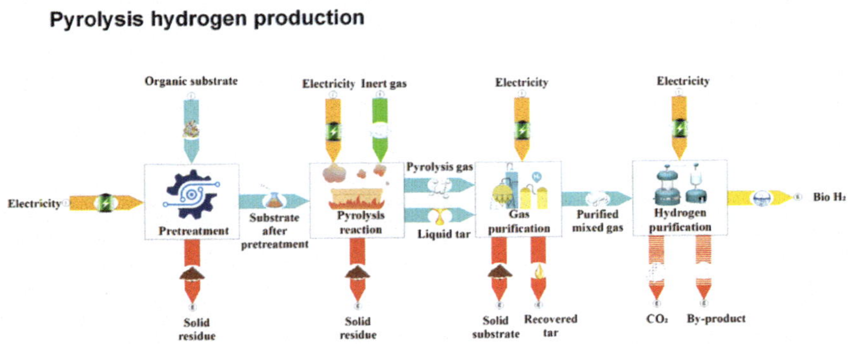

FIGURE 7.4 Material and energy flow diagram of pyrolysis biohydrogen production technology.

solid char residue). Gas purification inputs pyrolysis gas and electricity, adopting electrostatic precipitation or thermal cracking to recover tar and separate impurities (e.g., H_2S via amine absorption), producing mixed gas. Hydrogen purification further separates and purifies the mixed gas (e.g., pressure swing adsorption to isolate H_2 from CO/CH_4) to produce high purity H_2 (≥99.97%); CO_2 can be recycled for enhanced oil recovery, and waste liquids need advanced oxidation treatment to meet discharge standards (Li et al., 2020a). Despite challenges of high energy consumption and tar handling, it outperforms microbial fermentation in processing high-lignin rigid biomass (lignin is hard to degrade by microbes). Current studies address tar via catalytic cracking (using Ni-based catalysts to break tar into small molecule gases) or adsorption (activated carbon), which improve gas purification efficiency but increase process complexity and investment costs (Fig. 7.4).

7.2.5 Thermochemical gasification biohydrogen production technology

Thermochemical gasification biohydrogen production technology converts organic substrates into hydrogen via thermochemical gasification, featuring a four-stage process: pretreatment, gasification reaction, gas purification, and hydrogen purification (Wang et al., 2019). It boasts a fast reaction rate, can handle a wide range of biomass (unrestricted by microbial metabolism and applicable to various wastes), enables stable large-scale hydrogen production, which is favorable for industrial applications, and allows for easy regulation of process parameters such as temperature and pressure to optimize hydrogen yield. However, it demands high temperatures (600°C–1000°C) and pressures, relying on energy input that leads to high energy consumption; syngas contains impurities such as CO_2 and

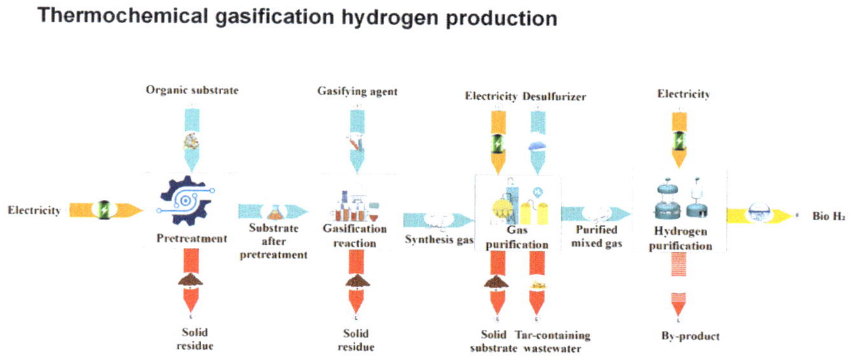

FIGURE 7.5 Material and energy flow diagram of thermochemical gasification biohydrogen production technology.

CH_4, increasing purification costs; it involves high equipment investment and maintenance expenses and is prone to generating pollutants like tar. Technological stages: Pretreatment uses physical and chemical methods with organic substrates and electricity to obtain pretreated substrates, producing solid residues (Singh et al., 2023). Gasification reaction inputs pretreated substrates and gasifying agents, where high-temperature cracking breaks organic bonds into syngas (Sarkar & Kumar, 2009), with solid char residue formed. Gas purification inputs syngas, electricity, and desulfurizing agents, recovering tar and separating impurities to produce mixed gas and separate waste (Materazzi et al., 2019). Hydrogen purification yields high-purity H_2 ($\geq$99.97%); byproducts like CO_2 can be recycled. Integrating into industrial energy systems, some projects use self-produced syngas for heating, creating a "gas-to-gas" cycle, though ensuring heating stability for gasification conditions is crucial (Fig. 7.5).

7.2.6 Microbial electrolysis cell (MEC) biohydrogen production technology

MEC biohydrogen production technology achieves hydrogen production via the synergistic decomposition of organic substrates by electrogenic microorganisms and electrolysis reactions (Ganguly et al., 2025). Its core is leveraging electrons from electrogenic microbial metabolism and combining them with externally applied low voltage to drive hydrogen production, unlike traditional electrolysis that depends on high electrical energy. The process has three stages: pretreatment, electrolysis reaction, and gas purification. It can generate high-purity hydrogen with low energy consumption and convert wastes like sewage, for instance, treating industrial or domestic sewage with high-concentration organic matter to

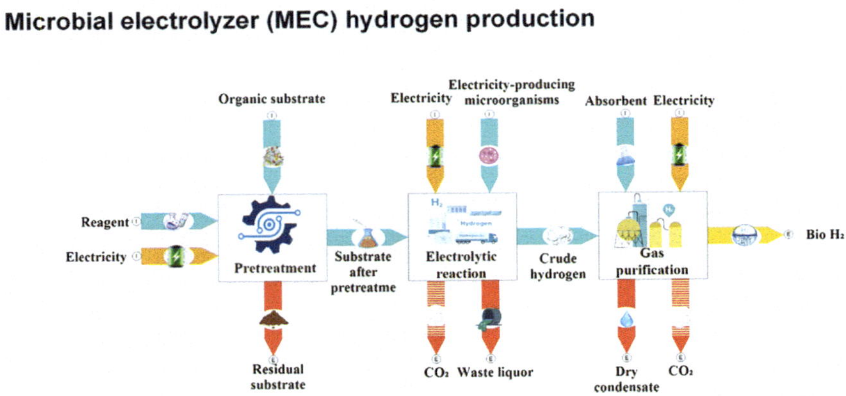

FIGURE 7.6 Material and energy flow diagram of microbial electrolysis cell (MEC) biohydrogen production technology.

produce hydrogen while reducing water pollution load. However, the system is complex (needing precise parameter control), the hydrogen production rate is slow (limiting large-scale use), and electrode materials are costly (prone to wear, with high maintenance costs). In pretreatment (Koul et al., 2022), organic substrates, reagents, and electricity are input; after treatment, pretreated substrates are obtained, with residual substrates produced. The electrolysis reaction involves inputting pretreated substrates, electricity, and electrogenic microorganisms, which, through a synergistic reaction yield crude hydrogen (containing CO_2 and waste liquid). For gas purification (Srivastava et al., 2020), crude hydrogen, absorbents, and electricity are input; after purification, high-purity hydrogen is produced, with byproducts recyclable and waste liquid requiring biochemical treatment. Valuable in coupling wastewater treatment and energy recovery, this technology enables "hydrogen production from waste" and wastewater purification, suitable for pilot applications in municipal wastewater treatment plants. Currently, research focuses on developing new electrode materials (e.g., low-cost carbon-based materials) and acclimatizing electrogenic microorganisms to boost hydrogen production rate and electrode service life (Fig. 7.6).

7.2.7 Photobiological hydrogen production technology

Photobiological hydrogen production technology decomposes water to produce hydrogen with light energy and photocatalysts (Olabi et al., 2023), featuring a three-stage process: pretreatment, photolysis reaction, and gas purification. It has advantages like a fast reaction rate (unrestricted by microbial metabolism), direct solar energy utilization (greatly

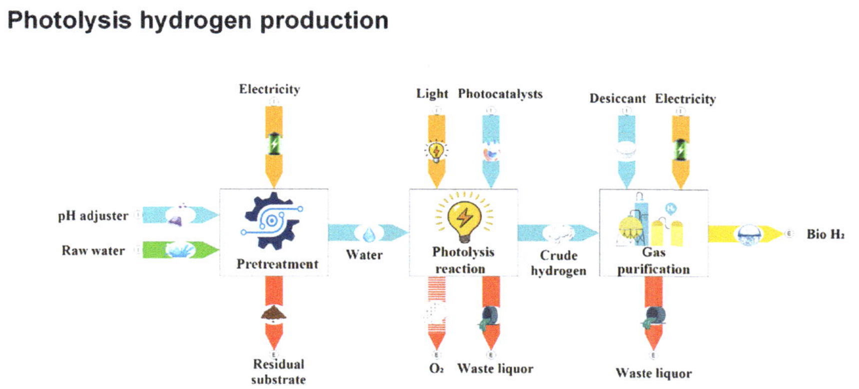

FIGURE 7.7 Material and energy flow diagram of photobiological hydrogen production technology.

reducing dependence on traditional electricity and enhancing low-carbon properties), and water as a clean, widely available substrate. However, it relies on light (efficiency plummets on cloudy days), uses costly and easily deactivated catalysts, requires sacrificial agents (hindering large-scale application), and lacks waste resource recovery benefits. In pretreatment (Sivaramakrishnan et al., 2021), pH regulators, raw water, and electricity are input; after treatment, water suitable for photolysis is obtained, with residual substrates produced. Strict control of water pH and impurity content is essential to avoid impacting photocatalyst activity and photolysis efficiency. The photolysis reaction involves inputting treated water, light energy, and photocatalysts, generating crude hydrogen (containing O_2 and waste liquid) via decomposition. For gas purification (Rawat et al., 2023), crude hydrogen, desiccants, and electricity are input; after purification, high purity hydrogen is produced, oxygen can be collected, and waste liquid needs treatment. Currently, research focuses on developing low-cost, high-stability photocatalysts to tackle the large-scale application bottleneck through material innovation. Moreover, integrating photolysis systems with energy storage devices is a direction to address hydrogen production interruptions at night or on cloudy days by storing excess daytime hydrogen energy (Fig. 7.7).

7.2.8 Electrolytic biohydrogen production technology

Electrolytic biohydrogen production technology uses electrical energy to decompose water for hydrogen production (Koroneos et al., 2008), relying on electrochemical principles to break water molecules' stable structure and convert electrical energy into storable hydrogen energy, a

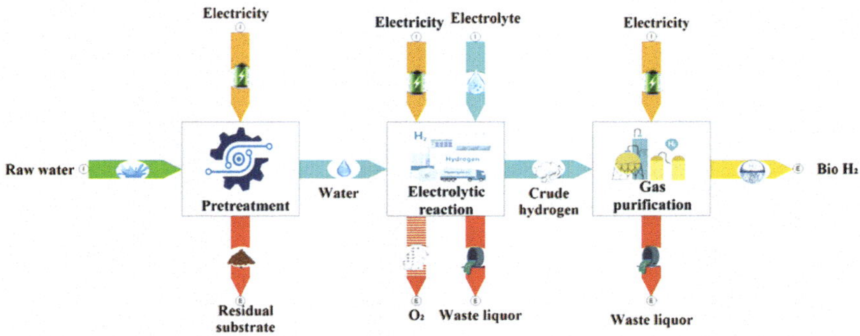

FIGURE 7.8 Material and energy flow diagram of electrolytic biohydrogen production technology.

mainstream green hydrogen production route. Its process has three stages: pretreatment, electrolysis reaction, and gas purification. It can cut carbon emissions and store electricity from renewable sources (Ghorbani et al., 2025), yet it depends on external electrical energy (losing low carbon merit if electricity is not clean), has low energy conversion efficiency, and is tightly linked to the power grid, leading to poor flexibility. In pretreatment (Liu et al., 2024), raw water and electricity are inputs; after purification, water fit for electrolysis is obtained, with residual substrates produced. For the electrolysis reaction (Xu et al., 2022), treated water, electricity, and electrolytes (such as potassium hydroxide and sulfuric acid, which boost water conductivity and speed up electrolysis) are input, generating crude hydrogen, O_2, and waste liquid. Gas purification then refines crude hydrogen into high-purity hydrogen, with proper waste liquid disposal. Often paired with intermittent renewable energy (RE) such as wind and photovoltaic power, this technology acts as an energy storage device to balance grid load, enabling "green electricity-green hydrogen" integrated development. To enhance economics, some projects explore the "electrolytic hydrogen production + industrial byproduct utilization" model—using industrial wastewater instead of pure water as electrolysis feedstock reduces water costs and realizes wastewater resource utilization (Fig. 7.8).

7.2.9 Biological conversion-chemical reforming hybrid biohydrogen production technology

The biological conversion chemical reforming hybrid biohydrogen production technology integrates biological conversion and chemical reforming processes, with workflows including pretreatment, biological conversion, mixed gas purification, chemical reforming, water–gas shift reaction, and gas separation and purification (Lefranc et al., 2024). This

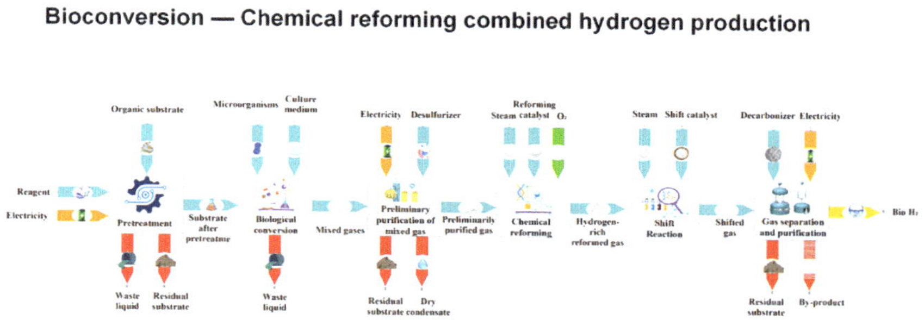

FIGURE 7.9 Material and energy flow diagram of biological conversion-chemical reforming hybrid biohydrogen production technology.

"biology + chemistry" synergistic model leverages biological conversion's raw material adaptability and chemical reforming's high-efficiency hydrogen production, achieving advantage complementarity. It breaks single technology efficiency bottlenecks and is compatible with various biomass types (Lefranc et al., 2024). However, the system is complex (high regulation difficulty), equipment and maintenance costs are high, and stability is poor. In pretreatment, organic substrates are processed to obtain pretreated products, generating waste liquid and residual substrates. Biological conversion uses microorganisms to decompose substrates, producing mixed gas and waste liquid. Mixed gas purification initially removes impurities, reducing catalyst poisoning risk in subsequent chemical reforming and ensuring reaction efficiency. Through catalytic reactions in secondary biological conversion and chemical reforming, hydrogen-rich gas is generated and component ratios adjusted. Finally, gas separation and purification yield high-purity hydrogen, with byproducts recyclable after treatment. As an innovative multitechnology integration path, commercial promotion requires solving parameter matching and coordinated control among subsystems. Currently in the laboratory to pilot transition, pilot studies by some teams, via modular equipment, enable independent regulation and rapid switching of each link, laying an experimental foundation for subsequent system optimization (Fig. 7.9).

7.3 Economic analysis for biohydrogen production technologies

TEA provides a rigorous, system-level framework for assessing both the commercialization potential and the economic feasibility of hydrogen production technologies. In contrast to assessments that are exclusively technical in nature, TEA explicitly integrates engineering design and process modeling with economic evaluation. This integration enables

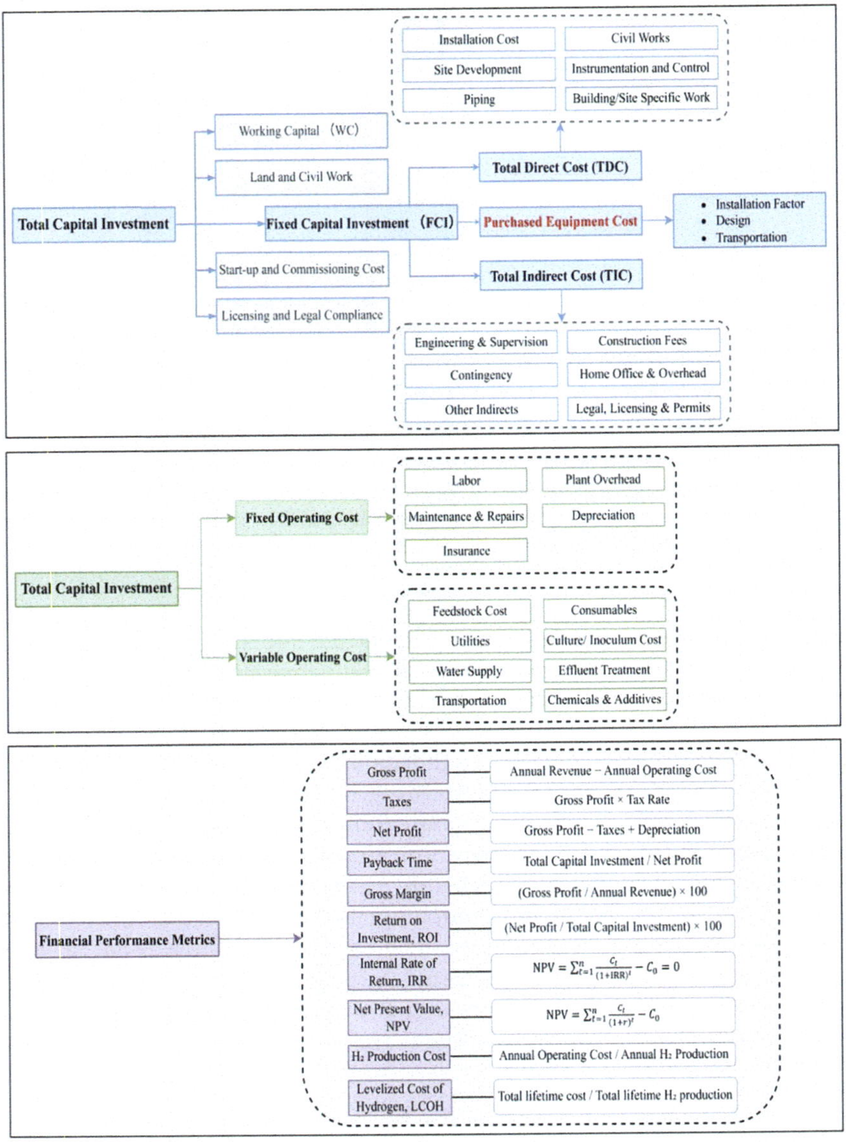

FIGURE 7.10 Framework for techno-economic analysis of hydrogen production pathways.

a comprehensive quantification of cost structures and investment risks across the life cycle of a project. The methodological foundation of TEA is generally organized into three interconnected modules: CAPEX, operating expenditure (OPEX), and financial performance indicators (Fig. 7.10).

CAPEX represents the initial investments required to establish a biohydrogen production facility. These expenditures are conventionally divided into two categories: direct costs, which include equipment procurement, installation, and site preparation; and indirect costs, which cover engineering design, licensing, permitting, and project management. The summation of these expenditures constitutes the fixed capital investment (FCI), a pivotal determinant of the long-term financial viability of biohydrogen projects. In TEA studies, equipment costs are frequently scaled using approaches such as the Lang factor method, factorial estimates, or parametric correlations. In addition, cost indices [e.g., chemical engineering plant cost index (CEPCI)] and regional adjustment factors are typically employed to ensure temporal and geographical comparability across different case studies. Conversely, OPEX encompasses the recurring annual expenses associated with plant operation. These expenditures are generally categorized into fixed costs, such as labor, insurance, and depreciation, and variable costs, which are directly dependent on production activity. In the case of biohydrogen systems, variable costs mainly include feedstock procurement, water usage, energy consumption, and waste treatment or byproduct handling. Because biological processes are inherently sensitive to feedstock composition and pretreatment requirements, feedstock procurement and processing costs often dominate the OPEX structure, alongside energy inputs for maintaining anaerobic or photobioreactor conditions. Consequently the sensitivity of OPEX to fluctuations in commodity prices, particularly biomass feedstock and electricity, emerges as a critical determinant of the economic robustness and scalability of biohydrogen pathways. The third analytical pillar comprises financial performance metrics, which translate engineering and cost inputs into investment-relevant indicators. Commonly adopted metrics include NPV, internal rate of return (IRR), and PBP, which collectively capture project profitability, capital efficiency, and investment risk. In addition, the LCOH serves as a standardized benchmark for cross-technology comparisons, expressing the average discounted cost per unit of hydrogen over the operational lifetime. Sensitivity and scenario analyses are frequently incorporated to quantify the effects of uncertainties in feedstock price, capital cost escalation, plant scale, and financing structures, thereby strengthening the robustness and policy relevance of the results. Building upon this framework, TEA has been extensively applied in the field of biohydrogen production to evaluate the feasibility of emerging technologies and to compare their performance against conventional fossil-based routes. Among the pathways currently at a higher level of technological maturity, thermochemical processes (e.g., pyrolysis and gasification), SMR, water electrolysis, and biochemical processes (e.g., DF and PF) continue to represent the dominant technological options (Lopes et al., 2019; Rawat et al., 2023; Sinnott & Towler, 2019; Zuo et al., 2025) (Fig. 7.10).

7.3.1 Pyrolysis-based hydrogen production

Pyrolysis is characterized by flexible system design, broad feedstock adaptability, and the potential for byproduct recovery, which collectively strengthen its competitiveness within biomass to hydrogen pathways (Senthil Rathi et al., 2023). Studies have shown that hybrid heating configurations combining solar energy with biochar combustion achieve an optimal balance between carbon efficiency (83%) and production cost (0.29 €/kg). In contrast, systems solely relying on solar input can raise carbon efficiency to 90%, but the limited capacity for heat storage increases costs (0.34 €/kg). Conventional pyrolysis without biochar recovery performs worst, with carbon efficiency dropping to 74% and unit costs reaching 0.27 €/kg, underscoring the dual role of biochar recovery in enhancing both economic and environmental outcomes (Amjed et al., 2024). Cross-technology comparisons further highlight the relative advantages of pyrolysis. In a 10 t/h bagasse system, liquefaction achieved the highest hydrogen yield but suffered from excessive CAPEX and OPEX, leading to an NPV loss of −113.7 million USD. Gasification showed higher feedstock conversion efficiency but incurred even larger losses (−128.3 million USD) because of high heat demand and limited Fischer–Tropsch synthesis efficiency. By contrast, pyrolysis provided moderate yields while requiring substantially lower capital investment (52 million USD) and annual operating costs (6.16 million USD), resulting in the smallest NPV loss (−65.7 million USD) and confirming its superior economic competitiveness under current conditions (Ramírez-Díaz et al., 2023). Additional case studies have validated the financial viability of pyrolysis in integrated systems. For instance a pyrolysis–gasification hybrid using empty fruit bunches with optimized heat integration achieved a 66% reduction in energy consumption and a 54% decrease in total costs, delivering an NPV of 250 million USD, an IRR of 22%, and a PBP of 5.98 years (Detchusananard et al., 2022). Similarly an Aspen Icarus-based analysis of biooil production indicated a minimum profitable scale of around 20 kt/year, with a product value (PV) of 0.47 USD/kg and an ROI of 21.9%. Sensitivity analysis revealed that PV is strongly dependent on plant scale, biooil yield, and capital costs, suggesting that decentralized pyrolysis facilities in resource-abundant regions could achieve PV levels as low as 0.27 USD/kg (Do et al., 2014). Overall, pyrolysis demonstrates considerable economic potential, particularly for high-moisture feedstocks and decentralized applications where system complexity needs to be minimized. Its economic advantages mainly derive from multiproduct valorization, modular design, and opportunities for heat integration. However, its competitiveness is still constrained by conversion efficiency and downstream upgrading requirements, indicating that further advances in process efficiency and system integration are essential for commercial deployment.

7.3.2 Gasification-based hydrogen production

Gasification, driven by high-temperature partial oxidation, achieves higher hydrogen yields and conversion efficiencies than pyrolysis, with **fluidized-bed (FB)** and **entrained-flow (EF)** reactors as the dominant configurations. **EF** systems deliver higher thermal efficiency [56% vs. 45%**lower heating value (LHV)** for **FB**] but require nearly double the CAPEX (USD 527 vs. 236 million) and result in higher **minimum selling price (MSP)** (3.40 vs. 3.10 USD/kg). When carbon capture and liquefaction are incorporated, efficiencies decline to 50% and 41% for **EF** and **FB**, respectively; nonetheless, **EF** systems can approach near zero CO_2 emissions, whereas **FB** remains constrained by elevated nitrogen content in the flue gas. Economic sensitivity analyses underscore the pivotal influence of feedstock costs, financing structures, and carbon pricing on project feasibility. At the current biomass price of 100 USD/t, **FB** systems achieve economic viability only if feedstock costs are reduced below 60 USD/t. By contrast, **EF** systems integrated with **carbon capture and storage (CCS)** become competitive at carbon prices of approximately 120 USD/t CO_2 (Salkuyeh et al., 2018). From a process design perspective, steam gasification achieves a favorable balance between syngas quality (15–20 MJ/Nm^3), moderate operating temperatures (700°C–1200°C), and compatibility with CO_2 capture, making it the most commercially promising subpathway (Parthasarathy & Narayanan, 2014). Regional case studies further support its feasibility; for instance, in Australia, coal gasification coupled with **CCS** yielded **LCOH** of 4.12–4.90 AUD/kg while reducing emissions from 16.3–20.9 to as low as 0.1–1.1 kg CO_2/kg H_2 (Hosseini et al., 2024). Comparative analyses also reveal trade-offs across reactor designs: although dual BD systems required the highest CAPEX, they achieved the lowest posttax hydrogen price, with the **NPV** comparable to other advanced pathways (Yao et al., 2017). Long-term projections further suggest that gasification costs could decline to 1.34–2.27 USD/kg, reinforcing its potential to reach commercial competitiveness (Nikolaidis & Poullikkas, 2017). Overall, gasification is widely recognized as one of the most commercially attractive thermochemical routes for hydrogen production. Nonetheless, its near-term competitiveness remains contingent on addressing persistent barriers, particularly tar removal, cost-effective CO_2 capture integration, and the substantial burden of CAPEXs.

7.3.3 Steam methane reforming (SMR)

SMR is widely recognized as the most industrially mature hydrogen production technology, underpinned by well-established infrastructure and proven operational reliability. It has consistently demonstrated strong cost competitiveness, particularly in distributed and medium-scale applications utilizing natural gas or biogas as feedstock. For biogas-based systems the hydrogen production cost is typically 2.3–2.5 USD/kg H_2 over an 8-year operating cycle, with extended annual operating hours markedly improving overall economic performance. Sensitivity analyses further indicate that SMR maintains stable performance even under ± 30% cost fluctuations, underscoring its resilience to market variability (Braga et al., 2013). With regard to sophisticated configurations the BioRobur system, which integrates SMR with autothermal reforming and biogas derived from municipal solid waste, achieves further cost reductions. At a capacity of 100 Nm^3/h, with 8030 h/year of operation and a 10-year amortization period, hydrogen production costs are reduced to 2.5 €/kg, well below the EU green hydrogen target of 5 €/kg. Cost structure analysis reveals that reactors account for approximately 38% of CAPEX, with the remainder largely attributable to heat exchangers, pressure swing adsorption units, and compressors. OPEX is dominated by feedstock and utilities, amounting to around 3% of CAPEX, highlighting the decisive role of feedstock costs and utilization rates in determining overall feasibility. Regional case studies further validate its economic viability (Montenegro Camacho et al., 2017). For instance an SMR system based on palm oil mill effluent processing 20,000 t/h of feedstock achieved an annual hydrogen output of 963 t with IRR ranging from 18.5% to 33% and PBP of 5–8.5 years (Wee et al., 2022). Overall, SMR remains the most cost-effective pathway in natural gas-rich regions, combining technical maturity with economic advantages, whereas its integration with waste valorization strategies offers additional sustainability benefits.

7.3.4 Water electrolysis

Water electrolysis, driven by renewable electricity, enables the production of hydrogen with a zero-carbon footprint and demonstrates strong compatibility with intermittent RE systems. Despite these advantages, its economic feasibility is currently constrained by a cost structure dominated by electricity, which typically accounts for nearly two-thirds of the total production cost. Small-scale plants (30 Nm^3/h) report production costs of 16.54–17.99 USD/kg for proton exchange membrane and alkaline water electrolysis, underscoring the substantial economic challenges that limit wider deployment (Lee et al., 2017). Although scale-up can partially mitigate costs, electrolysis continues to face significant disadvantages compared with more established hydrogen production pathways. Policy support is, therefore, regarded as indispensable for improving competitiveness. Capital subsidies alone are insufficient to close the cost gap between green hydrogen and fossil-based alternatives, and recent studies indicate that even under favorable policy frameworks, the LCOH for electrolysis remains substantially higher (Crespi et al., 2024). Operational subsidies, capacity-based incentives, and market pull mechanisms are thus critical to accelerate industrial-scale adoption. Techno-economic assessments further identify electricity price, stack durability, and capital intensity as the most decisive factors shaping the viability of electrolysis. Long-term improvements in stack lifetimes, advances in material innovation, and reductions in renewable power costs are expected to drive cost declines. Moreover, integrated system designs, such as biomass–electrolysis hybrid configurations producing hydrogen, electricity, and dimethyl ether, have demonstrated enhanced feasibility, delivering IRR exceeding 20% and PBP of less than 8 years in pilot studies (Xu et al., 2025). Overall, water electrolysis remains one of the most promising pathways for zero-carbon hydrogen production, but large-scale deployment will depend on technological advancements, declining renewable electricity costs, and well-designed policy frameworks that directly address its high sensitivity to energy prices and capital requirements.

7.3.5 Biochemical pathways

Biochemical pathways for hydrogen production mainly include DF and PF, each exhibiting distinct advantages and limitations. DF operates under mild conditions and accommodates a wide range of organic feedstocks such as agricultural residues, lignocellulosic biomass, wastewater, algae, and food waste with relatively low capital intensity, making it suitable for decentralized small- to medium-scale systems. Case studies highlight promising financial outcomes: A bread waste-fed continuous stirred-tank reactor (2 t/day) achieved the PBP in only 4.8 years; solid-state DF at 3 t/day reached an IRR of 24.1%; and hybrid DF systems at 10 t/day delivered an IRR of 20.2% (Han et al., 2016a, 2016b). Despite such potential the broader deployment of DF remains constrained by microbial instability and the absence of mature scale-up strategies, necessitating advances in process integration and microbial community control. PF, by contrast, offers theoretical carbon-neutral potential through light-driven conversion of volatile fatty acids but suffers from poor light conversion efficiency and prohibitively high photobioreactor costs. Even under optimistic assumptions, hydrogen production costs exceed 2.13 USD/kg, with >33% increases triggered by a 20% rise in CAPEX, whereas large-scale facilities producing over 1 M GJ/year demand initial investments above 40 million USD (Ahmed et al., 2021; Akkerman et al., 2002; Sathyaprakasan & Kannan, 2015). Consequently the commercial viability of PF will require substantial breakthroughs in photoreactor design and metabolic engineering (Touloupakis & Torzillo, 2019). Overall, DF currently demonstrates stronger economic potential in decentralized applications, whereas PF remains largely confined to the laboratory. Scalability, microbial community stability, and system-level integration are common bottlenecks across both pathways, and overcoming these barriers will be essential for advancing the commercial viability of biochemical hydrogen production.

7.3.6 Economic sensitivity and commercialization barriers of hydrogen production pathways

From an economic sensitivity perspective, CAPEX, feedstock price, and scale effects are the primary determinants influencing hydrogen production costs and investment returns. Thermochemical pathways exhibit strong dependency on capital optimization and economies of scale, with economic performance improving substantially through equipment cost amortization and energy efficiency gains. Electrochemical routes, by contrast, are highly sensitive to electricity price fluctuations, as power costs dominate their overall cost structure, making their economic viability heavily contingent on renewable electricity availability and pricing dynamics. Biochemical pathways show cost structures that are more constrained by substrate price, process stability, and scalability, conferring competitive advantages in small-scale, resource-proximate applications, but with significant limitations in large-scale deployment.

In terms of financial performance, DF demonstrates relatively high IRR and short PBP under small- to medium-scale configurations because of its low capital intensity and moderate operating costs, suggesting strong potential for decentralized energy solutions. However, its commercialization remains challenged by process instability, microbial community management complexity, and lack of scale-up strategies. PF, despite its theoretical advantages of carbon neutrality and zero-emission hydrogen production, suffers from low solar-to-hydrogen conversion efficiency and high reactor costs, confining its development to the laboratory stage in the near term. Water electrolysis, driven by policy incentives and renewable integration, is advancing rapidly; yet its commercialization remains hindered by high CAPEX, substantial operating costs, and acute sensitivity to electricity pricing. In comparison, SMR and biomass gasification maintain a competitive edge in terms of technological maturity and scale-driven cost reductions, although both require improvements in CO_2 capture integration and operational efficiency at high temperatures.

Across all pathways, commercialization faces dual challenges of cost reduction and system integration. For thermochemical technologies, breakthroughs in tar mitigation, CO_2 reduction integration, and catalyst lifetime management are essential to lower life cycle costs. Electrochemical pathways require substantial progress in reducing electrolyzer manufacturing

TABLE 7.1 Comparative analysis of hydrogen production pathways.

Technology pathway	Core advantages	Key bottlenecks	Critical sensitivity factors
Pyrolysis	Broad feedstock adaptability; moderate investment	Tar removal; energy efficiency improvement	Equipment investment; biochar market price
Gasification	High hydrogen yield potential; scalability	High energy consumption; costly gas purification	Capital cost; carbon capture integration
Steam reforming	High maturity; low production cost	CO_2 emissions; reliance on fossil fuels	Natural gas price; plant load factor
Water electrolysis	Zero-carbon process; integration with renewables	High electricity cost; large CAPEX	Electricity price; stack lifetime
Dark fermentation	Economic viability at small-to-medium scale; waste valorization	Scale-up limitations; process stability issues	Feedstock cost; hydrogen selling price

CAPEX, *capital expenditure*.

costs, extending stack lifetimes, and improving dynamic compatibility with variable RE inputs. Biochemical routes demand enhanced process stability, advanced microbial consortia, and cost-effective reactor scale-up solutions, supported by system-level integration to ensure operational robustness and economic feasibility (Table 7.1).

7.4 Life cycle analysis of biohydrogen production technologies

Biohydrogen production technology has now reached a relatively mature stage of development, with numerous studies conducting LCAs on biohydrogen production systems (Do et al., 2018). Currently, in 81.1% of published papers, the functional unit is defined as the production of 1 kg of biohydrogen. The most commonly used LCA software tools are openLCA and SimaPro, accounting for 11.5% and 19.4% of usage, respectively. The assessment methodologies employed include RECIPE and CML. The frequently used life cycle impact assessment indicators comprise global warming potential (GWP), acidification potential (AP), eutrophication potential (EP), human toxicity potential (HTP), and resource depletion (RD) (Nakamura et al., 2012).

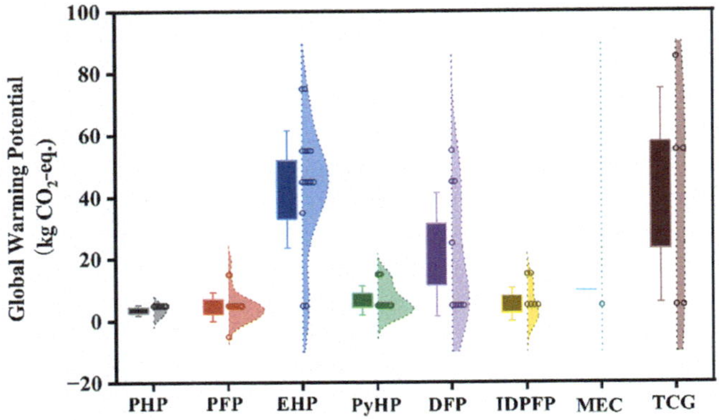

FIGURE 7.11 Net GWP values per functional unit of nine biological hydrogen production technologies. *GWP*, Global warming potential; *MEC*, microbial electrolysis cells.

7.4.1 Global warming potential

GWP serves as a standardized metric for assessing the climate impact of GHGs (Krinitzki et al., 2024), with carbon dioxide (CO_2) as the reference gas (GWP = 1). Key GHGs include CO_2, methane (CH_4), nitrous oxide (N_2O), and hydrofluorocarbons (HFCs) (Wu et al., 2021). These gases exhibit strong infrared radiation absorption, driving irreversible climatic changes such as glacial retreat and permafrost thaw. The high GWP of CO_2 is further exacerbated by its dissolution in seawater, forming carbonic acid that contributes to ocean acidification, ecosystem disruption, and biodiversity loss. Among nine biohydrogen production technologies the net GWP per functional unit ranges from −2.1 to 84.3 kg CO_2 eq (Fedele et al., 2023). As shown in Fig. 7.11, thermochemical gasification exhibits the highest net GWP (7.33–84.3 kg CO_2 eq), with a median of 55.36 kg CO_2 eq, surpassing other technologies by 34%–98%. Conversely, PF demonstrates the lowest GWP range (−1.6–12.1 kg CO_2 eq), with a median of 4.05 kg CO_2 eq, 13.8%–98% lower than alternatives. The main emission link of thermochemical biological hydrogen production technology is the fermentation reaction, which is the primary contributor to GWP emissions. The adoption of substrate type modification technology for emission reduction can achieve a 50% reduction in GWP. The gas separation and purification link represents a negative emission, primarily attributed to the production of hydrogen energy. For PF the fermentation reaction is the primary GWP contributor, though substrate modification can reduce emissions by 64.2%, whereas the gas separation/purification phase yields negative emissions because of hydrogen energy coproduction. In photolytic biohydrogen production the photocatalytic reaction dominates emissions, but medium recycling, nutrient recovery, and biophotocatalytic nitrogen/phosphorus removal can achieve a 10% GWP reduction. The remaining six technologies, ranked by descending GWP, are thermochemical gasification > electrolysis > DF > combined DF–PF > MECs. Major emission sources include the electrochemical reaction, pyrolysis reaction, fermentation reaction (DF), and PF phase (combined fermentation). Mitigation strategies such as RE integration [e.g., RE-powered electrolysis/MECs, solar/geothermal heated pyrolysis , and full RE adoption in fermentation)] coupled with process optimization (e.g., microbial consortia adjustment and MEC performance enhancement) can reduce GWP by 90%, 80%, 64.7%, 60.3%, and 86.2%, respectively. Meanwhile, specific links within certain hydrogen production technologies are associated with negative emissions. To elaborate, in the case of pyrolysis-based hydrogen production, the hydrogen purification link contributes to negative emissions primarily because of its role in generating a range of value-added products, including carbon-containing liquid fuels, various types of chemicals, fuel energy sources, biomass energy, and biofuels. When it comes to DF hydrogen production technology, the link that achieves negative emissions is identified as the gas separation and purification link; this negative emission effect is mainly driven by the production of multiple substances during the process, such as carbon-containing liquid fuels, different categories of chemicals, fuel energy, biomass energy, and biofuels. For DF–PF hydrogen production the specific link responsible for negative emissions is also the gas separation and purification link. Similar to the aforementioned DF technology, this link realizes negative emissions largely because it can produce a series of products, including carbon-containing liquid fuels, various chemicals, fuel energy sources, biomass energy, and biofuels. As for MEC-based hydrogen production the link that leads to negative emissions is the gas purification link. The fundamental reason for this negative emission phenomenon lies in the fact that this purification link is capable of generating multiple energy and chemical products, such as carbon-containing liquid fuels, different kinds of chemicals, fuel energy, biomass energy, and biofuels (Fig. 7.11).

7.4.2 Acidification potential and eutrophication potential

AP is a critical indicator in environmental science and LCA (Ge et al., 2017), quantifying the potential contribution of pollutants emitted by substances or processes to the acidification of the atmosphere, water bodies, or soil. It primarily evaluates chemicals such as CO_2 and N_2O that lower environmental pH, triggering acidification effects. The dissolution of CO_2 increases hydrogen ion (H^+) concentration in seawater (reducing pH), inhibiting shell and skeletal formation in calcifying organisms such as mollusks and corals, while also disrupting physiological functions such

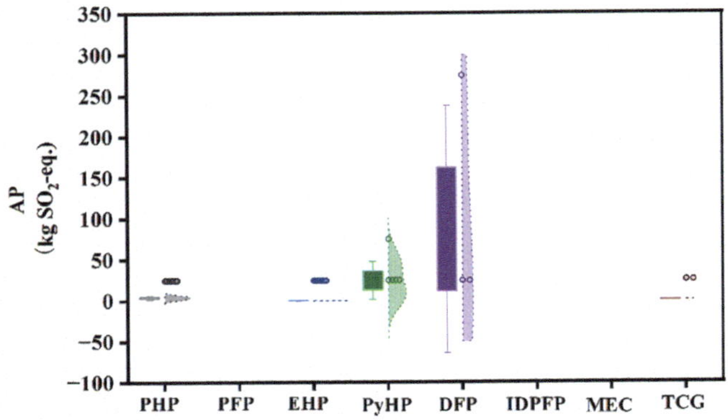

FIGURE 7.12 Net AP values per functional unit of nine biological hydrogen production technologies. *AP*, Acidification potential; *MEC*, microbial electrolysis cells.

as spawning and larval survival. Acidification stimulates denitrifying bacteria in estuarine sediments to release N_2O, a potent GHG, and accelerates N_2O emissions from acidified soils. Moreover, ocean acidification reduces CO_2 absorption capacity, creating a positive feedback loop that exacerbates climate change (Hall et al., 2020). This consequently causes climate zones such as temperate and subtropical zones to shift toward higher latitudes, which in turn alters crop cultivation zones (e.g., the latitudes suitable for crop cultivation in Northeast China have moved northward) and simultaneously affects the distribution range of natural vegetation. Among nine biohydrogen production technologies the net AP per functional unit ranges from 0.001 to 263.4 kg SO_2 eq. As depicted in Fig. 7.12, DF exhibits the highest AP (0.008–263.4 kg SO_2 eq), with a median of 0.216 kg SO_2 eq, surpassing other technologies by 78%–97%. In contrast, pyrolysis-based hydrogen production shows the lowest AP range (0.00038–0.005 kg SO_2 eq), with a median of 0.001 kg SO_2 eq, 6%–97% lower than alternatives. The remaining three technologies, ranked by descending AP, are photolytic biohydrogen production > electrolysis > thermochemical gasification. These technologies differ in their primary AP emission sources: electrolysis and photolysis are dominated by the electrochemical reaction and photocatalytic reaction, respectively, whereas pyrolysis and thermochemical gasification both derive emissions mainly from the pyrolysis reaction. DF primarily emits during the fermentation reaction. Targeted mitigation strategies can achieve significant AP reduction. Key approaches include RE integration, such as wind/solar-powered electrolysis; solar thermal/geothermal-assisted pyrolysis; solar-dried or geothermal-heated DF; RE-supported microbial cultivation in photolysis; and concentrated solar/nuclear-powered thermochemical gasification supplemented by equipment retrofits (e.g., fossil fuel combustion systems with desulfurization/denitrification) and process optimization (e.g., microbial strain selection and culture condition tuning). These measures yield AP reductions of 70.1%, 89%, 95.8%, 69.5%, and 80%, respectively. Meanwhile, specific links within certain hydrogen production technologies exhibit the characteristic of negative emissions. To provide concrete examples, in the context of pyrolysis-based hydrogen production technology, the hydrogen purification link, one of the key process links in this technology, is associated with negative emissions, and this link primarily generates two types of energy products that are environmentally friendly and renewable, namely, biomass-based energy and clean fuel energy. When it comes to DF hydrogen production technology, the specific link that achieves the effect of negative emissions is identified as the gas separation and purification link; this link plays a crucial role in the overall DF process, and its main output consists of two important energy sources, which are biomethane (a type of clean gaseous fuel with high energy efficiency) and biomass energy (a sustainable energy derived from organic materials). As for photolytic biological hydrogen production technology the link responsible for realizing negative emissions is the gas emission link. Unlike the negative emission links of the aforementioned two hydrogen production technologies, this gas emission link in photolytic biological hydrogen production mainly produces biomass energy, a kind of energy that can effectively reduce reliance on fossil fuels and mitigate environmental pollution (Fig. 7.12).

EP serves as a pivotal metric in environmental science and LCA (Sanderson & Coade, 2010), quantifying the capacity of pollutants released by substances or processes to induce aquatic eutrophication (Ge et al., 2017). It specifically evaluates nutrients such as nitrogen and phosphorus that stimulate excessive proliferation of algae and plankton, disrupting aquatic ecosystems. Nutrient enrichment triggers algal blooms (e.g., cyanobacteria) and red tides, which block sunlight and deplete dissolved oxygen, creating anoxic conditions that suffocate aquatic organisms like fish (Bernasová et al., 2024). This process shifts dominant species (e.g., diatoms to cyanobacteria), simplifying food webs as hypoxia-tolerant species (e.g., chironomids) replace sensitive ones (e.g., mayflies). Severe eutrophication further reduces species diversity, whereas anaerobic sediment decomposition releases N_2O, a potent GHG, and marine eutrophication weakens the carbon sequestration capacity of seagrass beds, exacerbating climate change. Among nine biohydrogen production technologies the net EP per functional unit ranges from 0.0004 to 0.4 kg phosphate equivalent (kg PO_4^{3-} eq) (Preisner et al., 2020). As illustrated in Fig. 7.13, photolytic biohydrogen production exhibits the highest EP range (0.03–0.93 kg PO_4^{3-} eq), peaking at 0.4 kg PO_4^{3-} eq and surpassing other technologies by 46.6%–98.0%. Conversely, pyrolysis-based hydrogen production demonstrates the lowest EP range (0.0004–0.004 kg PO_4^{3-} eq), with a minimum

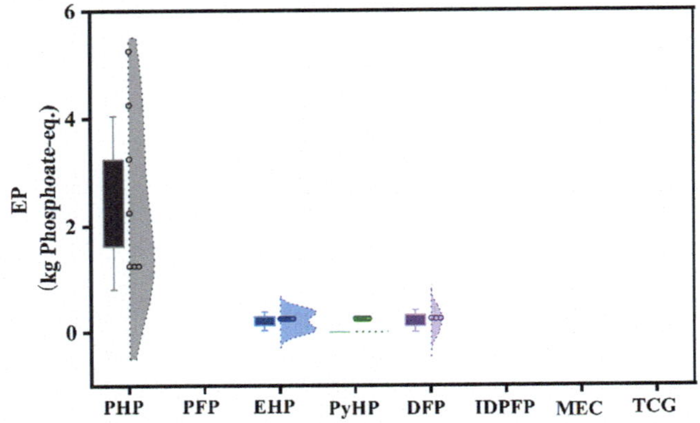

FIGURE 7.13 Net EP values per functional unit of nine biological hydrogen production technologies. *EP*, Eutrophication potential; *MEC*, microbial electrolysis cells.

of 0.002 kg PO_4^{3-}eq, 6%–99%. For the other two biological hydrogen production technologies, their EP (environmental pollution) values, when ranked in descending order of magnitude, are as follows: electrolytic biological hydrogen production technology > DF hydrogen production technology. These two distinct hydrogen production technologies differ significantly in the focus of their main EP emission-contributing links. Specifically, for electrolytic hydrogen production technology, its primary emission-generating link is the reaction link, which serves as the dominant contributor to the overall EP emissions of this technology. To mitigate such emissions effectively a series of targeted emission reduction technologies can be adopted, including replacing phosphorus-containing additives with phosphorus-free corrosion inhibitors (which helps reduce phosphorus-related pollution in wastewater), establishing a comprehensive wastewater recycling and reuse system (to minimize the volume of wastewater discharged into the environment), and implementing advanced nitrogen and phosphorus removal treatment on the finally discharged wastewater (to further lower the content of harmful nutrients in the effluent). Through the combined application of these technologies a remarkable 97.5% reduction in EP emissions can be achieved for electrolytic hydrogen production. In contrast, for DF hydrogen production technology, its main emission link is also the reaction link, and this link similarly acts as the key contributor to its EP emissions. For this technology the emission reduction is accomplished through the adoption of specific technical measures such as DF–PF (which optimizes the fermentation process to reduce pollutant generation), biological nitrogen and phosphorus removal (utilizing microbial activities to naturally decompose and remove nitrogen and phosphorus pollutants), and wastewater recycling and reuse (consistent with the water-saving and emission-reducing concept applied in electrolytic hydrogen production). The implementation of these measures enables an 80% reduction in EP emissions for DF hydrogen production. Meanwhile, there are differences in the negative emission links and their corresponding products between the two technologies. The negative emission link of electrolytic hydrogen production technology is the purification link, and this link primarily generates biomass-based energy, a type of RE derived from organic materials that is environmentally friendly. On the other hand the negative emission link of DF hydrogen production technology is the separation and purification link, which mainly produces microalgal biodiesel (a clean liquid fuel with good combustion performance) and biochar energy (a solid energy source with high carbon content and low pollutant emissions) (Fig. 7.13).

7.4.3 Human toxicity potential

HTP is a vital indicator in the fields of environmental science, risk assessment, and LCA (Chen et al., 2022). It is specifically designed to measure the potential capacity of pollutants emitted from a substance or a process to exert toxic impacts on human health. HTP focuses on various types of toxic substances, including heavy metals (such as lead, mercury, cadmium, and arsenic, which are highly stable and can accumulate in the human body over time); persistent organic pollutants (Lu et al., 2023) [POPs, with typical examples being dioxins, polychlorinated biphenyls (PCBs), organochlorine pesticides, etc. These substances are characterized by strong persistence, long-distance migration, and high bioaccumulation, making them pose long-term threats to human health]; and pesticides (including organophosphorus pesticides, pyrethroid pesticides, carbamate pesticides, and other categories). After these toxic substances are released into the environment, they can enter the human body through multiple exposure pathways. These pathways mainly include the air (for instance, toxic volatile organic compounds or particulate matter containing toxic substances are inhaled into the human respiratory system); water (such as toxic substances dissolving in drinking water sources or surface water, which are then ingested by humans through drinking or contact); soil (e.g., toxic substances in the soil are absorbed by crops and then enter the human body through the food chain, or humans come into direct contact with contaminated soil), or the food chain (toxic substances accumulate step by step from lower trophic levels to higher trophic levels in the ecosystem and finally reach the human body through the consumption of contaminated food) (Sarangi & Nanda, 2020). By evaluating the entire process from the release of these toxic substances to their exposure to humans, HTP further assesses the possibility and intensity of acute or chronic health hazards that may be caused to the human body. Acute health

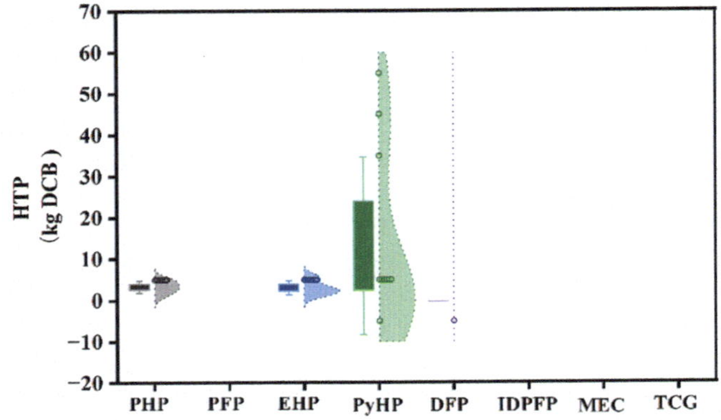

FIGURE 7.14 Net HTP values per functional unit of nine biological hydrogen production technologies. *HTP*, Human toxicity potential; *MEC*, microbial electrolysis cells.

hazards usually refer to severe adverse reactions that occur shortly after short-term exposure to high concentrations of toxic substances, such as acute poisoning symptoms like nausea, vomiting, dizziness, or even life-threatening conditions in severe cases, whereas chronic health hazards mainly involve long-term, subtle damage to the human body because of long-term exposure to low concentrations of toxic substances, which may manifest as chronic diseases such as cancer; damage to the nervous system, endocrine system, reproductive system, or immune system; and may even have intergenerational impacts in some cases. Among nine biohydrogen production technologies the net HTP per functional unit ranges from −2.1 to 84.3 kg 1,4-dichlorobenzene equivalent (kg DCB). As shown in Fig. 7.14, pyrolysis-based hydrogen production exhibits the highest HTP range (−0.00027–52.57 kg DCB), with a median of 0.26 kg DCB, surpassing the other eight technologies by 34%–98%. The main emission link of photocatalytic biological hydrogen production technology is the photocatalysis link, and this specific link is recognized as the primary contributor to HTP (a key indicator related to environmental or energy-related emissions) emissions of the entire technology. To reduce such HTP emissions effectively, photocatalytic biological hydrogen production technology can adopt targeted emission reduction technologies, among which the core ones include RE replacement (specifically replacing traditional fossil energy with clean, sustainable energy sources such as solar energy, wind energy, or hydropower to cut down on emissions generated from energy consumption) and integrated optimization of system energy efficiency (optimizing the coordination and operation efficiency of various components within the hydrogen production system, such as the photocatalytic reactor, energy conversion devices, and auxiliary equipment, to minimize energy waste and the corresponding HTP emissions). Through the scientific application of these emission reduction technologies, photocatalytic biological hydrogen production technology is capable of achieving a 60% reduction in HTP emissions. For electrolytic biological hydrogen production technology, its main emission link is the electrolysis reaction link, and this link also plays a dominant role in contributing to the technology's HTP emissions (Zhao et al., 2015). In contrast to photocatalytic biological hydrogen production, electrolytic biological hydrogen production realizes emission reduction through a set of more comprehensive emission reduction technologies. These technologies mainly cover RE power supply replacement (using electricity generated from RE sources to power the electrolysis reaction, thereby avoiding emissions from electricity produced by fossil fuels), system energy efficiency optimization (improving the energy utilization rate of the electrolysis cell, power supply system, and other key equipment to reduce unnecessary energy consumption and associated emissions), and nutrient recovery from biological residual liquids (recovering valuable nutrients such as nitrogen, phosphorus, and organic matter from the residual liquids generated during the biological reaction process, not only reducing the pollutant content in the discharged waste but also realizing the recycling of resources, which indirectly contributes to HTP emission reduction). With the combined implementation of these technologies, electrolytic biological hydrogen production technology can achieve a significant 87.1% reduction in HTP emissions. Meanwhile, there is a certain degree of similarity between the two hydrogen production technologies in terms of their negative emission links and the corresponding products generated. Specifically the negative emission link of photocatalytic biological hydrogen production technology is the gas purification link, and the main outcome of this link is the production of hydrogen energy, a clean, high-efficiency energy source that does not produce harmful emissions during its utilization process. In the same vein the negative emission link of electrolytic biological hydrogen production technology is also the gas purification link, and this link similarly focuses on the production of hydrogen energy, further highlighting the environmental advantages of both technologies in terms of generating clean energy while achieving negative emissions (Kumar et al., 2019) (Fig. 7.14).

7.4.4 Abundance depletion

AD (abundance depletion), which stands for resource depletion, is a crucial indicator in the fields of environmental science, risk assessment, and LCA (Islam et al., 2019). At present the range of net AD values per functional unit for the nine biological hydrogen production technologies is −0.0000182–3.5 kg SB, as illustrated in Fig. 7.15. Among these nine technologies, pyrolysis-based hydrogen production technology has the highest

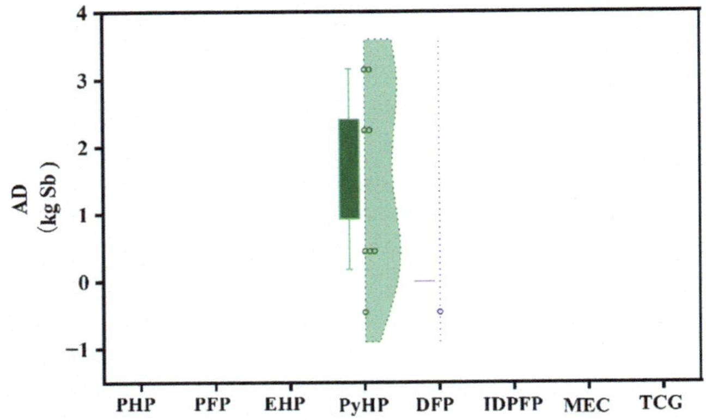

FIGURE 7.15 Net AD values per functional unit of nine biological hydrogen production technologies. *MEC,* Microbial electrolysis cells.

net AD value, with a median of 0.7 kg SB eq, this median value is 60%–98% higher than that of the other eight biological hydrogen production technologies. When it comes to DF hydrogen production technology, its main emission link at present is the fermentation reaction link (Karim et al., 2025), and this specific fermentation reaction link is recognized as the primary contributor to AD emissions of the entire DF hydrogen production process. In contrast the gas treatment link of DF hydrogen production technology exhibits the characteristic of negative emissions (Lourinho & Brito, 2021). The fundamental reason behind this negative emission phenomenon lies in the fact that the gas treatment link is mainly responsible for the production of hydrogen energy, a clean and RE source. The generation of hydrogen energy not only provides a valuable energy output but also helps offset part of the resource depletion associated with the technology, thereby resulting in the negative AD emission effect of the gas treatment link (Mahidhara et al., 2019) (Fig. 7.15).

7.5 Major challenges and future prospects

Current research on the economic and environmental analysis of biohydrogen production confronts significant methodological constraints that impede cross-technology comparability. Economically the absence of standardized assessment frameworks perpetuates analytical fragmentation, as studies examining isolated pathways, such as DF or gasification, fail to harmonize critical parameters. This is exemplified by inconsistent LCOH calculations, where the inclusion of pyrolysis byproduct credits contrasts with the exclusion of water electrolysis' grid synergies, thereby distorting comparative validity. Furthermore, insufficient integration of technological maturity, scale effects, and regional resource endowments obstructs context optimized technology selection.

These methodological limitations extend to environmental evaluations, where compartmentalized metrics undermine holistic analysis. Although LCAs quantify indicators including GWP and AP, emission mitigation strategies remain siloed within technological domains. PF relies on photon-conversion efficiency for GWP reduction, whereas pyrolysis necessitates RE substitution, yet their respective cost-benefit tradeoffs lack systematic comparative examination. Crucially, integrated economic–environmental assessments remain underdeveloped, particularly regarding carbon capture's consequential tradeoff of 20% operational cost escalation against >80% emission reduction in gasification systems.

Compounding these issues are persistent data robustness concerns. Economic analyses exhibit divergent assumptions, notably in discount rates and feedstock price volatility ranges, that undermine cross-study comparability, evidenced by > 30% cost variations for natural gas steam reforming across literature. Concurrently, environmental datasets derived predominantly from laboratory-scale studies demonstrate limited industrial relevance, with scaled emission factors such as DF's N_2O exhibiting 15%–20% deviations from pilot-scale measurements.

Transitioning to industrial deployment reveals three interconnected challenges. Economic viability remains constrained across technological pathways: thermochemical systems suffer acute capital intensity (> 60% for gasification reactors) and 15%–25% operational cost inflation from tar management, whereas electrochemical alternatives exhibit critical electricity price sensitivity (> \$0.05/kWh erodes competitiveness against conventional steam reforming) exacerbated by inadequate electrolyzer longevity (5–10 years) and prohibitive capital expenditures (40%–50% of CAPEX). Environmental performance tradeoffs further complicate adoption, as median gasification emissions (55.36 kg CO_2 eq/kg H_2) necessitate carbon capture implementation that reduces system efficiency by ~10%, whereas DF's elevated eutrophication potential increases wastewater remediation costs. Systemic deficiencies in cross-technology integration, manifested through unquantified synergies like DF–PF's 20% substrate utilization improvement and absent dynamic models for feedstock price volatility or carbon policy shifts, complete this triad of industrial barriers.

Addressing these multidimensional challenges necessitates coordinated research advancements. Economic methodology requires unified frameworks integrating CAPEX/operational expenditures, byproduct revenues, and carbon valuation with standardized boundaries (8%–10% discount rates and regionally calibrated feedstock pricing) to enable cross-pathway comparability, supplemented by dynamic regionalized assessment tools quantifying context-specific opportunities like 15%–20% cost reduction in Southeast Asian palm-waste gasification. Environmental evaluation must prioritize life cycle-integrated mitigation strategies—advancing catalytic tar conversion (30% cost reduction target), microbial consortium optimization (40% EP reduction via coupled fermentation), and renewable integration (sub-5 kg CO_2 eq/kg H_2 GWP target)—whereas developing carbon cost-abatement equilibrium models to evaluate critical tradeoffs such as carbon capture breakeven thresholds at \$120/t CO_2. Concurrent technological innovation should focus on modular bioreactor development (25% CAPEX reduction), hybrid process intensification (10%–15% energy savings), and electrolyzer durability enhancement (20-year operational target), supported by industrial databases documenting operational parameters across scales and open-source platforms for dynamic techno-economic environmental modeling.

Collectively, transcending fragmented analytical paradigms through methodological harmonization, cross pathway integration, and empirical data synthesis will resolve current limitations. Future endeavors must converge on scalable cost environment synergies, leveraging region-specific resource configurations and policy mechanisms to transition biohydrogen from fundamental research to industrial feasibility, thereby establishing the quantitative foundations required for a sustainable hydrogen economy.

7.6 Conclusion

This chapter systematically analyzes nine biohydrogen production pathways, including DF/PF, pyrolysis, gasification, microbial electrolysis, photolysis, electrolysis, and hybrid systems, detailing their mechanisms, process flows, and energy/material exchanges. TEA reveals the cost structures and bottlenecks of pyrolysis, gasification, and steam reforming, whereas LCA quantifies environmental impacts, highlighting the decarbonization potential of PF/DF. Current limitations include fragmented economic/environmental frameworks and data gaps. Industrialization is challenged by economic viability thresholds, environmental trade-offs, and insufficient system integration. Future progress hinges on standardizing assessments, advancing cross-pathway synergies, and establishing scalable industrial data infrastructure to enable biohydrogen's transition from lab-scale innovation to a sustainable hydrogen economy pillar.

References

Ahmad, A. A., Zawawi, N. A., Kasim, F. H., Inayat, A., & Khasri, A. (2016). Assessing the gasification performance of biomass: A review on biomass gasification process conditions, optimization and economic evaluation. *Renewable and Sustainable Energy Reviews, 53*, 1333–1347. https://www.journals.elsevier.com/renewable-and-sustainable-energy-reviews. doi:10.1016/j.rser.2015.09.030.

Ahmed, S. F., Rafa, N., Mofijur, M., Badruddin, I. A., Inayat, A., Ali, M. S., Farrok, O., & Yunus Khan, T. M. (2021). Biohydrogen production from biomass sources: Metabolic pathways and economic analysis. *Frontiers in Energy Research, 9*. http://www.frontiersin.org/Energy_Research/. doi:10.3389/fenrg.2021.753878.

Akkerman, I., Janssen, M., Rocha, J., & Wijffels, R. H. (2002). Photobiological hydrogen production: Photochemical efficiency and bioreactor design. *International Journal of Hydrogen Energy, 27*. doi:10.1016/S0360-3199(02)00071-X.

Amjed, M. A., Sobic, F., Romano, M. C., Faravelli, T., & Binotti, M. (2024). Techno-economic analysis of a solar-driven biomass pyrolysis plant for bio-oil and biochar production. *Sustainable Energy & Fuels, 8*(18), 4243–4262. doi:10.1039/d4se00450g.

Arfan, M., Eriksson, O., Wang, Z., & Soam, S. (2023). Life cycle assessment and life cycle costing of hydrogen production from biowaste and biomass in Sweden. *Energy Conversion and Management, 291*, 117262. doi:10.1016/j.enconman.2023.117262.

Bernasová, T., Nedbal, V., Ghorbani, M., Brom, J., Amirahmadi, E., & Bernas, J. (2024). Eutrophication risk potential assessment between forest and agricultural sub-catchments using LCIA principles. *Land, 13*(8). Www.mdpi.com/journal/land/. doi:10.3390/land13081150.

Bhatia, S. K., Jagtap, S. S., Bedekar, A. A., Bhatia, R. K., Rajendran, K., Pugazhendhi, A., Rao, C. V., Atabani, A. E., Kumar, G., & Yang, Y. H. (2021). Renewable biohydrogen production from lignocellulosic biomass using fermentation and integration of systems with other energy generation technologies. *Science of the Total Environment, 765*. Www.elsevier.com/locate/scitotenv. doi:10.1016/j.scitotenv.2020.144429.

Braga, L. B., Silveira, J. L., Da Silva, M. E., Tuna, C. E., Machin, E. B., & Pedroso, D. T. (2013). Hydrogen production by biogas steam reforming: A technical, economic and ecological analysis. *Renewable and Sustainable Energy Reviews, 28*, 166–173. doi:10.1016/j.rser.2013.07.060.

Chen, D., Kuang, Y., Wang, H., Liang, J., & Zhao, J. (2022). Insights into the mechanism of naproxen inhibiting biohydrogen production from sludge dark fermentation. *Process Safety and Environmental Protection, 167*, 390–397. http://www.elsevier.com/wps/find/journaldescription.cws_home/713889/description#description. doi:10.1016/j.psep.2022.09.015.

Chezeau, B., & Vial, C. (2019). Combined effects of digestate viscosity and agitation conditions on the fermentative biohydrogen production. *Biochemical Engineering Journal, 142*, 105–116. doi:10.1016/j.bej.2018.11.016.

Crespi, E., Luca, G., Testi, M., Maggi, C., Bona, V., Barone, M. B., Staffetti, G., & Crema, L. (2024). Renewable hydrogen production through electrolysis: An analysis of the cost gap for its economic competitiveness in Italy. *International Journal of Hydrogen Energy, 68*, 1163–1177. https://www.sciencedirect.com/science/journal/03603199. doi:10.1016/j.ijhydene.2024.04.303.

Detchusananard, T., Wuttipisan, N., Limleamthong, P., Prasertcharoensuk, P., Maréchal, F., & Arpornwichanop, A. (2022). Pyrolysis and gasification integrated process of empty fruit bunch for multi-biofuels production: Technical and economic analyses. *Energy Conversion and Management, 258*, 115465. doi:10.1016/j.enconman.2022.115465.

Do, J. Y., Choi, S., Nahm, K., Kim, S. K., & Kang, M. (2018). Reliable hydrogen production from methanol photolysis in aqueous solution by a harmony between In and Zn in bimetallic zinc indium sulfide. *Materials Research Bulletin, 100*, 234–242. doi:10.1016/j.materresbull.2017.12.034.

Do, T. X., Lim, Y. I., & Yeo, H. (2014). Techno-economic analysis of biooil production process from palm empty fruit bunches. *Energy Conversion and Management, 80*, 525–534. doi:10.1016/j.enconman.2014.01.024.

Ebrahimian, F., Denayer, J. F. M., & Karimi, K. (2022). Efficient coproduction of butanol, ethanol, and biohydrogen from municipal solid waste through a cocultivated biorefinery. *Energy Conversion and Management, 255*. https://www.journals.elsevier.com/energy-conversion-and-management. doi:10.1016/j.enconman.2022.115303.

Fedele, L., Lombardo, G., Greselin, I., Menegazzo, D., & Bobbo, S. (2023). Thermophysical properties of low GWP refrigerants: An update. *International Journal of Thermophysics, 44*(5). doi:10.1007/s10765-023-03191-5.

Ganeshan, P., Vigneswaran, V. S., Gowd, S. C., Kondusamy, D., Sanjay kumar, C., Krishnamoorthy, N., Kumar, D., Juneja, A., Paramasivan, B., Raju, N. N., Rajendran, K., & Pugazhendhi, A. (2023). How does techno-economic analysis and lifecycle assessment help in commercializing the biohydrogen supply chain? *Fuel, 341*. http://www.journals.elsevier.com/fuel/. doi:10.1016/j.fuel.2023.127601.

Ganguly, A., Sun, P., Liu, X., Delgado, H. E., Sun, L., & Elgowainy, A. (2025). Techno-economic and life cycle analysis of bio-hydrogen production using bio-based waste streams through the integration of dark fermentation and microbial electrolysis. *Green Chemistry, 27*(21), 6213–6231. http://pubs.rsc.org/en/journals/journal/gc. doi:10.1039/d4gc05020g.

Ge, C., Chai, Y., Wang, H., & Kan, M. (2017). Ocean acidification: One potential driver of phosphorus eutrophication. *Marine Pollution Bulletin, 115*(1–2), 149–153. Www.elsevier.com/locate/marpolbul. doi:10.1016/j.marpolbul.2016.12.016.

Ghorbani, B., Zendehboudi, S., Saady, N. M. C., & Naterer, G. F. (2025). Hydrogen production and status in Canada: Foundations for a low-carbon future. *Canadian Journal of Chemical Engineering, 103*(9), 4125–4171. http://onlinelibrary.wiley.com/journal/10.1002/(ISSN)1939-019X. doi:10.1002/cjce.25620.

Hall, E. R., Wickes, L., Burnett, L. E., Scott, G. I., Hernandez, D., Yates, K. K., Barbero, L., Reimer, J. J., Baalousha, M., Mintz, J., Cai, W.-J., Craig, J. K., DeVoe, M. R., Fisher, W. S., Hathaway, T. K., Jewett, E. B., Johnson, Z., Keener, P., Mordecai, R. S., ... Styron, J. (2020). Acidification in the U.S. Southeast: Causes, potential consequences and the role of the southeast ocean and coastal acidification network. *Frontiers in Marine Science, 7*. doi:10.3389/fmars.2020.00548.

Han, W., Fang, J., Liu, Z., & Tang, J. (2016a). Techno-economic evaluation of a combined bioprocess for fermentative hydrogen production from food waste. *Bioresource Technology, 202*, 107–112. Www.elsevier.com/locate/biortech. doi:10.1016/j.biortech.2015.11.072.

Han, W., Hu, Y. Y., Li, S. Y., Li, F. F., & Tang, J. H. (2016b). Biohydrogen production from waste bread in a continuous stirred tank reactor: A techno-economic analysis. *Bioresource Technology, 221*, 318–323. Www.elsevier.com/locate/biortech. doi:10.1016/j.biortech.2016.09.055.

Hassan, G. K., Massanet-Nicolau, J., Dinsdale, R., Jones, R. J., Abo-Aly, M. M., El-Gohary, F. A., & Guwy, A. (2019). A novel method for increasing biohydrogen production from food waste using electrodialysis. *International Journal of Hydrogen Energy, 44*(29), 14715–14720. http://www.journals.elsevier.com/international-journal-of-hydrogen-energy/. doi:10.1016/j.ijhydene.2019.04.176.

Hossain, S., Saha, C. K., Ismail, M., Reza, T., Kabir, K. B., & Kirtania, K. (2023). Prospects and challenges of renewable hydrogen generation in Bangladesh. *International Journal of Hydrogen Energy, 48*(54), 20588–20612. http://www.journals.elsevier.com/international-journal-of-hydrogen-energy/. doi:10.1016/j.ijhydene.2023.03.059.

Hosseini, T., Culley, S. A., Zecchin, A., Maier, H. R., & Ashman, P. J. (2024). Impact of co-digestion and degree of centralization on the yield and viability of biomethane production: A case study in regional Australia. *Energy Conversion and Management: X, 22*. https://www.sciencedirect.com/science/journal/25901745. doi:10.1016/j.ecmx.2024.100585.

Islam, M. H., Burheim, O. S., & Pollet, B. G. (2019). Sonochemical and sonoelectrochemical production of hydrogen. *Ultrasonics Sonochemistry, 51*, 533–555. doi:10.1016/j.ultsonch.2018.08.024.

Jamali, N. S., Jahim, J. M., O-Thong, S., & Jehlee, A. (2019). Hydrodynamic characteristics and model of fluidized bed reactor with immobilised cells on activated carbon for biohydrogen production. *International Journal of Hydrogen Energy, 44*(18), 9256–9271. http://www.journals.elsevier.com/international-journal-of-hydrogen-energy/. doi:10.1016/j.ijhydene.2019.02.116.

Jin, M., Wei, X., Mu, X., Ren, W., Zhang, S., Tang, C., & Cao, W. (2024). Life-cycle analysis of biohydrogen production via dark-photo fermentation from wheat straw. *Bioresource Technology, 396*, 130429. doi:10.1016/j.biortech.2024.130429.

Karim, S., Tanwar, N., Das, S., Ranjit, R., Banerjee, A., Gulafshan, N., Gupta, A., Kumar, A., & Dutta, A. (2025). Shaping the future of green hydrogen production: overcoming conventional challenges with molecular catalysts, immobilization, and scalable electrolyzers. *ACS Catalysis, 15*(2), 1073–1096. http://pubs.acs.org/page/accacs/about.html. doi:10.1021/acscatal.4c05986.

Koroneos, C., Dompros, A., & Roumbas, G. (2008). Hydrogen production via biomass gasification—A life cycle assessment approach. *Chemical Engineering and Processing: Process Intensification, 47*(8), 1261–1268. doi:10.1016/j.cep.2007.04.003.

Koul, Y., Devda, V., Varjani, S., Guo, W., Ngo, H. H., Taherzadeh, M. J., Chang, J. S., Wong, J. W. C., Bilal, M., Kim, S. H., Bui, X. T., & Parra-Saldívar, R. (2022). Microbial electrolysis: A promising approach for treatment and resource recovery from industrial wastewater. *Bioengineered, 13*(4), 8115–8134. http://www.tandfonline.com/loi/kbie20. doi:10.1080/21655979.2022.2051842.

Krinitzki, C., Kaczorowski, M., & Hartz, C. (2024). Zum globalen Erwärmungspotenzial (GWP) von zentrisch belasteten Stützen im Konstruktiven Ingenieurbau. *Bautechnik, 101*(5), 309–319. http://onlinelibrary.wiley.com/journal/10.1002/(ISSN)1437-0999. doi:10.1002/bate.202300048.

Kucharska, K., Rybarczyk, P., Hołowacz, I., Łukajtis, R., Glinka, M., & Kamiński, M. (2018). Pretreatment of lignocellulosic materials as substrates for fermentation processes. *Molecules, 23*(11), 2937. doi:10.3390/molecules23112937.

Kumar, G., Mathimani, T., Rene, E. R., & Pugazhendhi, A. (2019). Application of nanotechnology in dark fermentation for enhanced biohydrogen production using inorganic nanoparticles. *International Journal of Hydrogen Energy, 44*(26), 13106–13113. http://www.journals.elsevier.com/international-journal-of-hydrogen-energy/. doi:10.1016/j.ijhydene.2019.03.131.

Lee, B., Chae, H., Choi, N. H., Moon, C., Moon, S., & Lim, H. (2017). Economic evaluation with sensitivity and profitability analysis for hydrogen production from water electrolysis in Korea. *International Journal of Hydrogen Energy, 42*(10), 6462–6471. http://www.journals.elsevier.com/international-journal-of-hydrogen-energy/. doi:10.1016/j.ijhydene.2016.12.153.

Lefranc, L., Linares, J. I., Santos, A. M., Arenas, E., Martín, C., & Moratilla, Y. (2024). Biohydrogen with negative CO_2 emissions from municipal solid waste for Decarbonising the public bus fleet. *Journal of Environmental Management, 371*. https://www.sciencedirect.com/science/journal/03014797. doi:10.1016/j.jenvman.2024.123258.

Li, G., Cui, P., Wang, Y., Liu, Z., Zhu, Z., & Yang, S. (2020a). Life cycle energy consumption and GHG emissions of biomass-to-hydrogen process in comparison with coal-to-hydrogen process. *Energy, 191*, 116588. doi:10.1016/j.energy.2019.116588.

Li, G., Wang, S., Zhao, J., Qi, H., Ma, Z., Cui, P., Zhu, Z., Gao, J., & Wang, Y. (2020b). Life cycle assessment and techno-economic analysis of biomass-to-hydrogen production with methane tri-reforming. *Energy, 199*, 117488. doi:10.1016/j.energy.2020.117488.

Li, W., Zhang, J., Yang, J., Zhang, J., Li, Z., Yang, Y., & Zang, L. (2022). Comparison of copper and aluminum doped cobalt ferrate nanoparticles for improving biohydrogen production. *Bioresource Technology, 343*, 126078. doi:10.1016/j.biortech.2021.126078.

Liu, M., Zeng, J., Huang, G., Liu, X., He, G., Yao, S., Shang, N., Zheng, L., & Wang, P. (2024). Assessing energy consumption, carbon emissions, and costs in biomass-to-gas processes: A life-cycle assessment approach. *Sustainability, 16*(12), 5209. doi:10.3390/su16125209.

Lopes, T. F., Carvalheiro, F., Duarte, L. C., Gírio, F., Quintero, J. A., & Aroca, G. (2019). Techno-economic and life-cycle assessments of small-scale biorefineries for isobutene and xylo-oligosaccharides production: A comparative study in Portugal and Chile. *Biofuels, Bioproducts and Biorefining, 13*(5), 1321–1332. http://onlinelibrary.wiley.com/journal/10.1002/(ISSN)1932-1031. doi:10.1002/bbb.2036.

López-Hernández, B. N., Escamilla-Alvarado, C., Guadalupe Paredes, M., Albalate-Ramírez, A., Rodríguez-Valderrama, S., Magnin, J.-P., & Rivas-García, P. (2024). Environmental impact assessment of biohydrogen production from orange peel waste by lab-scale dark and photofermentation processes. *Revista Internacional de Contaminación Ambiental, 40,* 27–40. doi:10.20937/RICA.55063.

Lourinho, G., & Brito, P. S. D. (2021). Electrolytic treatment of swine wastewater: Recent progress and challenges. *Waste and Biomass Valorization, 12*(2), 553–576. http://www.springer.com/engineering/journal/12649. doi:10.1007/s12649-020-00951-4.

Lu, C., Wang, G., Zhang, Q., Yang, X., Yu, J., Liu, T., Petracchini, F., Zhang, Z., Sun, Y., Jiang, D., Liang, X., Li, Y., Zhang, Y., Zhang, T., & Zhang, H. (2023). Comparison of biorefinery characteristics: Photo-fermentation biohydrogen, dark fermentation biohydrogen, biomethane, and bioethanol production. *Applied Energy, 347*. https://www.journals.elsevier.com/applied-energy. doi:10.1016/j.apenergy.2023.121463.

Mahidhara, G., Burrow, H., Sasikala, C., & Ramana, C. V. (2019). Biological hydrogen production: Molecular and electrolytic perspectives. *World Journal of Microbiology and Biotechnology, 35*(8). Www.wkap.nl/journalhome.htm/0959-3993. doi:10.1007/s11274-019-2692-z.

Materazzi, M., Taylor, R., & Cairns-Terry, M. (2019). Production of biohydrogen from gasification of waste fuels: Pilot plant results and deployment prospects. *Waste Management, 94,* 95–106. Www.elsevier.com/locate/wasman. doi:10.1016/j.wasman.2019.05.038.

Mohanakrishna, G., Sneha, N. P., Rafi, S. M., & Sarkar, O. (2023). Dark fermentative hydrogen production: Potential of food waste as future energy needs. *Science of the Total Environment, 888*. Www.elsevier.com/locate/scitotenv. doi:10.1016/j.scitotenv.2023.163801.

Montenegro Camacho, Y. S., Bensaid, S., Piras, G., Antonini, M., & Fino, D. (2017). Techno-economic analysis of green hydrogen production from biogas autothermal reforming. *Clean Technologies and Environmental Policy, 19*(5), 1437–1447. doi:10.1007/s10098-017-1341-1.

Muntasir Shovon, S., Ahamed Akash, F., Abdur Rahman, M., Rahman, W., Chakraborty, P., Monir, M. U., Sarkar, S. M., Abd Aziz, A., & Chowdhury, S. (2024). Advancements in hydrogen generation, storage, and utilizations: A comprehensive review of current trends in Bangladesh. *Energy, 292*. https://www.sciencedirect.com/science/journal/03605442. doi:10.1016/j.energy.2024.130477.

Nakamura, K., Kanno, T., Mokudai, T., Iwasawa, A., Niwano, Y., & Kohno, M. (2012). Microbial resistance in relation to catalase activity to oxidative stress induced by photolysis of hydrogen peroxide. *Microbiology and Immunology, 56*(1), 48–55. doi:10.1111/j.1348-0421.2011.00400.x.

Nikolaidis, P., & Poullikkas, A. (2017). A comparative overview of hydrogen production processes. *Renewable and Sustainable Energy Reviews, 67,* 597–611. https://www.journals.elsevier.com/renewable-and-sustainable-energy-reviews. doi:10.1016/j.rser.2016.09.044.

Olabi, A. G., Abdelkareem, M. A., Mahmoud, M. S., Elsaid, K., Obaideen, K., Rezk, H., Wilberforce, T., Eisa, T., Chae, K. J., & Sayed, E. T. (2023). Green hydrogen: Pathways, roadmap, and role in achieving sustainable development goals. *Process Safety and Environmental Protection, 177,* 664–687. http://www.elsevier.com/wps/find/journaldescription.cws_home/713889/description#description. doi:10.1016/j.psep.2023.06.069.

Pachapur, V. L., Kutty, P., Pachapur, P., Brar, S. K., Le Bihan, Y., Galvez-Cloutier, R., & Buelna, G. (2019). Seed pretreatment for increased hydrogen production using mixed-culture systems with advantages over pure-culture systems. *Energies, 12*(3). https://www.mdpi.com/1996-1073/12/3. doi:10.3390/en12030530.

Parthasarathy, P., & Narayanan, K. S. (2014). Hydrogen production from steam gasification of biomass: Influence of process parameters on hydrogen yield—A review. *Renewable Energy, 66*, 570–579. http://www.journals.elsevier.com/renewable-and-sustainable-energy-reviews/. doi:10.1016/j.renene.2013.12.025.

Preisner, M., Neverova-Dziopak, E., & Kowalewski, Z. (2020). Analysis of eutrophication potential of municipal wastewater. *Water Science and Technology, 81*(9), 1994–2003. doi:10.2166/wst.2020.254.

Ramírez-Díaz, R. C., Prato-Garcia, D., & Vasquez-Medrano, R. (2023). How sustainable is the biohydrogen produced from sugarcane vinasse? An approach based on life cycle assessment. *Biomass Conversion and Biorefinery, 13*(16), 14755–14775. https://www.springer.com/journal/13399. doi:10.1007/s13399-022-03242-1.

Rani, P., Yadav, D. K., Yadav, A., Bishnoi, N. R., Kumar, V., Ram, C., Pugazhendhi, A., & Kumar, S. S. (2024). Frontier in dark fermentative biohydrogen production from lignocellulosic biomass: Challenges and future prospects. *Fuel, 366*, 131187. doi:10.1016/j.fuel.2024.131187.

Rawat, S., Rautela, A., Yadav, I., Misra, S., & Kumar, S. (2023). A comprehensive review on enhanced biohydrogen production: Pretreatment, applied strategies, techno-economic assessment, and future perspective. *BioEnergy Research, 16*(4), 2131–2154. doi:10.1007/s12155-023-10598-3.

Rocha-Meneses, L., Hari, A., Inayat, A., Shanableh, A., Abdallah, M., Ghenai, C., Shanmugam, S., & Kikas, T. (2022). Application of nanomaterials in anaerobic digestion processes: A new strategy towards sustainable methane production. *Biochemical Engineering Journal, 188*, 108694. doi:10.1016/j.bej.2022.108694.

Safari, M., Tondro, H., & Zilouei, H. (2022). Biohydrogen production from diluted-acid hydrolysate of rice straw in a continuous anaerobic packed bed biofilm reactor. *International Journal of Hydrogen Energy, 47*(9), 5879–5890. http://www.journals.elsevier.com/international-journal-of-hydrogen-energy/. doi:10.1016/j.ijhydene.2021.11.222.

Salkuyeh, Y. K., Saville, B. A., & MacLean, H. L. (2018). Techno-economic analysis and life cycle assessment of hydrogen production from different biomass gasification processes. *International Journal of Hydrogen Energy, 43*(20), 9514–9528. http://www.journals.elsevier.com/international-journal-of-hydrogen-energy/. doi:10.1016/j.ijhydene.2018.04.024.

Sanderson, B. G., & Coade, G. (2010). Scaling the potential for eutrophication and ecosystem state in lagoons. *Environmental Modelling & Software, 25*(6), 724–736. doi:10.1016/j.envsoft.2009.12.006.

Sarangi, P. K., & Nanda, S. (2020). Biohydrogen production through dark fermentation. *Chemical Engineering & Technology, 43*(4), 601–612. doi:10.1002/ceat.201900452.

Sarkar, S., & Kumar, A. (2009). Techno-economic assessment of biohydrogen production from forest biomass in Western Canada. *Transactions of the ASABE, 52*(2), 519–530. doi:10.13031/2013.26809.

Sathyaprakasan, P., & Kannan, G. (2015). Economics of bio-hydrogen production. *International Journal of Environmental Science and Development, 6*(4), 352–356. doi:10.7763/IJESD.2015.V6.617.

Senthil Rathi, B., Senthil Kumar, P., & Rangasamy, G. (2023). A sustainable approach on thermal and catalytic conversion of waste plastics into fuels. *Fuel, 339*, 126977. doi:10.1016/j.fuel.2022.126977.

Shahbaz, M., yusup, S., Inayat, A., Patrick, D. O., & Ammar, M. (2017). The influence of catalysts in biomass steam gasification and catalytic potential of coal bottom ash in biomass steam gasification: A review. *Renewable and Sustainable Energy Reviews, 73*, 468–476. https://www.journals.elsevier.com/renewable-and-sustainable-energy-reviews. doi:10.1016/j.rser.2017.01.153.

Singh, R., Kumar, R., Sarangi, P. K., Kovalev, A. A., & Vivekanand, V. (2023). Effect of physical and thermal pretreatment of lignocellulosic biomass on biohydrogen production by thermochemical route: A critical review. *Bioresource Technology, 369*. Www.elsevier.com/locate/biortech. doi:10.1016/j.biortech.2022.128458.

Sinnott, R., & Towler, G. (2019). *Chemical engineering design: SI edition*. Butterworth-Heinemann.

Sivaramakrishnan, R., Shanmugam, S., Sekar, M., Mathimani, T., Incharoensakdi, A., Kim, S.-H., Parthiban, A., Edwin Geo, V., Brindhadevi, K., & Pugazhendhi, A. (2021). Insights on biological hydrogen production routes and potential microorganisms for high hydrogen yield. *Fuel, 291*, 120136. doi:10.1016/j.fuel.2021.120136.

Song, H., Luo, S., Huang, H., Deng, B., & Ye, J. (2022). Solar-driven hydrogen production: Recent advances, challenges, and future perspectives. *ACS Energy Letters, 7*(3), 1043–1065. http://pubs.acs.org/journal/aelccp. doi:10.1021/acsenergylett.1c02591.

Srivastava, R. K., Shetti, N. P., Reddy, K. R., & Aminabhavi, T. M. (2020). Biofuels, biodiesel and biohydrogen production using bioprocesses. A review. *Environmental Chemistry Letters, 18*(4), 1049–1072. https://www.springer.com/journal/10311. doi:10.1007/s10311-020-00999-7.

Sun, H., Shen, J., Hu, M., Zhang, J., Cai, Z., Zang, L., Zhang, F., & Ji, D. (2021). Manganese ferrite nanoparticles enhanced biohydrogen production from mesophilic and thermophilic dark fermentation. *Energy Reports, 7*, 6234–6245. http://www.journals.elsevier.com/energy-reports/. doi:10.1016/j.egyr.2021.09.070.

Touloupakis, E., & Torzillo, G. (2019). Chapter 14—Photobiological hydrogen production. In F. Calise, M. D. D'Accadia, M. Santarelli, A. Lanzini, & D. Ferrero (Eds.), *Solar hydrogen production* (pp. 511–525). Academic Press. doi:10.1016/B978-0-12-814853-2.00014-X.

Wang, Y., Li, G., Liu, Z., Cui, P., Zhu, Z., & Yang, S. (2019). Techno-economic analysis of biomass-to-hydrogen process in comparison with coal-to-hydrogen process. *Energy, 185*, 1063–1075. https://www.journals.elsevier.com/energy. doi:10.1016/j.energy.2019.07.119.

Wee, A. N. C. H., Erison, A. E., Edward Anyek, E. H., Pakpahan, G. R., Lim, J. R., & Tiong, A. N. T. (2022). Techno-economic assessment of hydrogen production via steam reforming of palm oil mill effluent. *Sustainable Energy Technologies and Assessments, 53*. http://www.journals.elsevier.com/sustainable-energy-technologies-and-assessments. doi:10.1016/j.seta.2022.102575.

Wu, D., Hu, B., & Wang, R. Z. (2021). Vapor compression heat pumps with pure low-GWP refrigerants. *Renewable and Sustainable Energy Reviews, 138*. https://www.journals.elsevier.com/renewable-and-sustainable-energy-reviews. doi:10.1016/j.rser.2020.110571.

Xu, W., Yang, L., Niu, Z., Wang, S., Wang, Y., Zhu, Z., & Cui, P. (2025). Thermodynamic and economic analysis of a novel DME-power polygeneration system based on the integration of biomass gasification and alkaline electrolysis of water for hydrogen production. *Energy, 314*, 134185. doi:10.1016/j.energy.2024.134185.

Xu, X., Zhou, Q., & Yu, D. (2022). The future of hydrogen energy: Bio-hydrogen production technology. *International Journal of Hydrogen Energy, 47*(79), 33677–33698. http://www.journals.elsevier.com/international-journal-of-hydrogen-energy/. doi:10.1016/j.ijhydene.2022.07.261.

Yaashikaa, P. R., Keerthana Devi, M., & Senthil Kumar, P. (2022). Biohydrogen production: An outlook on methods, constraints, economic analysis and future prospect. *International Journal of Hydrogen Energy, 47*(98), 41488–41506. doi:10.1016/j.ijhydene.2022.07.082.

Yao, J., Kraussler, M., Benedikt, F., & Hofbauer, H. (2017). Techno-economic assessment of hydrogen production based on dual fluidized bed biomass steam gasification, biogas steam reforming, and alkaline water electrolysis processes. *Energy Conversion and Management, 145*, 278–292. doi:10.1016/j.enconman.2017.04.084.

Yusuf, M., Salam Farooqi, A., Kok Keong, L., Hellgardt, K., & Abdullah, B. (2021). Contemporary trends in composite Ni-based catalysts for CO_2 reforming of methane. *Chemical Engineering Science, 229*, 116072. doi:10.1016/j.ces.2020.116072.

Zhang, Q., Jiao, Y., He, C., Ruan, R., Hu, J., Ren, J., Toniolo, S., Jiang, D., Lu, C., Li, Y., Man, Y., Zhang, H., Zhang, Z., Xia, C., Wang, Y., Jing, Y., Zhang, X., Lin, R., Li, G., … Tahir, N. (2024). Biological fermentation pilot-scale systems and evaluation for commercial viability towards sustainable biohydrogen production. *Nature Communications, 15*(1). https://www.nature.com/ncomms/. doi:10.1038/s41467-024-48790-4.

Zhao, Y., Liang, X., Mu, H., & Zhang, X. (2015). Biohydrogen production from sewage sludge by sequential dark and photo fermentation. *Journal of Biobased Materials and Bioenergy, 9*(1), 95–100. http://docserver.ingentaconnect.com/deliver/connect/asp/15566560/v9n1/s18.pdf?expires=1431485357&id=81776710&titleid=72010009&accname=Elsevier+BV&checksum=EDE082491F66304C2E1CE20EB07673AF. doi:10.1166/jbmb.2015.1496.

Zuo, X., Toam, Q., & Zhong, Y. (2025). Comparative analysis of hydrogen production methods: Environmental impact and efficiency of electrochemical and thermochemical processes. *International Journal of Hydrogen Energy, 118*, 426–440. https://www.sciencedirect.com/science/journal/03603199. doi:10.1016/j.ijhydene.2025.03.171.

CHAPTER

8

Prospects of biohydrogen

Chaoyang Lu[a], *Zhipeng Song*[a], *Xueting Zhang*[b], *Quanguo Zhang*[a,b], *Mujeeb Alam*[a], *and Danping Jiang*[a,*]

[a]Henan Agricultural University, Zhengzhou, China, [b]Huanghe Science and Technology University, Zhengzhou, China

*Corresponding author. E-mail address: jiangdanping@henau.edu.cn

8.1 Sustainability outlook of biohydrogen production

The way the world uses energy is changing because countries want to cut carbon emissions and use energy sources that are better for the environment. For this purpose, hydrogen, a way to store energy, has become into people's view because of its nature of being nonharmful to the environment when combusted, especially performing outstandingly in areas where it is hard to make electricity directly, such as heavy industry, transportation, and storing energy for a long time.

The methods used to produce hydrogen, such as thermochemical method, over the past few decades may not be able to meet the future demand, which urges people to find more cost-effective and environment-friendly alternatives. Therefore biohydrogen production technology stands out among different technologies. During the biohydrogen production process the hydrogen-producing microorganisms can utilize food waste, lignocellulosic biomass, wastewater, and other natural materials as substrate to produce hydrogen and, at the same time, sequester the carbon in environment. As a consequence, our reliance on fossil energy has also been reduced, and the related emission reduction issues have also been alleviated, which makes it possible to achieve low-carbon or near-neutral during the life cycle of energy production and conversion. Furthermore the feasibility of the combination of biohydrogen production with agricultural and forestry affairs, municipal administration, and environmental management would lead to a circular economy.

Biohydrogen: Advances in Theory and Production
DOI: https://doi.org/10.1016/B978-0-443-27686-6.00006-2

This section will give a full summary of the role of biohydrogen in the future to help boom the environment and the economy. The contributions of renewable raw materials to biohydrogen production and biohydrogen production for carbon footprint are highlighted. Moreover, the different available technological pathways and the positive impact of favorable policies are summarized. In addition, significant issues of contestation, such as low technical efficiency, high cost, and the scale-up challenge, will be further discussed; and the possible solution for future study will be proposed as well.

It is hopeful to explain how a pivotal role would be played by biohydrogen in the development of green energy and how a more robust and low-emission energy system would be created in the future.

8.1.1 Renewable raw materials

The ability of biohydrogen technology to truly deliver on its sustainability promise depends largely on the renewability and environmental friendliness of its feedstocks. These factors directly shape the technology's carbon cycling characteristics and determine how well it aligns with natural ecosystems and societal systems. Unlike fossil fuels, which rely on geological carbon storage, biohydrogen draws on the contemporary carbon cycle using biomass that grows continuously through photosynthesis, and during the growth stage, this biomass could absorb CO_2 from the atmosphere. Under ideal conditions, this mechanism allows hydrogen production to achieve carbon balance rather than net emissions, which is the heart of its sustainability. Moreover, the category of feedstocks for biohydrogen production demonstrates remarkable diversity, skillfully avoiding the ethical dilemmas and development bottlenecks associated with early biofuels that "competed with food for land." Its main resources have shifted from food crops to more sustainable second-generation biomass. The core advantage of biohydrogen production is the renewable nature of its feedstocks, including agricultural waste, organic wastewater, algal biomass, and other biomass resources. In addition, the digestate or residues remaining after biohydrogen production are rich in nutrients, such as nitrogen, phosphorus, and potassium; therefore it might be possible to return them to farmland as high-quality organic fertilizers so that parts of the chemical fertilizers will be replaced. Hence the technology will provide a closed-loop cycle of biomass utilization of "cultivation–harvest–waste–energy and fertilizer–recultivation." Moreover, it will reduce agricultural pollution owning to huge amounts of chemical fertilizers used and improve the organic matter content of soil.

Agricultural wastes, such as crop straw, bagasse, and corn cobs, are widely available feedstocks that do not require additional cultivation and have no competition with food production (Jiang et al., 2022). Within the global agricultural production system, agricultural wastes could represent a substantial and stable resource for biohydrogen production, serving as a typical "stock" resource. In addition, a range of organic wastewater, including residual streams from food processing, livestock operations, and municipal sludge (Chaves et al., 2025), can also be used to produce hydrogen via anaerobic fermentation, as it contains abundant carbohydrates and proteins. The key advantage using these waste streams is that no extra pretreatment steps are needed, making the processing easier and cheaper. Therefore the wastewater treatment facilities will evolve into energy-recovery platforms and possess the capability to transform the materials that have been viewed as disposal burdens into renewable energy resources.

Another viable pathway for biohydrogen generation is provided by microalgae, such as green algae and cyanobacteria. These organisms not only can produce biohydrogen by using sunlight as energy-driven but also can be changed into feedstocks that are used for fermentation (Liu et al., 2022). In addition, growing microalgae does not take land away from other types of farming, because it can grow well in areas that are not very good for farming, such as saline soils, deserts, and coastal regions. Moreover, microalgae can be cultivated in open ponds or closed photobioreactors using seawater, brackish water, or nutrient-rich wastewater. These characteristics make them a very sustainable biomass source, because they can significantly reduce freshwater use as well as fertilizer use.

Renewable biomass used for biohydrogen production can be get sustainably through natural cycles or standardized cultivation, which provides a robust and secure resource supply link for bioenergy generation. Consideration of technological evolution and economic feasibility, the combination of biohydrogen systems with more efficient pretreatment processes, microbial metabolic regulation technologies, and comprehensive storage and transportation facilities will become inevitable. As the biohydrogen production industry expands, the integration among agriculture, manufacturing, and the energy sector will become increasingly close, thereby promoting the formation of a more coordinated resource utilization model. This development path is conducive to building a more resilient and sustainable regional energy system, providing a guarantee for long-term stable energy supply.

8.1.2 Low-carbon emissions

Compared with traditional fossil-fuel-based hydrogen production, biohydrogen offers clear advantages in terms of carbon emissions. This is particularly true of photobiological hydrogen. Gray hydrogen is usually made by changing methane using heat and pressure. Blue hydrogen also uses a process called carbon capture and storage (CCS) to capture carbon emissions, but it still depends on fossil fuels and cannot fully get rid of emissions. Biohydrogen uses tiny living things like algae to change food into hydrogen. It also uses solar energy to make hydrogen, which means there is much less pollution (Thajuddin et al., 2026). This drop is not only because of how the products are made but also because of the special carbon cycle of the things that are used to make the products. These things take in the air's CO_2 when they are growing through photosynthesis. If everything goes well, the carbon released when making biohydrogen can be thought of as having been fixed recently. This makes the carbon loop almost closed, and it means that the total amount of carbon released is lower over the whole lifecycle.

Dark fermentation relies on agricultural residues or organic wastewater as substrates, and most of the associated carbon emissions arise from transporting these materials and carrying out necessary pretreatment. Nevertheless, utilizing such waste-derived carbon streams—materials that would otherwise decompose or be burned and release greenhouse gases—offers clear environmental benefits. By contrast the process of photobiological fermentation generates hydrogen through photosynthetic activity. This uses sunlight to convert organic substrates into energy. The CO_2 taken up during this process can offset part of the emissions, forming an almost self-contained carbon cycle and enabling hydrogen production with very low net carbon impact (Zhang et al., 2024).

More importantly the low-carbon potential of biohydrogen goes beyond achieving carbon neutrality, pointing toward the possibility of future negative-carbon development. By combining biohydrogen production with biomass CCS—capturing and permanently sequestering CO_2 generated during hydrogen production—it is possible to achieve net removal of atmospheric CO_2 through the biomass energy cycle. This makes biohydrogen a potential "negative-carbon technology," with significant strategic value for global emission reduction and carbon management.

8.1.3 Technical diversity

Biohydrogen production technology has diversified technological pathways, representing a core advantage in adapting to different regional resource endowments and application scenarios. Currently, several major technological routes [such as dark fermentation, photo-fermentation, combined dark–photo fermentation, direct/indirect biophotolysis, and microbial electrolysis cells (MECs)] for biohydrogen production have been established, and each technology has distinct characteristics and complementary roles.

Dark fermentation produces hydrogen under light-free conditions by utilizing anaerobic microbes to decompose organic matter. It can achieve a theoretical hydrogen yield up to 4 mol H_2/mol glucose, operate at a relatively high rate, and adapt to a wide range of substrates, making it particularly suitable for treating high-strength organic wastewater or agricultural waste. However, this process generates CO_2 along with a large number of organic acids that cannot be further utilized, which limits complete substrate conversion and overall energy efficiency. However, photo-fermentation demonstrates its unique value that can employ photosynthetic bacteria to further convert the organic acids produced in dark fermentation into hydrogen, and the theoretical yield is 12 mol H_2/mol glucose. Therefore combining dark fermentation with photo-fermentation would significantly enhance the overall hydrogen yield, and the theoretical values may even reach 6.31 mol H_2/mol carbohydrates, enabling a deep resource recovery from waste streams (Das & Basak, 2025).

Beyond these three fermentation pathways that rely on organic substrates, the biohydrogen technology categories also include more routes. Direct/indirect biophotolysis uses algae or cyanobacteria to split water directly into hydrogen, requiring in theory only water and sunlight, which represent an ultimate clean pathway, although current challenges remain, such as low hydrogen production rate and oxygen inhibition. MECs ingeniously integrate electrochemistry and biocatalysis together with a low external voltage (0.6–1.0 V), and the microorganisms oxidize organic matter, whereas protons are reduced at the cathode to produce hydrogen.

Current research of biohydrogen production focus extends beyond optimizing individual processes and is increasingly interesting directly toward breakthroughs via genetic engineering of microbial strains (such as enhancing oxygen tolerance of hydrogenases), novel photobioreactor designs, and the coupling of multiple processes (such as combined dark–photo fermentation, biophotolysis–photo-fermentation, and dark fermentation–MECs). Such technological diversity enables biohydrogen production to be flexibly adapted and combined on the basis of the specific feedstock properties, site constraints, and energy demands.

8.1.4 Policy support

With the steady advancement of low-carbon development, biohydrogen is playing an increasingly important role in the national energy strategy. The European Union pointed out the key role of biohydrogen in renewable hydrogen production and has included it in its core strategy for addressing climate change. The European Union has set a target of achieving an annual production capacity of 10 million tons of biohydrogen by 2030 in its "hydrogen strategy." Moreover, they also provide financial support for technological research, development, and demonstration applications via measures including innovation funds and the European Horizon. China's "14th 5-Year Plan for Hydrogen Development" emphasizes the potential of hydrogen production from biomass and encourages the implementation of demonstration projects in provinces such as Guangdong and Jiangsu by using agricultural and forestry waste as feedstocks. The US "Inflation Reduction Act (IRA)" provides incentives for biohydrogen production, which can receive a subsidy of up to $3/km. In addition, Japan and South Korea have established demonstration facilities for biohydrogen production and also collaborated with related industries. Japan's "Green Innovation Fund" and South Korea's "Hydrogen Economy Roadmap" both list biohydrogen technology as a key development area aiming to support the low-carbon transformation of regional energy systems (Neumann et al., 2023).

Government support for biohydrogen typically involves research grants, industrial incentives, and carbon-pricing mechanisms. These aim to reduce production costs and facilitate market entry. Importantly the way policies are set is changing. In the past, too much focus was placed on R&D. Now the focus is shifting toward other areas, such as planning supply chains, selling products in new markets, and developing infrastructure. As the global push for carbon neutrality gathers pace, the drive for biohydrogen is set to gain even more traction. A stable market environment can be cultivated by expanding flexible financial instruments. This can also be achieved by refining regulatory standards. Another way to achieve this is to strengthen international collaboration. In this environment the long-term growth of the biohydrogen sector is sustained by the reinforcement of policy measures and market forces.

8.2 Significant issues of contestation

8.2.1 Low technical efficiency

The low efficiency of biological hydrogen production technology is mainly reflected in the slow hydrogen production rate of microorganisms and insufficient light energy conversion efficiency.

Biohydrogen production has the characteristics of slow hydrogen production rate and environmental sensitivity. The metabolic pathways of hydrogen production in microorganisms are complex, involving key enzymes such as hydrogenase or nitrogenase, whose activity is easily affected by environmental factors. The optimal pH range for most hydrogen production bacteria is 5.5–7.5, and deviations from this range would inhibit the enzyme activities. Mesophilic bacteria prefer the temperature range of 30°C–40°C, and extreme temperatures would halt the metabolic activity because the essence of enzymes is protein. Moreover, hydrogenase and nitrogenase are deactivated in the presence of oxygen, necessitating strict anaerobic conditions and thereby increasing the complexity of system control. In addition, there is a competition between microbial growth and hydrogen production; essentially the microbial fermentation process serves their own growth and energy acquisition rather than being intended for "hydrogen production" for human use. Consequently the metabolic pathways spontaneously proceed in the direction most beneficial for the microorganisms themselves. Therefore the substrate energy is diverted toward cell biomass proliferation rather than hydrogen generation, resulting in lower hydrogen production rate.

Photo-fermentation is regarded as a "follow-up and complementary" process to dark fermentation, aiming to enhance the overall energy conversion efficiency by further using the volatile fatty acids (VFAs). The efficiency of light energy transmission and conversion has a significant impact on photo-fermentation biohydrogen production. However, the bacterial broth within the reactor has a certain density, making it difficult for light to penetrate to deeper layers. This results in uneven light distribution inside the reactor, creating illuminated and dark zones, leading to "light source wastage," as the cells near the surface of reactor may experience light saturation or even photodamage because of excessive light, whereas the cells in the interior may suffer from insufficient light and consequently low efficiency. Hence achieving uniform illumination becomes a significant engineering challenge for large-scale photobioreactors to obtain higher light energy conversion efficiency. The same problem may be observed in the biophotolysis hydrogen production.

MECs are an advanced technology that combines microbial electrochemistry with applied electrical energy, and the core components are electrogenic microorganisms attached to the anode. However, electrical resistances are present throughout the MEC system, including in the electrolyte, membrane, electrodes, and external connections. As current flows through these components, ohmic losses occur, resulting in energy dissipation in the form of heat and consequently reducing the overall energy efficiency of the system. Especially proton exchange membranes are expensive and prone to contamination and blockage, which can lead to an increase in internal resistance. Although membrane-free design can reduce the overall system cost, it may allow hydrogen to escape to the anode. Hydrogen is consumed by microorganisms at the anode, which reduces the purity of hydrogen.

8.2.2 High cost

Biohydrogen production has made some breakthroughs, but its cost remains a key constraint on its development. The main limiting factors at present are as follows:

1. The construction cost of bioreactors is very high. Strict anaerobic environment and lighting conditions are required in the process of biohydrogen production. The devices that must be equipped in reactors, such as temperature control, pH adjustment, and gas collection systems, result in higher initial investment.
2. Because of the high costs of strain cultivation and maintenance the most efficient hydrogen-producing strains (*Clostridium* or photosynthetic bacteria) need some special culture medium, which is prone to contamination or being deactivated; they need constant cultivation conditions to keep strains viable, increasing the operation cost (Lu et al., 2016).

3. In separating products and purifying them, many challenges exist. The biohydrogen production usually has several by-products such as carbon dioxide and methane, which means that additional gas separation technology such as pressure swing adsorption or membrane separation is required, which leads to the increase in cost.
4. Low hydrogen production efficiency exists because of low microbial metabolic efficiency, and biohydrogen yield usually cannot meet the theoretical value so that the feedstock utilization is not high enough (Lu et al., 2023).
5. Economies of scale are lacking. At present, large-scale biohydrogen production technology is lacking because most hydrogen production is carried out in laboratories, which results in fixed costs that cannot be shared.

In addition to common challenges, different biohydrogen production methods face different difficulties. Dark fermentation hydrogen production produces a large number of metabolites, which reduces the amount of hydrogen. Biophotolysis and photosynthetic biohydrogen production have a strict dependence on light, which limits their large-scale production. In addition, the efficiency of light energy conversion inside the reactor remains a key issue that needs to be addressed. Precious metal catalysts and external power supplies have constrained the large-scale development of MECs.

In general, biohydrogen production has not yet progressed from being mainly tested in the laboratory to large-scale industrial production. Its technology itself is still immature, so it has led to high costs. To reduce future costs, we need more efforts in terms of technological breakthroughs, process optimization, economies of scale, and policy support, among other factors. As long as the world calls for green hydrogen, biohydrogen will have strong economic competitiveness in terms of continuous innovation and ongoing optimization despite carbon neutralization policies becoming more prevalent in countries globally.

8.2.3 The scale-up challenge

The commercialization of biohydrogen production goes through challenges as far as increasing its scale from lab to industry, which takes many forms.

1. Reactor size-up and process steadiness: Because of the issue of nonuniform mass transfer and heat transfer during the scale-up from lab-scale bioreactors (such as batch culture), the generation amount of hydrogen is obviously weakened (Zheng et al., 2022). For example, improper light distribution and the substrate diffusion difficulty caused by insufficient density gradient could be very dangerous for the activities of microbes

inside photobioreactors. More importantly, continuous operations of microbial strains make it possible that these microbial strains would degenerate continuously, especially metabolically inhibited strains with heavy accumulation of organic acids or contaminated strains, which are prone to appear after a long period of operation.

2. Hydrogen collection and purification are often generated with mixed gas containing impurities like CO_2, so there is a necessity for effective separation and purification. Although mature methods have been developed to purify hydrogen, namely, conventional technologies using membrane separation or pressure swing adsorption are not easy to be available at lower cost. Besides, because of its very small molecular weight, hydrogen is much more likely to leak out of the reaction system compared to other flammable gases, which can lead to problems with operation.
3. Storage and transportation: Low density and high flammability mean that hydrogen must be stored under pressure or cryogenic conditions. However, biohydrogen is usually generated under low pressure; therefore energy is needed to compress it. As distributed production facilities can alleviate transportation requirements, the necessary infrastructure, such as pipelines or liquid-H_2 transport trucks, represent an upfront cost.
4. System integration and economic feasibility: Raw material pretreatment, automation control system, waste liquid treatment, etc. are all issues that need to be considered in large-scale production, which will increase the complexity and cost of the system.

Further sorting out technologies, such as dark fermentation, light fermentation, MECs, and biomass thermochemical conversion, will reveal more technical issues. In large-scale dark fermentation biohydrogen production facilities the mixing dead zone can result in insufficient mixing of substrates and colonies, which further leads to imbalanced reactor operation. The accumulation of metabolites such as butyric acid and ethanol can severely inhibit the activity of enzymes and hydrogen-producing microorganisms. In addition, in large-scale biohydrogen production processes, the microbial colony structure in large reactors may change because of small environmental changes, which can easily lead to the loss of advantages for hydrogen-producing microorganisms in survival competition, resulting in the collapse of the hydrogen production system.

To solve the problem of scaling up biophotolysis and photofermentation, one must overcome two major issues—obtaining uniform light distribution and enough light penetration depth—to achieve uniform irradiation on all the microorganisms. In laboratory conditions, one can simply use external light sources to provide uniform irradiation onto the thin layer of liquid culture; however, when scaled to industrial levels, because of the attenuation effect of cells on surrounding areas, cells far from the center are illuminated insufficiently, which drastically reduces the irradiance of the central core and forms a dark region. Not only does it bring big trouble for the whole process because of the low irradiance, but it also induces uneven growth of the microbes. To address this difficulty, we have to use complex internal illumination systems or adopt tubing or flat-panel reactors with thin layers, both of which cause extremely high investment and maintenance costs and also more challenges for cleaning and controlling contamination.

MECs have impressive performance when applied in laboratories, but scaling them up toward real-world applications is affected by both electrochemical and biological constrains. First, internal resistances increase dramatically with reactor sizes, resulting in higher ohmic losses and lower energy efficiencies. Second the ratio between the electrode surface area and the reactor volume would decrease drastically during scaling-up process, which would lead to less total biomass and lower activities of electroactive biofilms. Moreover, the proton exchange membrane, which is costly already, would be too expensive for large areas and even more likely to suffer from fouling and mechanical degradation.

8.3 Way forward of biohydrogen production

8.3.1 Genetic engineering interventions

The rapid development of genetic engineering has provided new opportunities and challenges for biohydrogen production technology. With the help of precise gene-editing tools such as CRISPR the metabolic source code of hydrogen-producing microorganisms can be rewritten to overcome the physiological limitations shaped by natural evolution. Moreover, the microbial community structure can be adjusted by genetic engineering via modifying the key functional bacteria within it, thereby enhancing the hydrogen yield.

A major reason for the low efficiency of dark fermentation is the high sensitivity of hydrogenases to their own product, hydrogen, along with metabolic shifts toward by-products such as butyric acid and ethanol. Therefore knocking out or downregulating key enzyme genes in these competing metabolic pathways seems to be a possible method to redirect carbon flux toward the acetate pathway, which offers the highest theoretical hydrogen yield, thereby enhancing production. However, the chemical energy and reductant produced from photosynthesis are primarily used for carbon fixation, aiming for cell growth rather than hydrogen evolution; hence it is expected to further enhance the hydrogen production efficiency by regulating the functions of related metabolic genes through gens engineering to weaken the carbon fixation pathways, allowing more energy to flow into the hydrogen production pathway. Moreover, the utilization of nonfood biomass resources, including forestry processing residues, agricultural waste, and microalgae with high cellulose content for biohydrogen production, is predicted to become mainstream; hence the microbes that can directly decompose and utilize lignocellulosic hydrolysates because of introducing foreign genes encoding cellulases, xylanases, and similar enzymes into common hydrogen producers are required. This means that previously required costly pretreatment to become accessible to microbes can be directly converted by these super microbes, eliminating complex enzymatic saccharification steps and substantially reducing feedstock processing costs (Jiang et al., 2025; Wang et al., 2025).

8.3.2 Machine learning association

The interactive effect of various variables on the responses in the biohydrogen process is nonlinear and complicated inside the bioreactor, resulting in inaccurate predictions of the conventional statistical computational models, such as response surface methodology, while machine learning is a viable option for precisely predicting the process of biohydrogen production (Kumar Sharma et al., 2022).

The machine learning methods include artificial neural networks (ANNs), adaptive neuro-fuzzy interference systems (ANFIS), random forests (RF), and support vector machines (SVMs). It can dig out complex relationships from massive and multidimensional data, thereby accelerating the industrial application process of biohydrogen production. Hence the correlation between machine learning methods and biohydrogen production process parameters is necessary to be further studied. For example, in-depth research of machine learning in bioreactor design, feedstock selection, catalyst screening, determinant operational parameter identification (such as pH, temperature, illumination, C/N ratio, and organic loading), microbial community structure changes, and hydrogen yield, hydrogen production rate, and by-products prediction are needed. Moreover, machine learning is expected to be used to predict and optimize the performance of hydrogen-producing enzymes and to carry out design modifications, thereby obtaining enzyme variants with superior performance. However, data collection remains a key issue that needs to be addressed for the further application of machine learning in biohydrogen production, and a deeper integration with artificial intelligence (AI) might be a better option.

8.3.3 Hybrid approach integration

For future biohydrogen production, one destination is to transcend the limitations of individual technologies and achieve the cascading use of materials and energy through process coupling, thereby establishing a nearly closed-loop resource cycling system. The integration of different biohydrogen production approaches has been considered a promising strategy to achieve the goal.

Recently, dark fermentation integrated with photo-fermentation is popular, whereas there are also some reports on combined dark fermentation, photo-fermentation, and anaerobic digestion. However, the research mainly focuses on optimizing process parameters (e.g., pH and hydraulic retention time) and microbial community control to address challenges such as product inhibition and system stability (Lu et al., 2018, 2020). Further strengthening of research based on engineering scaling-up and industrial application is necessary. Moreover, the combination of biophotolysis and photo-fermentation and the integration of MECs and fermentation still need further exploration to achieve a highly efficient system. Another significant integration direction is the deep coupling of wastewater treatment with biohydrogen production and developing high-value animal feed or fertilizers based on the digestate of biohydrogen production.

In addition, simulating natural photosynthetic systems to develop efficient and stable bioabiotic hybrid systems provides new approaches for biohydrogen development. Photosynthetic microorganisms (such as cyanobacteria and green algae) or isolated photosystem proteins with nonbiological components such as semiconductor materials and nanocatalysts are integrated to achieve the conversion of light energy into hydrogen energy through biomimetic design (Alam et al., 2024). However, the current research mainly focuses on optimizing energy transfer pathways, enhancing light absorption efficiency, and addressing the stability issues of biological components. Carrying out large-scale applications still encounters many challenges.

8.3.4 Novel bioreactor system

Bioreactor system plays a key role in industrialization of biohydrogen production technology. Excellent mass transfer, light transmission, and heat transfer properties are the issues that future reactors need to address. The development of intelligent, digital, and automated bioreactor systems is of utmost urgency.

Recently the researche on bioreactors mainly concentrated on the respects, such as the shape of the bioreactor, the blender installation, material selection, and light source arrangement. With the rapid development of AI technology the biohydrogen production reactor system has ushered in new opportunities. For example, the hydrogen production process in the bioreactor can be intelligently controlled by creating a highly realistic virtual model (such as bioreactor digital twin) for the real bioreactor with the assistance of AI so that realize monitoring the real data. Furthermore the biohydrogen yield can be enhanced by automatically adjusting control strategy according to the analysis by AI algorithms.

Low-cost and high-performance manufacturing materials have always been the goal that bioreactor strive to. The development of antipollution, highly selective, and low-resistance nonmetallic or composite membranes for MECs has come to be necessary to decrease the cost and energy consumption. Moreover, the application of light-conducting materials and intelligent-responsive materials (such as response for temperature and illumination changes) in bioreactors is also expected to further enhance the transmission and uniform distribution of light, heat, and mass within the bioreactor.

To serve the integration of different biohydrogen production routines the corresponding bioreactor system integration also needs to achieve matching. However, the bioreactor system integration still encounters many challenges, such as the connection issue between dark fermentation and photo-fermentation as well as MECs, microbial community conflicts and ecological imbalance, separation and purification of the product gas as well as high investment and operating costs.

8.3.5 Policy and industrial chain synergy

Coordinated advancement of policies and industrial chains is required to achieve industrialization of biohydrogen production.

In terms of policy support, measures such as financial subsidies, tax incentives, research funding, and the introduction of carbon pricing mechanisms are expected to reduce the cost of biohydrogen and enhance its economic competitiveness compared to fossil fuels. Moreover, standards of biohydrogen purity, carbon accounting systems, and the development of regional hydrogen pipeline networks between resource production sites and demand centers also need further support from the policy, thereby driving biohydrogen production from the laboratory to large-scale markets and supporting the goals of carbon neutrality (Jiang et al., 2023).

In terms of industrial chain collaboration, integrations like biohydrogen production with chemical sector as a raw material for green chemical products such as ammonia and methanol (Jiang et al., 2023), biohydrogen production with hydrogen fuel cell vehicle industry to establish a full-scenario demonstration; and biohydrogen production with feedstock supply chain to form a circular economy model. In addition, establishment of a precise accounting and certification system covering the entire lifecycle carbon footprint of biohydrogen and formulation of international hydrogen standards and rules are impetrated.

8.4 Conclusion

This chapter provides a sustainability outlook of biohydrogen production, including the availability of renewable feedstocks, the characteristic low-carbon emission, the diversity of production technologies, and the recent policies and governmental support. Concurrently the significant issues of contestation, such as low technical efficiency, high cost, and the scale-up challenges, are also discussed. In addition, to solve the problems mentioned and boom the biohydrogen industry all over the word, genetic engineering interventions, machine learning association, hybrid approach integration, novel bioreactor systems, policy, and industrial chain synergy need to be further investigated and developed.

References

Alam, M., Jiang, D., Zhang, H., Zhang, L., Nadeem, F., Tahir, N., Lu, C., & Zhang, Q. (2024). Regulating the metabolites through SiC, ZnO, and SnO_2 nano-photocatalysts for elevating efficiency of photo fermentative bioH_2 from *Arundo donax* L. *Industrial Crops and Products, 221*. https://www.sciencedirect.com/science/journal/09266690. doi:10.1016/j.indcrop.2024.119395.

Chaves, T. C., Peiter, F. S., Gois, G. N. S. B., Telles, N. L. d. S., Almeida, R. M. R. G., & de Amorim, E. L. C. (2025). Synergistic effects on biohydrogen production from vinasse and molasses co-digestion: Influence of mixture composition on process stability. *Energy, 329*. https://www.sciencedirect.com/science/journal/03605442. doi:10.1016/j.energy.2025.136677.

Das, S. R., & Basak, N. (2025). Optimization of process parameters in microbial bio-hydrogen production from waste wheat starch as substrate by dark and photo fermentation. *International Journal of Hydrogen Energy, 98*, 500–513. https://www.sciencedirect.com/science/journal/03603199. doi:10.1016/j.ijhydene.2024.12.118.

Jiang, D., Ge, X., Lin, L., Chen, Z., Zhang, Q., & Li, Y. (2023). Biological conversion of methane to methanol at high H_2S concentrations with an H_2S-tolerant methanotrophic consortium. *Renewable Energy, 204*, 475–484. http://www.journals.elsevier.com/renewable-and-sustainable-energy-reviews/. doi:10.1016/j.renene.2022.12.106.

Jiang, D., Zhang, X., Jing, Y., Zhang, T., Shui, X., Yang, J., Lu, C., Chen, Z., Lei, T., & Zhang, Q. (2022). Towards high light conversion efficiency from photo-fermentative hydrogen production of *Arundo donax* L. By light–dark duration alternation strategy. *Bioresource Technology, 344*. Www.elsevier.com/locate/biortech. doi:10.1016/j.biortech.2021.126302.

Jiang, D., Zhang, L., Pei, W., Song, Z., Zhang, T., Jing, Y., Chen, Z., Li, W., Zhang, Q., & Lu, C. (2025). Correlation between ethanol addition and photo-fermentative biohydrogen production of *Arundo donax* L. with a consortium. *Renewable Energy, 248*, 123077. doi:10.1016/j.renene.2025.123077.

Kumar Sharma, A., Kumar Ghodke, P., Goyal, N., Nethaji, S., & Chen, W. H. (2022). Machine learning technology in biohydrogen production from agriculture waste: Recent advances and future perspectives. *Bioresource Technology, 364*. Www.elsevier.com/locate/biortech. doi:10.1016/j.biortech.2022.128076.

Liu, H., Zhang, Z., Lu, C., Wang, J., Wang, K., Guo, S., & Zhang, Q. (2022). Effects of enzymatic hydrolysis and alkalization pretreatment on biohydrogen production by chlorella photosynthesis. *Bioresource Technology, 349*, 126859. doi:10.1016/j.biortech.2022.126859.

Lu, C., Zhang, H., Zhang, Q., Chu, C. y., Tahir, N., Ge, X., Jing, Y., Hu, J., Li, Y., Zhang, Y., & Zhang, T. (2020). An automated control system for pilot-scale biohydrogen production: Design, operation and validation. *International Journal of Hydrogen Energy, 45*(6), 3795–3806. http://www.journals.elsevier.com/international-journal-of-hydrogen-energy/. doi:10.1016/j.ijhydene.2019.04.288.

Lu, C., Zhang, Z., Ge, X., Wang, Y., Zhou, X., You, X., Liu, H., & Zhang, Q. (2016). Biohydrogen production from apple waste by photosynthetic bacteria HAU-M1. *International Journal of Hydrogen Energy, 41*(31), 13399–13407. http://www.journals.elsevier.com/international-journal-of-hydrogen-energy/. doi:10.1016/j.ijhydene.2016.06.101.

Lu, C., Zhang, Z., Zhou, X., Hu, J., Ge, X., Xia, C., Zhao, J., Wang, Y., Jing, Y., Li, Y., & Zhang, Q. (2018). Effect of substrate concentration on hydrogen production by photo-fermentation in the pilot-scale baffled bioreactor. *Bioresource Technology, 247*, 1173–1176. Www.elsevier.com/locate/biortech. doi:10.1016/j.biortech.2017.07.122.

Lu, C., Wang, G., Zhang, Q., Yang, X., Yu, J., Liu, T., Petracchini, F., Zhang, Z., Sun, Y., Jiang, D., Liang, X., Li, Y., Zhang, Y., Zhang, T., & Zhang, H. (2023). Comparison of biorefinery characteristics: Photo-fermentation biohydrogen, dark fermentation biohydrogen, biomethane, and bioethanol production. *Applied Energy, 347*, 121463. doi:10.1016/j.apenergy.2023.121463.

Neumann, F., Zeyen, E., Victoria, M., & Brown, T. (2023). The potential role of a hydrogen network in Europe. *Joule, 7*(8), 1793–1817. https://www.journals.elsevier.com/joule. doi:10.1016/j.joule.2023.06.016.

Thajuddin, F., Rasheed, A. S., Elumalai, A., Palanivel, P., Nooruddin, T., & Dharumadurai, D. (2026). Harnessing microalgae and cyanobacteria for sustainable bioenergy production: Integration of molecular engineering and AI-driven approaches. *Renewable Energy, 256*. https://www.sciencedirect.com/science/journal/09601481. doi:10.1016/j.renene.2025.123961.

Wang, G., Lu, C., Sun, P., Song, C., Zhang, Q., Jing, Y., Jiang, D., Sun, J., Lai, M., He, X., Liu, J., & Liang, X. (2025). Enhanced energy conversion efficiency through photo-fermentation biohydrogen and biomethane co-production. *International Journal of Hydrogen Energy, 151*, 150221. doi:10.1016/j.ijhydene.2025.150221.

Zhang, Q., Jiao, Y., He, C., Ruan, R., Hu, J., Ren, J., Toniolo, S., Jiang, D., Lu, C., Li, Y., Man, Y., Zhang, H., Zhang, Z., Xia, C., Wang, Y., Jing, Y., Zhang, X., Lin, R., Li, G., … Tahir, N. (2024). Biological fermentation pilot-scale systems and evaluation for commercial viability towards sustainable biohydrogen production. *Nature Communications, 15*(1). https://www.nature.com/ncomms/. doi:10.1038/s41467-024-48790-4.

Zheng, Y., Zhang, Q., Zhang, Z., Jing, Y., Hu, J., He, C., & Lu, C. (2022). A review on biological recycling in agricultural waste-based biohydrogen production: Recent developments. *Bioresource Technology, 347*, 126595. doi:10.1016/j.biortech.2021.126595.

Index

Page numbers followed by "*f*" and "*t*" indicate, figures and tables respectively.